S0-BKW-222

# Phenolic Substances in Grapes and Wine, and Their Significance

# ADVANCES IN FOOD RESEARCH

*Edited by*

C. O. CHICHESTER E. M. MRAK G. F. STEWART

*University of California*
*Davis, California*

*Editorial Board*



*Supplement 1*

Phenolic Substances in Grapes and Wine, and Their Significance. 1969
VERNON L. SINGLETON AND PAUL ESAU

# Phenolic Substances in Grapes and Wine, and Their Significance

VERNON L. SINGLETON

PAUL ESAU

DEPARTMENT OF VITICULTURE AND ENOLOGY
UNIVERSITY OF CALIFORNIA
DAVIS, CALIFORNIA

1969

Academic Press New York and London

ACADEMIC PRESS, INC.
111 Fifth Avenue, New York, New York 10003

*United Kingdom Edition published by*
ACADEMIC PRESS, INC. (LONDON) LTD.
Berkeley Square House, London W1X6BA

LIBRARY OF CONGRESS CATALOG CARD NUMBER: 75-84257

PRINTED IN THE UNITED STATES OF AMERICA

# Preface

We believe that this work will be of interest to all with a serious interest in the practical technology or in the fundamental science of grapes and wine, since it is widely recognized that the phenols of these products have great importance to enologists, wine consumers, and winemakers. Many properties and reactions during growing, processing, and storage involve the anthocyanins, tannins, and other phenols of grapes.

We hope also to serve those interested in other products importantly involved with phenols: leather, tea, beer, other fruits, etc. The breadth, depth, and importance of studies of phenols in grapes and wine seem to us greater than in other foods. Although we have attempted to be complete as well as critical in coverage of the grape or wine literature, we have also brought in other phenolic studies when they seemed to complement or clarify.

In addition, we hope to encourage fundamental geneticists, organic chemists, and biochemists to recognize and apply themselves to the solution of additional problems of commercial importance as well as of scientific interest. Equally, we have tried to point out fundamental meanings behind practical observations in the winery or vineyard and to synthesize the subject into as coherent a whole as possible. We have speculated on the interpretation of various findings, and therefore have tried to bring in evidence from as many directions as possible. About 1225 references have been cited, and at least three times that number were considered. While revisions have been made as new publications were noted, it is probable that some worthy ones have been missed. The literature review was completed about May, 1968.

It is a pleasure to express our gratitude to colleagues who encouraged and advised us. Professors H. W. Berg and A. D. Webb of the University of California at Davis read and made suggestions on the entire manuscript, and R. E. Kunkee and C. S. Ough did so on specific sections. Kelvin Deming helped greatly in editing. Mrs. Kathryn Singleton aided materially in literature collection, checking, and proofreading.

*Davis, California*
*June, 1969*

Vernon L. Singleton
Paul Esau

# Contents

# Phenolic Substances in Grapes and Wine, and Their Significance

# I

# Introduction

The fruit produced in largest tonnage worldwide is the grape, with a major part of the harvest used in producing wine. Major factors in making wine something more than a colorless, tart, alcoholic solution are the phenolic substances it contains. The changes that occur during the processing and aging of wine are intimately tied to reactions of the phenolic substances. Thus, the study of phenolics would appear to be particularly necessary to understanding grapes and wine, and these fruit products would perhaps be preferred for the study of plant phenols and their reactions. Nevertheless, the food scientist may be less aware of research on phenolics in grapes than in tea, cacao, persimmons, citrus, peaches, pears, or apples (Bate-Smith, 1954a; Hulme, 1958a; Kefford, 1959; Joslyn and Goldstein, 1964a; Roelofsen, 1958; Stahl, 1962). There are several reasons why this may be so.

A great amount of the early research on phenolics was done with grapes and wine, preceding similar studies with other plant material. In the dawn of the Scientific Revolution such men as Pasteur, Berthellot, Büchner, Berzelius, and Fresenius used grapes or wine as a preferred medium of study. Thus, enology could be considered the parent of microbiology, biochemistry, stereochemistry (the word "racemize" stems from the Latin for grape cluster), and food technology. The very success of this early golden period occasioned, on the one hand, a period of complacency and digestion, and on the other, a turning to simpler systems for the development of details.

Research on grape phenolics has not made as coherent an advance as it might have because, aside from its complexity, a series of recurring crises have interrupted and disorganized research on grape and wine.

The world wars were disruptive to all research programs, but especially to enology since the wars overran the European centers of such research. Other crises peculiar to grape and wine production included the devastation of vineyards of European grapes by pests, particularly pests introduced from the New World. The worst plague of this sort was the root louse phylloxera, which killed large portions of the vineyards of Europe during the latter half of the 19th century. Owing to the four-year delay between planting and appreciable grape yield, plus poor forecasting of supply and demand, viticulture has been more than usually plagued by boom-and-bust economic cycles. Finally, a very disruptive influence was the prohibition of alcoholic beverages in the United States, from before 1918 until 1933.

Since World War II, however, we have been in a period of relative stability, and progress has been accelerated tremendously by this as well as by the general resurgence in interest and new techniques in phenolic biochemistry. Much of the worthwhile research on the phenols of grapes and wine has been published in journals devoted more or less exclusively to viticulture or enology, most of them in foreign languages. The consequence has been that the scientist and technologist working with grapes or wine have not been in close enough communication with the rest of the scientific community.

We hope to improve awareness in both directions by reviewing as critically and completely as we can the entire subject at hand. Several good reviews have already appeared on some aspects of the subject. Within this present series (Advances in Food Research) Amerine (1954) has included tannins and pigments in his review of the organic composition of wine. Mackinney and Chichester (1954) have discussed color in wines. Bate-Smith (1954a) has reviewed flavonoids in foods, including some examples from grapes. Joslyn and Goldstein (1964a) have reviewed astringency, which relates importantly to wine phenols. Durmishidze (1955) has summarized Russian work, particularly that of his own group. Pascal Ribéreau-Gayon (1964a, 1965a) has reviewed the occurrence of phenolics in grapes and wine, emphasizing the work of his group. Webb (1964) has critically reviewed the anthocyanins of grapes, and Herrmann (1963a,b) has written a series of reviews on phenolics in foods, including grapes and wines. We hope to cover the subject in a more integrated fashion and bring it up to date. We propose to give an introductory background, to list the specific compounds identified, and to evaluate the rigor of identification. Then, we intend to discuss the amounts present and the significance of these compounds in the production and attributes of grapes or wines.

## A. Importance of Grapes and Wine as Foods

Grapes form a larger part of the human diet than any other single fruit. The production in the principal producing countries of the world in 1963 was 43,236,000 tons of grapes, 19,260,000 tons of all types of citrus fruits, and 34,340,000 tons of apples, apricots, cherries, peaches, pears, and plums combined (Anon., 1965). Vineyards total approximately 24,640,000 acres, a significant portion of the estimated 3.5 billion world's total of cropped acres (Protin, 1966; Guidry, 1964). About 76% of this grape acreage is in Europe, including Russia, with other important plantings in Iran, Turkey, Algeria, Morocco, South Africa, Argentina, Chile, Australia, and the United States, among others. In 1966, the United States produced 2,875,600 tons of apples, 1,702,500 tons of peaches, and 8,177,200 tons of oranges. On roughly 500,000 acres, 3,733,600 tons of grapes were produced, 3,400,000 tons of this in California (Anon., 1968b).

Grapes probably enter the diet in more varied forms than any other fruit. Like most fruit, they are consumed fresh, canned (mainly in fruit cocktail), as juices, concentrates, jellies, and as dried raisins. However, in contrast to many other fruits, they also go into wines and brandies consumed in very large total gallonage. Not only are these beverages of many types, but they may be consumed alone as well as with meals. Wines, brandies, grape concentrate, and raisins are used in home cooking, and are also sources of special flavors and other qualities in commercial food preparations and, sometimes, in other products, such as tobacco. The world production of wines in 1965 was about 7.3 billion gallons (Protin, 1966). In California in 1965 about 14.3% of the grapes were sold as table grapes, 32.7% as raisins, 1.4% canned, and the remaining 51.6% crushed for wine (Anon., 1967a).

The form in which the grape is consumed affects the phenolic substance which enters the diet. The total phenolic content of the fleshy part of a grape berry is relatively low, as with most fruits. With grapes, however, it is common to consume the peel and often the seeds as well, and these tissues are rich in phenolic substances. Many other fruits are peeled, and the seeds are not eaten or, if eaten, are low in tannins and other phenols. Still more to the point is the effect of wine making. The phenolic substances as a group are extracted from the tissue much more easily into aqueous alcohol than into juice or water. Depending upon the methods used, more or less of the phenolics of the solid parts of grapes will be extracted into the wine and may produce considerably different effects on the appearance, flavor, and other properties of the wine.

The annual consumption of wine is only about 1 gallon per capita in the United States, but in many countries it is over 10 gallons and in some even more than 30 gallons. In such high-consumption countries it is not considered unusual for a working man to consume a liter of robust red wine per day. This would represent a daily consumption of about 2 grams of phenolic substances calculated as tannic acid. It would appear that grapes and wine can be a major source of phenolic substances in the diet. It is difficult to visualize as large an intake from other regularly consumed products, with the possible exception of tea.

## B. Botanical Relationships

Grapes are dicotyledonous Spermatophytes of the order Rhamnales, family Vitaceae, genus *Vitis*. They are the only plants in the order that are of economic significance as food. The members of the other family in the Rhamnales, the Rhamnaceae, are mostly shrubs or small trees and include *Zizyphus jujuba* and a few similar species with fruit that is edible but not widely commercialized. *Rhamnus cathartica* and *Rhamnus purshiana* are sources of phenolic (emodin) purgatives—the latter gives the drug Cascara Sagrada (Bailey, 1935; Anon., 1968a).

The Vitaceae are mostly climbing shrubs with tendrils and include 11 genera, among them *Ampelopsis*, *Cissus* (Kangaroo vine), *Parthenocissus* (Virginia creeper, Boston Ivy), and *Vitis* (grapes). Except for grapes, few are in cultivation, and those that are are used primarily as ornaments. In a few interesting cases, plant parts other than the fruit are eaten, though these are comparatively unusual. The green shoot-tips of *Cissus* (*Vitis*) *quadrangularis* are eaten as a vegetable in Ceylon and India (Bhaskara *et al.*, 1939; Bailey, 1935). *Cissus* (*Vitis*) *capensis*, of southern California, and *Vitis* (*Cissus?*) *opaca*, of Australia, produce tubers which can be eaten or fed to animals (Baker and Smith, 1906). Leaves of *Ampelopsis meliaefolia* are eaten in Japan (Kotake and Kubota, 1940). Of course, grape leaves are an occasional item in Mediterranean and other cuisine.

In its stricter limitations, the genus *Vitis* includes nearly 60 species and is widely distributed, particularly in the north temperate zone (Winkler, 1962; Bailey, 1935). By far the most important from a commercial viewpoint is *Vitis vinifera*, the European grape. Familiar varieties include most of the table grapes, such as "Thompson Seedless," "Flame Tokay," "Emperor," "Malaga," and "Ribier," as well as prestigious wine varieties such as "Pinot noir," "Cabernet-Sauvignon," "White Riesling," "Chardonnay," "Sémillon," and "White Muscat."

Other species which are important as fruit producers include *Vitis labrusca*, the fox grape, native to the Allegheny region, with its typical variety, "Concord." The muscadine grape, or "Scuppernong" is the best-known variety of *Vitis rotundifolia*. They are native to the southeastern United States (and not to be confused with muscat or muscadet). They are classed by some as a separate or subgenus.

Many other species of *Vitis* have been used in crosses, primarily with *Vitis vinifera*, in efforts to produce good fruiting properties and viticultural advantages such as pest resistance. In fact, many of the varieties which are considered typical of *Vitis labrusca*, particularly "Delaware," perhaps "Catawba," and even "Concord," may well carry some *Vitis vinifera* "blood." Owing partly to the antiquity of the cultivation of the European grape and partly to the economic pressures produced by a series of disasters and disruptions in viticulture, there are perhaps 10,000 named cultivars of grapes. Some are widely dispersed, and are often known by different names in different areas. The description and differentiation of these varieties, ampelography, has produced a voluminous specialized literature (Galet, 1962; Turković, 1967). Varietal distinctions relate to compositional differences (anthocyanin-bearing versus white grapes, for example) and are of great importance to grape and wine quality. Fortunately, relatively few cultivars (about 100) account for a very large portion of the world's production.

## C. Other Aspects

The fact that the grape is relatively isolated from other food plants botanically, and therefore worthy of study has already been discussed. The importance of the grape as a major fruit crop, and one consumed in many ways which may cause it to be the major source of phenolic substances in the human diet, has been outlined. It is clearly important, therefore, to learn the specific phenols present and the amounts and variations produced during the growing and processing of grapes and grape products.

In addition, phenolics play very active roles, both positive and negative, in plant products in general and grapes and wines in particular. The phenolic composition of the vine and its fruit reflects biogenesis, which in turn reflects the usefulness to the plant of the phenolic substances and the interplay of stimulus and response to the environment. In fact, as secondary substances that are not uniformly present in or generally essential to the life of plant cells but that occur widely in many forms, phenols appear uniquely capable of great variation in plants in response to genetic and environmental influences. We are beginning to discover and understand the specific participation of phenols in important regulatory mechanisms in plants. For example, some natural phenols stimulate and others inhibit enzymatic oxidation of indoleacetic acid and thus may participate in regulation of the effective level of this important plant hormone and, through it, plant growth rate, etc. (Gortner and Kent, 1958; Galston, 1967). Phenol content can indicate a sufficiency or deficiency of soil water (Gortner, 1963). Phenol content is also markedly affected by the plant's response to light (Harborne, 1964a). Twining of tendrils seems closely related to specific variation in phenolic composition (Galston, 1967). Phenols appear to be involved in mineral deficiency effects in plants—boron, for example (Lee and Aronoff, 1967)—and are sometimes key substances in disease and pest resistance (Goodman *et al.*, 1967). Again, the great potential value of the study of grape phenols is apparent not only for the viticulturist and vineyardist, but also for the whole food grower-processor community.

The qualitative and quantitative makeup of the phenols in wine reflects not only the grape but also the fermentation practice and the further processing. As in other fruit products, phenols are importantly involved in astringent and bitter flavors, natural colors, and browning and other reactions with oxygen. With wines, however, since storage is long and aging may improve quality, the changes associated with fruit phenols and time are especially important. Enzymic and nonenzymic browning reactions, which are undesirable in grapes and other fresh fruit, are produced deliberately in certain wines and minimized in others. Some wines are highly oxidized, and others are kept as free of oxidation as possible. A very wide range of phenolic pigmentation is sought for different wines, from the nearly colorless through all shades of yellow to brown, and of anthocyanin colors, pink to nearly "black." Wines produced over the world also encompass levels of astringency and bitterness from undetectable to strongly offensive for the uninitiated consumer. Standardization of phenol content and control of reactions involving phenols have long been major concerns of enologists, who are now sharing, with other food scientists, an interest in the development of more fundamental and less "artful" understanding of these important compounds.

# II

# Qualitative Phenolic Composition of Grapes and Wines

There has been a tendency for substances in natural products to be identified on the basis of evidence that is limited in some way or another. In the better older work, more or less heroic efforts were made to isolate sizable amounts of a substance in order to prove its identity by crystallization, degradation, synthesis, etc. Such studies were generally most effective in the detection of relatively major and unique substances. They tended to err in producing artifacts or incomplete separations from complex mixtures of similar compounds. With the advent of chromatographic methods coupled with spectrophotometry, specific reagents, and so on, it has been possible to elucidate much more complex situations, but there has been a tendency to assume identity with a known substance when mobility or reaction is similar in only a few systems.

No doubt, considering probable biochemical unity, most of these identifications will prove correct, but we feel that reports of substances identified should include critical details of the procedures employed. This seems particularly necessary with phenolic substances, since they usually occur in plants in complex arrays of potentially interfering substances.

## A. Phenols and Related Substances Other Than Flavonoids

### 1. Derivatives of Hydroxylated Benzoic Acid

In the late 1800's, chemical preservatives were added to foods (including wine). These preservatives included salicylic acid, *p*-hydroxybenzoic acid and its esters, and some other nonphenolic substances such as *p*-chlorobenzoic acid and chloroacetic acid. Ash (1952) states that 90% of the wine made in 1900 in California had salicylic acid added, a worldwide practice. Most such additions were prohibited when pure food laws were adopted in the major wine-producing and wine-importing countries. Legal actions against makers of wines alleged to contain salicylic acid and other additions led to contests over procedures of detection and to the claim that a certain low level of salicylic acid was present naturally. Pellet (1906) has given an extensive summary of the early reports. The common detection procedure was to extract acidic solutions of wine or saponified wine with nonpolar solvents or to steam distill to separate the putative salicylic acid. This was then "identified" by the production of a violet color with ferric chloride. By such procedures, claims were made of identification of small amounts of salicylic acid in natural grapes and untreated wine. Traphagen and Burke (1903) reported the equivalent of at least 0.32 mg of salicylic acid per kg of fresh "Concord" grapes, and similar amounts in most fruits. Chelle (1925) concluded from his own work and a critical review that small amounts of salicylic acid can occur naturally in wine. For effective preservation, added salicylic acid must be present at concentrations of about 500 mg/liter (Fonzes-Diacon and Laforce, 1926; Gentilini, 1960). Gentilini (1960) showed that the official Italian method was capable of detecting salicylic acid or its methyl ester added to wine at levels of 1 mg/liter or more. This qualitative chemical test, and paper chromatography of a concentrated ether-petroleum ether extract of 50 ml of 24 different red and white table wines from the Veneto district failed to show the presence of salicylic acid or extractable esters of it. Butanol/water was the system used, and ferric chloride gave a violet spot with known salicylic acid at $R_f$ 0.38. Guimberteau and Portal (1961) used paper chromatography to study acids in an ether extract of juices of 2 red ("Cabernet-Sauvignon," "Merlot") and 2 white ("Sauvignon blanc" and "Sémillon") varieties of grapes, plus 9 red and 9 white table wines known to be free of added salicylic acid. The extracts were prepared after refluxing the sample for 2 hours with 1 ml 5 *N* NaOH/25 ml wine, and acidifying. Of 4 chromatographic solvents tested,

BuOH-1.5 *N* $NH_4OH$ (35:65 v/v) was preferred and gave an $R_f$ of 0.52 with salicylic acid and of 0.56 with cinnamic acid. In comparison with known compounds, salicylic acid was located by fluorescence enhanced by $NH_3$; also used were bromphenol blue spray for acids and diazotized *p*-nitroaniline. Salicylic acid was not found in the juices or in 6 white and 5 red wines. Four of the red wines gave tests for 0.5, 0.5, 1.0, and 1.5 mg salicylic acid per liter; 3 white wines showed 0.5 mg/liter or less.

Ribéreau-Gayon (1963a) reports finding very small amounts of salicylic acid in certain wines by paper chromatography. However, this acid was not found in his careful studies with juices and wines from two grape varieties and is not reported in several other studies on the chromatographic separation of phenolic acids from grapes or wines. It appears certain from available data that natural salicylic acid can be no more than about 1 mg/liter, either free or readily liberated by alkali, in grape must or wine. We consider that the natural presence of any salicylic acid in grapes or wine is not proved conclusively, and suspect that the occasional traces reported in commercial wines may represent contamination from storage in wooden cooperage used before the addition of salicylic acid was outlawed, or other such artifacts. With long storage it would be expected that ethyl salicylate would form in wine. Bayer (1957) tentatively identified salicylic esters in wines by paper chromatography. Volatile salicylates, however, have generally not been found in very sensitive gas chromatographic studies of grape and wine volatiles (Webb, 1962, 1967; Stevens *et al.*, 1967). However, methyl salicylate has been tentatively identified in a commercial sample of muscat raisin fusel oil (Webb *et al.*, 1952), ethyl salicylate in a "Sylvaner" wine (Webb, 1962), and small amounts of salicylic acid in "Sauvignon blanc" wine (Chaudhary *et al.*, 1968).

Guimberteau and Portal (1961) reported 0.2 to 0.5 mg per liter of *p*-hydroxybenzoic acid ($R_f$ 0.14) in saponified samples of all 18 wines and 4 juices tested, and tentatively identified syringic acid in most of the saponified samples. Ribéreau-Gayon (1963a) reported *p*-hydroxybenzoic acid ($R_f$ 0.06) on paper chromatograms irrigated with toluene-acetic acid-water (4:1:5). This acid was found free in wines from "Cabernet-Sauvignon" and "Sémillon" grapes, although more was detected (0.5 to 1 mg) after saponification of both wines. It was not detected free in extracts of grape skins from the same varieties. After saponification it was detected only in Sémillon skin extracts, and in smaller amounts than in the wines. Esters of *p*-hydroxybenzoic acid may, of course, be added as preservatives in wines (Amati and Formaglini, 1965).

In the same study (Ribéreau-Gayon, 1963a), vanillic acid ($R_f$ 0.34), syringic acid ($R_f$ 0.24), and gentisic acid ($R_f$ 0.02) were detected, and the use of benzene-acetic acid-water (6:7:3) revealed protocatechuic acid ($R_f$ 0.08) and gallic acid ($R_f$ 0.005). In general, these substances were not detected, or were noted only in traces (vanillic, syringic), until after saponification of the skin extracts, and although detected free in one or both of the two wines, they were usually increased considerably by saponification. Upon saponification, the red "Cabernet-Sauvignon" gave much greater amounts of vanillic (15 mg/liter), syringic (30 mg/liter), protocatechuic (8 mg/liter), and gallic (12 mg/liter) acids in wine, and skins as well, than did the white "Sémillon" (about 1 mg/liter total). In the red samples, the increase upon saponification was as much as 30-fold. It seems possible that some degradation of flavonoids of the alkaline-fusion and cleavage type was occurring during saponification, so that the syringic acid, for example, may come from malvidin residues. Similarly, vanillic acid could come from peonidin, protocatechuic acid from cyanidin, and gallic acid from delphinidin. The origin of the gentisic acid is not clear. Ibrahim and Towers (1960), however, found gentisic, vanillic, and protocatechuic acids in acidic and alkaline hydrolyzates from 9 of 10 plant species examined. Tomaszewski (1960) found gentisic acid in bound form in all of 122 species examined including grapes. On the other hand, studies of grape phenolics without deliberate hydrolysis have failed to detect several of these substances. For example, Pifferi (1966) has found a total of 33 phenolics on two-dimensional chromatograms of phenols other than anthocyanins from 4 red wines. The wines were concentrated at low temperature under $CO_2$, and the phenols separated from sugars, anthocyanins, and other interfering substances by preliminary retention and elution on a cation-exchange resin column. Known substances, eight spot-detection procedures, and relative mobility were compared for the identifications reported. Although $\gamma$-resorcylic acid was reported only in Merlot wine, and gallic, protocatechuic, and *m*-digallic acid were reported in 3 or 4 of the wines, no vanillic, syringic, or gentisic acid was noted. In thin-layer chromatographic studies (Pifferi, 1965), vanillic acid was not found in similar wines although gallic and protocatechuic acids were.

Protocatechuic acid has been reported in the water-soluble substances from grape leaves (Boettinger, 1901). Of course, pyrocatechol or protocatechuic acid have long been known as degradation products from grape phenols. Carles and co-workers (1960) report, without experimental details, that protocatechuic acid and gallic acid were absent in grapes or young wines but appeared as the wine aged. In two-dimensional paper chromatograms comparing a red wine with tea, Colagrande

(1961) found gallic acid. Levels are so low that Kloss and Seifert (1927) could not reliably demonstrate their presence in wine with the techniques available at the time. Ribéreau-Gayon (1963a) reported 0–0.5 mg/liter of gallic acid in skin extracts or white wine, and 12 mg/liter before or after saponification of a red wine. Hennig and Burkhardt (1958) found free gallic acid in wines just after fermentation, especially if fermented on the skins, but not in Riesling grapes or in aged wine. Free gallic acid was not found in grape seed extracts containing depside gallates (Singleton *et al.*, 1966a). Durmishidze (1955) found gallic acid only in esterified form in phenol fractions from grapes and grapevines.

## 2. Derivatives of Hydroxycinnamic Acid

Simple observation shows that grapevines produce anthocyanins and that the vines are lignified. It is therefore obvious from our knowledge of the biosynthesis of such compounds that shikimic acid and cinnamates should occur in grapes. Shikimic acid and quinic acid have been found in grapes or wine (Kohman and Sanborn, 1931; Hennig and Burkhardt, 1957a,b, 1958; Carles and Lamazou-Betbeder, 1958; Carles and Alquier-Bouffard, 1962, Delmas *et al.*, 1963; Ribéreau-Gayon and Ribéreau-Gayon, 1963; Kliewer, 1966). Caffeic acid is very common in plants, often occurring as a caffeoyl quinate or chlorogenic acid. The phenolic cinnamic acids and related cinnamate-derived depsides are usually brightly fluorescent and relatively mobile on paper in both butanol-based and aqueous solvents. Their fluorescence is generally enhanced and changed in hue by fuming with ammonia, and, owing to cis-trans isomerism, they usually give two spots upon paper chromatography with aqueous solvents.

Substances meeting these criteria are readily demonstrated in grapes and wine. Specific identification of the native and derived substances poses more of a problem. Caffeic acid and *p*-coumaric acid or their derivatives have been reported in grapes and wine by several researchers, and there can be little doubt of their general presence (Hennig and Burkhardt, 1957a,b, 1958). Ribéreau-Gayon (1963a) reports about 1–15 mg/liter of caffeic acid and 0.3–30 mg/liter of *p*-coumaric acid in crushed grapes or wine. Ferulic acid has been reported in lesser amounts (0.2–1 mg/liter) and less frequently (Genevois, 1952; Ribéreau-Gayon, 1963a; Pifferi, 1965). Sinapic acid does not appear to have been found in grapes or wine, although syringic acid has, as previously stated. In this connection it is worthy of note that Bastianutti and Romani (1961) found that cinnamic acid, during recovery from wine for paper chromatography, was converted to benzoic acid. Ibrahim and Towers (1960) were able to recover only small amounts of ferulic or sinapic acids from plant extracts hydrolyzed by either acid or alkali, owing to destruction during heating. These facts suggest that the smaller acids, notably *p*-hydroxybenzoic, protocatechuic, vanillic, and syringic acids, may have been produced from, respectively, *p*-coumaric, caffeic, ferulic, and sinapic acids, at least in part. This would help explain why some of these compounds have been detected in some instances and not in others. Although Pifferi (1965) searched for *o*-coumaric acid in wine, he was unable to find it, which would add doubt to the presence of salicylic acid.

It appears generally true that the hydroxycinnamic acids may occur partly free in wines, but in the grape berry occur in appreciable amounts only as saponifiable forms (Hennig and Burkhardt; 1957a,b, 1958; Ribéreau-Gayon, 1963a). Smit (1953) and Joslyn and Smit (1954) classified one group of polyphenols from grape juice as caffetannins or chlorogenic-acidlike, based upon absorption spectra of fractions isolated by column chromatography on silica gel. Hennig and Burkhardt (1957a,b, 1958), based upon two-dimensional paper chromatography and several color tests, concluded that Riesling grapes, and wines, including fruit wines, contained chlorogenic and isochlorogenic acids. Using column chromatography on silicic acid, Sondheimer (1958) isolated from "Steuben" (non-*vinifera*) grapes four bands which, in comparison by spectra and other properties with known substances isolated from other natural products, corresponded to chlorogenic acid (140 mg/100 g dry weight), 20 mg of "band 510," and tentatively (slight differences in spectral maxima) 20 mg/100 g isochlorogenic acid plus 2 mg/100 g of neochlorogenic acid. Free caffeic acid was absent, as expected. Weurman and de Rooij (1958) precipitated the phenols from pressed juice from "Black Alicante" grapes with lead acetate, and chromatographed the recovered phenols on paper. The separated bands were rechromatographed in two dimensions in comparison with known chlorogenic, *p*-coumaric, and gallic acids. Identified as caffeic acid derivatives were neochlorogenic acid (probably), an apparently new chlorogenic isomer, and *p*-coumaroyl quinic acid. Jurics (1967a) lists the chlorogenic acid content of grapes as 125 mg/liter.

Tanner and Rentschler (1956; Rentschler and Tanner, 1956) found a small amount of chlorogenic acid in "Chasselas" wine but not in the juice of this grape, and attributed this to extraction from Chasselas berry solids in which they found chlorogenic acid. They did not find chlorogenic acid in three other varieties, "Räuschling," "Müller-Thurgau," and "Blue Burgundy," and concluded that chlorogenic, caffeic, and quinic acid do not occur in grape juice. Singleton *et al.* (1966a) reported that chlorogenic-acidlike substances were the major phenols of juice. Mehlitz and Matzik (1957) reported chlorogenic-like substances but not chlorogenic acid itself. Pifferi (1966) indicated that chlorogenic and isochlorogenic acid occurred in wine. Colagrande (1961) reported chlorogenic acid. Masquelier and Ricci (1964) reported chlorogenic acid (50–150 mg/liter plus 3-20 mg/liter of caffeic acid) in wine, based on paper chromatography in 4 solvents.

Since some isomers of chlorogenic acid have been clarified only fairly recently (Corse *et al.*, 1965, 1966), some of the earlier confusion is understandable. Work by Ribéreau-Gayon (1965c) appears to have cleared up the confusion in grapes, however, by discovering that caffeic acid occurred in "Cabernet-Sauvignon" grapes as monocaffeoyl tartaric acid and not as a caffeoyl quinate. Repeated paper chromatography of an ethyl acetate extract of ground leaves or berry skins gave a substance which, when saponified, gave products identified by spectra, chromatography, and color tests as caffeic acid and tartaric acid, in equal proportions. Similarly, monoferuloyl and mono-*p*-coumaroyl tartaric acids were identified. These had not been reported previously, and chicory is the only other known source of caffeoyl tartrates (Scarpati and Oriente, 1958; Scarpati and D'Amico, 1960). Considering the relatively rare accumulation of tartaric acid in plants, this is not surprising.

Feruloyl quinic acids, on the other hand, have been reported in grapes or wine by Masquelier and Ricci (1964). *p*-Coumaroyl quinic acids have been reported in wines by Pifferi (1966), Masquelier and Ricci (1964), and Burkhardt (1965). In the last case, *p*-coumaroyl quinate was found in all stages of the developing grape berry and wines of several types. The primary means of identification were paper chromatography with characteristic fluorescence, cochromatography, and isolation of a calcium salt. These spots appeared in addition to those designated as chlorogenic and caffeic acids.

It is not unequivocally clear at this point whether the hydroxycinnamates occur naturally as both tartaric acid derivatives and quinic acid derivatives in grapes or wine. It appears probable, pending further study, that both forms occur. Quinic acid is evidently present. Color reactions supposedly specific for chlorogenic acids have been obtained. *p*-Coumaric acid and caffeic acid are known to occur in grapes, as discussed later, as an acyl-sugar-anthocyanidin combination, so that there seems to be no biochemical reason to exclude alternative forms. It may be significant that Ribéreau-Gayon (1965c) found the tartaric acid derivatives in skin extracts and that most of the other workers have worked with high-fluid parts, predominantly juice or wine. The tartaric acid metabolism (concentration compared with sugar, etc.) is known to be considerably different in different tissues of the berry. Tartaric acid may reach levels high enough to cause intracellular crystallization in grape skin (Alwood, 1914).

### 3. Other Nonflavonoid Phenols

Chaudhary *et al.* (1968) have reported gas chromatographic detection of a trace of phenol itself in extracts of wines from normal and from botrytised "Sauvignon blanc" grapes. They reported *m*-cresol also, in only the botrytised sample. Aldehydes corresponding to the benzoic acid derivatives might be expected in wines. Vanillin has been reported in wine (Hennig and Villforth, 1942), but over 0.5 mg/liter has been considered evidence of sophistication (i.e., addition)(Grohmann and Mühlberger, 1954). Extraction from wooden cooperage, or conversion from lignin extracted from wooden cooperage, appears to be the source of vanillin in wine and, to our knowledge, it has not been found as such in grapes. Skurikhin (1967) reports the presence of vanillin, sinapaldehyde, coniferaldehyde, and syringaldehyde in sherries aged in oak barrels, but not in sherries produced by submerged culture and out of contact with wood.

Lignin occurs in grapevines, but few chemical studies of grape lignin have been made. Fernández and Sanchez-Gavito (1949) compared lignin preparations from dry grape skin and defatted seed powders. The Klason-lignin from skins gave no test for methoxyl groups, whereas that from seeds showed 4% methoxyl content. It was concluded that the lignins of the two tissues were similar but that lignin from seeds was more highly polymerized. Ethanol-soluble lignin was studied, but tannin contamination would appear to have been a problem. Defatted dry seed powder from "Almeria" and "Villanueva" grapes analyzed 35 and 38% Freudenberg-lignin, and the Klason-lignin content was about 50% in all samples.

Ellagic acid was apparently reported first in leaves of grapes, by Bate-Smith (1956) as part of taxonomic screening studies. Hennig and Burkhardt (1957a,b, 1958) reported ellagic acid and ellagitannins in red and white wines. The ellagic acid was identified by two-dimensional paper chromatographic comparison with known substances and various color tests. They found ellagic acid in large amounts in juice from green "Riesling" grapes, but noted that it could have been produced by polyphenoloxidase action on natural tannins during the analysis. Ellagic acid was also reported in large amounts in young wines fermented in the presence of grape solids, but it was removed by aging and processing. Pifferi (1966) also reported ellagic acid on chromatograms of red wines from "Clinton," "Baco," and "Merlot," but not "Raboso" grapes. Singleton *et al.*

(1966b) rigorously identified ellagic acid as the cause of a troublesome crystalline precipitate in loganberry wine. In spite of a great deal of study of precipitates in grape wines, no such ellagic acid precipitate in wine had been reported previously. This suggests that there is no appreciable reservoir of soluble ellagitannin in grapes to gradually release ellagic acid and cause precipitation in grape wine. The suggestion of Hennig and Burkhardt (1958) appears valid that the ellagic acid arises in grape products by oxidative coupling, and a source is suggested by the fact that gallic esters (which occur in grapes and wine) are converted to ellagic acid more readily than is gallic acid (Hathway, 1957 a,b). The possibility is discussed later that ellagic acid can enter wine as an extractive from wooden cooperage (Hennig and Burkhardt, 1962).

Hennig and Burkhardt (1958) reported three spots each from caffeic acid and *p*-coumaric acid on their chromatograms of wine. These were assigned the identities of the cis-trans isomers plus esculetin and umbelliferone. Pifferi (1966) assigned spots on his chromatograms from wine the identity of esculetin, methylesculetin (scopoletin?), and, tentatively, *p*-coumaric lactone (umbelliferone). It has been shown (Butler and Siegelman, 1959) that, catalyzed by metal ions such as ferric ion and ultraviolet light (which also produces cis-trans isomerism), caffeic acid converts *reversibly* to esculetin (and daphnetin in lesser amount). Other hydroxycinnamates such as ferulic acid give analogous coumarins. On a single two-dimensional paper chromatogram with normal procedures (no pretreatment to remove iron from the paper, or ultraviolet irradiation to locate spots) pure caffeic acid gave five spots on a chromatogram developed in two dimensions with 5% acetic acid—three of cis or trans caffeic acid, and two of esculetin. Butler and Siegelman (1959), however, showed that prewashing the paper to remove iron prevented the production of coumarin derivatives from the usual cinnamic acid derivatives on the chromatograms. Pifferi (1966) did pretreat his paper, so that this reaction should have been eliminated, yet the coumarins were still detected in wine. It seems reasonable that, with the traces of iron in wine and possibilities for oxidation, the same reaction can occur in wine as on paper, with the coumarins produced by hydroxylation, isomerization, and cyclization of the hydroxycinnamates. The identification of esculetin and similar coumarins as occurring naturally in grapes, however, must be discounted, owing to the possibility of their formation on paper in the presence of traces of iron. Daphnetin also can arise from caffeic acid, and under reductive conditions the reaction can be reversed (van Sumere *et al.*, 1959).

The foregoing illustrates beautifully the pitfalls of identifying substances by limited paper chromatography. One must conclude that more detailed corroborative evidence will be necessary to prove the existence of such coumarins in grapes; however, the oxidation and isomerization described should occur in wine, and substituted coumarins corresponding to the existing hydroxycinnamic acids can be expected to arise. Such hydroxyl-inserting reactions during analysis or in wine might also account for the *o*-hydroxy derivatives, such as γ-resorcylic acid and gentisic acid, which have been reported in grapes or wines in some instances and not in others. Both heat and activity by yeasts and bacteria have been found to convert *p*-coumaric acid and ferulic acid in whiskey mash to the respective 4-ethyl and 4-vinyl phenols and guaiacols (Steinke and Paulson, 1964).

As would be expected, tyrosine is found in grapes, and at a low level in wine (Castor, 1953a,b; Beridze and Sirbiladze, 1964; Nassar and Kliewer, 1966). Free tyrosine occurs in grapes at about 20–60 mg/kg fresh weight, but 75–90% of this disappears during fermentation to wine. Tyrosine, as an amino acid, is not considered to be a part of this review; however, it can be converted to tyrosol by yeasts. The conversion of nonphenolic substances to phenolic substances by yeasts has been demonstrated most conclusively with tyrosol. Ehrlich (1907) showed the conversion of tyrosine to tyrosol by yeast. Other microorganisms, such as the lactic acid bacteria, are capable of similar transformations, and tyramine, *p*-hydroxyphenylacetic acid, or 3-*p*-hydroxyphenyllactic acid may also be produced (Grimmer and Wauschkuhn, 1939). However, like other "fusel oil" alcohols, tyrosol may be produced from synthetic media, free of tyrosine or other preformed sources of phenol (Stevens, 1961). Of course, it is well known that yeast does not require an external source of phenylalanine, tyrosine, or other benzenoid substance, but is capable of synthesizing these compounds (Lingens *et al.*, 1967).

The occurrence of tyrosol in alcoholic beverages has been studied most in beer and sake. Takahashi and Abe reported 400 mg/liter in sake as early as 1913, but, more recently (Otsuka and Hara, 1967), 89–136 mg/liter were found. McFarlane and Millingen (1964) found that 73% of the aromatic amino acids of wort appeared as the corresponding alcohols in beer. Levels of tyrosol are about 18–40 mg/liter in beer, and 3–17 mg/liter in ale (McFarlane and Thompson, 1964; McFarlane, 1965). Wine has been studied only recently (Nykänen *et al.*, 1966; Ribéreau-Gayon and Sapis, 1965) and has been shown by gas and paper chromatographic methods to contain 10–45 mg/liter. Yamamoto *et al.* (1966) reported that yeast fermentation of a prepared medium containing L-tyrosine produced, in addition to tyrosol, *p*-hydroxybenzoic acid, *p*-hydroxyphenyllactic acid, *p*-hydroxyphenylacetic acid, *p*-hydroxybenzaldehyde or *p*-hydroxyphenylacetaldehyde, and *p*-hydroxyphenylpyruvic acid. The first three acids named were isolated and identified by melting point, elemental analysis, and infrared spectra, in addition to the paper chromatographic techniques employed for all.

## B. Flavonoids

### 1. Catechins (Flavan-3-ols)

The type compound of the natural flavan-3-ol series is *d*-catechin itself:

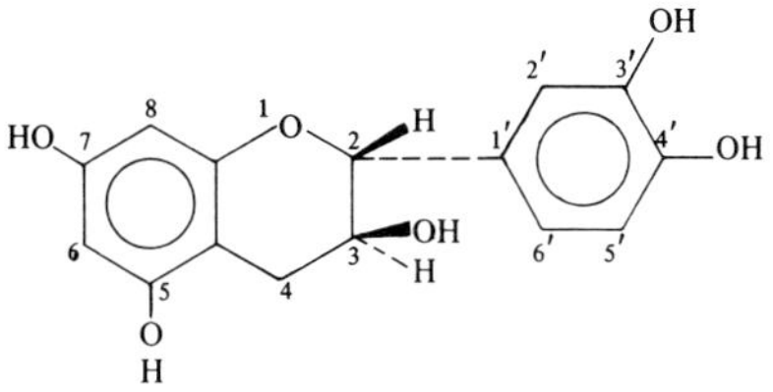

Members of this family of substances are sometimes referred to as catechols rather than catechins, especially in the German literature. Since 1,2-dihydroxybenzene is commonly called catechol in English, rather than pyrocatechol, this is confusing, a point that is mentioned only because it is poorly recognized even by major abstracting services. Catechin is the preferred trivial name. Although catechins had been known for many years (Freudenberg, 1956) their structural relationships to other flavonoids became clear only after hydrogenation of the known structure of cyanidin to *dl*-epicatechin (Freudenberg *et al.*, 1925). Thus, catechin and its epimer epicatechin are 5,7,3′,4′-tetrahydroxyflavan-3-ols, and gallocatechin and its isomer epigallocatechin have an additional 5′-hydroxyl. Other catechins are relatively rare botanically and have not been found in grapes, so they need not concern us here.

Carbons 2 and 3 in the catechin structure are asymmetric, and therefore four stereoisomers occur, *d*- and *l*-catechin representing one pair, and *d*- and *l*-epicatechin the other. The racemic mixtures *dl*-catechin and *dl*-epicatechin may have somewhat different properties in crystalline form than would be predicted from the properties of the respective antipodes. In fact, the individual isomers may crystallize in more than one form (Keller and Berger, 1946), and the formation of different hydrates also makes identification more difficult.

The configuration at the two asymmetric centers has been shown to be the trans arrangement with respect to the hydrogens on carbons 2 and 3 for the catechin pair, and cis for the epicatechin pair. The configuration and conformation of these and related compounds have been reviewed by Whalley (1962) and Clark-Lewis (1962). The absolute configuration of *d*-catechin is 2R:3S, by the nomenclature of Cohn *et al.* (1956), and *l*-epicatechin has the 2R:3R configuration. Obviously, *l*-catechin would be the 2S:3R form, and *d*-epicatechin the 2S:3S form. The absolute configurations in the gallocatechin–epigallocatechin set are analogous. For visualization, refer again to the absolute structure shown for *d*-catechin, and note that the C-2 phenyl substituent and the C-3 hydrogen are behind the plane of the paper, and the C-2 hydrogen and C-3 hydroxyl in front.

The specific structure has importance from a chemical and biochemical viewpoint. Epicatechin or epigallocatechin has the hydrogen on carbon 2 situated trans to the hydroxyl on carbon 3. They are readily dehydrated, a trans-elimination reaction, to flav-2,3-ens, but catechin or gallocatechin will break down in other ways before undergoing this reaction. Differences in sensitivity to oxidation would also be expected (Clark-Lewis, 1958). Epimerization in the catechins occurs rather readily in solution (Freudenberg, 1924). It involves inversion at carbon 2 because *d*-catechin (2R:3S) epimerizes to *d*-epicatechin (2S:3S), and *l*-epicatechin (2R:3R) to *l*-catechin (2S:3R). Racemization at carbon 3 can also occur under conditions permitting oxidation–reduction reactions (Clark-Lewis, 1958).

As a result of these facts it is possible for all four of the catechins to arise from inversions of any one, and the same would be true for the gallocatechins. Roberts and Wood (1953), in studies on tea, demonstrated that these catechin isomers can be separated by paper chromatography, and Mayer and Merger (1961) used column chromatography to separate the isomers from racemates. Although it is generally possible to separate optical isomers in chromatographic systems involving optically active stationary systems (cellulose in the case of paper), the separations in the catechin series seem to be completed unusually readily. This is attributed to the conformations of the molecules. The heterocyclic ring is not planar, as drawn above, but is puckered into a "halfchair" or "sofa" form (Roberts, 1956a; Whalley, 1962). Although the energy differences are apparently small and an equilibrium between alternative conformations no doubt occurs in a normal thermal range, there are believed to be enough differences in the most prevalent species to increase the differences in properties of the various catechin isomers (Clark-Lewis, 1962; Whalley, 1962; Roberts, 1956b; Haslam, 1966). The evidence indicates that the B-ring of the *d*- and *l*-catechins is predominantly at an angle (axial) and in the epicatechins is predominantly equatorial, in a plane parallel with the other two rings. This more angular, less compact shape of catechin and gallocatechin evidently gives these molecules a greater mobility or higher partition coefficient in butanol-type paper chromatographic systems than epicatechin or epigallocatechin. Although little or no difference occurs between the *d*- or *l*-forms of a single epimer (e.g., *dl*-catechin) in the usual partition systems, there is appreciable difference in single-phase adsorption systems such as 2% acetic acid in paper chromatography, although perhaps not enough for identification in mixtures (Haslam, 1966). The forms predominant (if not exclusive) in nature, *d*-catechin, *d*-gallocatechin, *l*-epicatechin, and *l*-epigallocatechin, adsorb less tightly (move faster) on cellulose than do their respective antipodes, *l*-catechin, *l*-gallocatechin, *d*-epicatechin, and *d*-epigallocatechin (Roberts, 1956b; Roux *et al.*, 1961).

The most extensive study of the catechins in grapes and wine has been that of Durmishidze and co-workers, in Russia, much of which has been summarized in one of only two known published books devoted exclusively to the tannins and anthocyanins of grapes and wine (Durmishidze, 1955). (The other is by Averna Sacca, 1906.) Most of the Russian studies on grape phenols were based upon the initial preparation of an amorphous solid containing the phenolic components. This amorphous "tannin" was prepared by steaming the plant material 20–30 minutes, air-drying it, grinding, removing chlorophyll, etc., with chloroform, and macerating the remaining solids with water-saturated ethyl acetate for 15–20 days at room temperature. The ethyl acetate was concentrated under vacuum and $CO_2$ at 40°C, and the tannins were precipitated by adding chloroform. The resultant colorless to pinkish astringent solid was freed from solvents and used for further tests. Prior studies indicated that ethyl acetate was capable of extracting nearly all the tannins from fruit parts, so that only a few percent more could be extracted by water, 95% ethanol, or 1% KOH.

From such preparations Durmishidze isolated *d*-catechin by procedures including lead acetate precipitation, and proved its presence unequivocally by optical rotation and derivative-degradation studies. He was able to isolate crystalline *d*-catechin from leaves, roots, and all cluster parts of "Saperavi" (red) and seeds and stems of "Mtsvane" and "Rkatsiteli" (white) grapes.

It was not possible in the studies reported (Durmishidze, 1955) to crystallize it, but by fractional precipitation of lead salts and extraction with ether a second substance was isolated from grape seeds. This material was free of gallic acid or *d*-catechin before tannase hydrolysis, but they were reported in the products of tannase hydrolysis and the material was considered to be *d*-catechin gallate. All the bound gallic acid was assumed to occur in this form, and bound gallic acid (liberated with tannase) was demonstrated in leaves, stems, skins, and seeds of "Saperavi" and seeds of "Rkatsiteli" and "Mtsvane" varieties.

A crystalline substance was isolated from a neutral lead acetate fraction of grape tannins which gave tests for pyrogallol moieties but did not liberate gallic acid upon hydrolysis. It oxidized readily, and most of the study of it was devoted to the methyl and acetyl derivatives. Both the original substance and the acetyl derivative were levorotatory. The substance was therefore identified as *l*-gallocatechin, the same as had been found in tea. It was identified in leaves and seeds of the "Saperavi" variety, and, based on less detailed studies, it appeared to be present generally in the plant.

Other studies, such as unsuccessful efforts to quantitatively combine the optical rotations of the substances separated and obtain that of the original preparation, convinced Durmishidze and co-workers that there were other substances present. Paper chromatographic studies were begun in an effort to identify them. It is stated that the standards were *d*-catechin, *l*-gallocatechin, and *d*-epicatechin gallate recovered from grape seeds. The apparent shift to *d*-epicatechin gallate in the naming of *d*-catechin gallate is not explained. The other standards used were *dl*-gallocatechin isolated from tea and "*dl*-catechin" obtained by epimerization of *d*-catechin according to the method of Freudenberg (1924). The major substance expected from this epimerization would be *d*-epicatechin. This and other apparent discrepancies make it difficult in the light of presently known absolute structures to interpret some of the identity assignments made by the Russian workers.

Durmishidze (1955) reports, from paper chromatography of grape berries and various vine parts, *l*-gallocatechin, *dl*-gallocatechin, *d*-catechin, *dl*-catechin, and *d*-epicatechin gallate. In wine (Nutsubidze, 1958a,b; Gerasimov and Smirnova, 1964) the same substances were reported except that *l*-gallocatechin was not reported in all instances. In a summary report (Durmishidze, 1966), the complete list of catechins in wine is given as *d*- and *dl*-catechin, *d*- and *l*-gallocatechin, *d*-catechin gallate, and *d*-epicatechin gallate. In the light of other work it is possible to reinterpret some of those assignments. The original work was concurrent with early work being done on tea catechins (Stahl, 1962). Bradfield and Bate-Smith (1950) showed that the substance previously called *l*-gallocatechin from tea should be designated *l*-epigallocatechin. Durmishidze (1951, 1955) stated that his *l*-gallocatechin was the same as that in tea, and the data reported (for example, specific rotation in acetone

−50°, acetate melting point 195°–200°C with specific rotation of −14°, ultraviolet absorption maximum 271 m$\mu$) compare well with those reported for *l*-epigallocatechin and not with the corresponding gallocatechin (Herrmann, 1959). It therefore seems clear that the "*l*-gallocatechin" of Durmishidze, and presumably that of later Russian work, is *l*-epigallocatechin.

Steam treatment, prolonged extraction, and lead precipitation are all conducive to isomerization and might explain reports of some of the unexpected isomers; however, it appears that, once paper chromatography came into use, the actual optical properties were not checked either in the later Russian work or in that by others. Dependence has been placed on comparison with previously identified standards. Chromatograms drawn by Durmishidze (1955) and Nutsubidze (1958a) show single discrete spots designated, in order of increasing mobility in *n*-butanol-acetic acid-water (4:1:3), to be *l*-gallocatechin, *dl*-gallocatechin, *dl*-catechin, *d*-catechin, and *d*-epicatechin gallate. The major isomerization which would be expected would be epimerization at carbon 2. If the forms designated *dl* were truly the less likely racemate at carbon 3, the spots for *d*- and *dl*-catechin should coincide or the *dl* should show one similar and one additional spot, regardless of the system. The same argument holds for the *l*- and *dl*-gallocatechin set. It therefore appears probable that the reported "*dl*-catechin" is epicatechin. It has the expected lower $R_f$ than *d*-catechin in this solvent on the chromatograms drawn. Durmishidze (1955) notes that Rodopulo (1950) had reported isolation of epicatechin from wine, but correctly noted that the melting point reported was low for epicatechin, although it was high for *d*-catechin.

We have already explained the redesignation of "*l*-gallocatechin" to *l*-epigallocatechin. On the Russian paper chromatograms, "*dl*-gallocatechin" ran faster and was shown as a separate discrete spot. It therefore seems that this was gallocatechin, but the optical isomeric form could not be considered proven by the data available. The designation of the gallates as *d*-catechin gallate or *d*-epicatechin gallate is consistent with the expected epimerization but inconsistent with the forms expected from work with other plants, especially tea.

Other workers have reported flavan-3-ols in wine. Lavollay *et al.* (1944; Lavollay and Sevestre, 1944) reported catechin and epicatechin in grapes or red wine. The predominant test given, however, involved reaction of ether or ethyl acetate extracts with concentrated sulfuric acid in an acetone solution containing a small amount of potassium persulfate or other oxidant. The production of a red-violet substance was concluded to be the result of oxidation of the catechins to cyanidin. With the possibility of extractable leucocyanidins, which would readily convert to cyanidin under the same conditions, this cannot be considered definite proof. Hennig and Burkhardt (1957a,b, 1958) presented rather detailed paper chromatographic evidence that *d*-catechin, *l*-epicatechin, and *l*-epigallocatechin occurred in the seeds, stems, and skins of "Riesling" grapes and young wines. Herrmann (1958, 1963a,b) reported *d*-catechin and *l*-epicatechin in wine grapes based on paper chromatography. Pifferi (1966) identified gallocatechin gallate, *l*-gallocatechin, *d*-catechin, and *l*-epicatechin in red wines by paper chromatography. Ribéreau-Gayon (1964a) states that *d*-catechin and *l*-gallocatechin were identified with certainty in his research with wine, but Ribéreau-Gayon and Stonestreet (1966a) report paper chromatographic identification of *d*-catechin, *l*-epicatechin, and *dl*-gallocatechin. Singleton *et al.* (1966a), using cochromatography, comparisons of properties, and, in the case of the gallate, hydrolysis to the constituent moieties, tentatively identified epicatechin, catechin, and epicatechin gallate in extracts of seeds of many varieties of grapes. Joslyn and Dittmar (1967) reported *d*-catechin and *l*-epicatechin in extracts of "Pinot blanc" skins and seeds, and gallocatechin in the seeds.

Jurics (1967a,b) boiled crushed grapes 15–30 min and examined them paper chromatographically with 10 solvents and with 14 spot-detecting agents to identify both *d*-catechin (101 mg/kg) and *l*-epicatechin (40 mg/kg). In 16 other fruits, not more than half as much *d*-catechin was found although several had similar *l*-epicatechin content. Iankov (1966) examined stems of 4 grape varieties chromatographically and reported *l*-catechin in 2, *l*-epicatechin gallate in 3, and *dl*-catechin, *l*-gallocatechin, *dl*-gallocatechin, and *l*-epigallocatechin gallate in all. Segal and Grager (1967) report paper chromatographic indentification of *d*-catechin, *dl*-catechin, *l*-gallocatechin, *dl*-gallocatechin, and *d*-epicatechin gallate. It would appear that these identifications require reinterpretation as previously described.

It is very probable that the 2R configuration (therefore *d*-catechin, *l*-epicatechin, *d*-gallocatechin, and *l*-epigallocatechin) is the only configuration occurring naturally (Weinges, 1964). The naturally occurring gallic acid depsides of catechins appear to be limited to the 3-galloyl esters of the *epi* forms (Roberts, 1956a). In view of the contrary reports with respect to grapes and the fact that the optical rotation has actually been determined in relatively few cases, it seemed necessary to prove the optical form present in grapes. Continuation of previous studies (Singleton *et al.*, 1966a; Su and Singleton, 1968, unpublished data) has shown by specific rotations on the isolated crystalline substances and derivatives that the expected forms *d*-catechin, *l*-epicatechin, and *l*-epicatechin gallate were the major flavan-3-ols occurring in extracts from seeds of many varieties of grapes. This confirmation of Durmishidze's identification of *d*-catechin, and the convincing evidence that *l*-epigallocatechin was isolated by Durmishidze, suggests that in grapes also the gallocatechin series should be the normal series *d*-gallocatechin and *l*-epigallocatechin. We therefore believe that, for example, the substances identified by Iankov (1966) are actually, respectively *l*-epicatechin, *l*-epicatechin gallate, *d*-catechin, *l*-epigallocatechin, *d*-gallocatechin, and *l*-epigallocatechin gallate. Similarly those of Segal and Grager (1967) appear to be, respectively, *l*-epicatechin, *d*-catechin, *l*-epigallocatechin, *d*-gallocatechin, and *l*-epicatechin gallate. There is evidence (Durmishidze, 1955) for dimeric or larger polymers of catechins in grapes. These will be discussed as tannins.

### 2. Flavan-3,4-diols, Anthocyanogens, and Condensed Tannins

The nomenclature in use for flavonoids, and for this group of flavonoids in particular, is still in a state of flux in spite of recommendations for a more systematic approach (Freudenberg and Weinges, 1960). Rosenheim (1920b), who worked with grape leaves, first used the name "leuco-anthocyanins." Since they are not glycosides, at least in most cases, the name has been modified to leucoanthocyanidins. The name proanthocyanidins has been proposed, and anthocyanogens, a term widely used in the scientific literature pertaining to beer, seems to us the preferable trivial name for the general class of substances readily convertible to anthocyanidins. It appears in indexes near anthocyanin; it implies a chemically masked form rather than a metabolic precursor or formation promoter. Leucoanthocyanidin is probably still the most prevalent usage, and appears likely to remain so for some time. More specific chemical names are to be preferred, of course, where possible.

Most (but not all) of the anthocyanogens studied in detail have turned out to be, or to include, flavan-3,4-diols as a major structural unit. The flavan-3,4-diols are equivalent to the catechin structures, but with one more alcoholic hydroxyl on carbon 4. Thus, they have 3 asymmetric carbons, and each substitution pattern, such as a cyanidin-forming (leucocyanidin, procyanidin?, cyanidinogen?) 5,7,3′,4′-tetrahydroxyflavan-3,4-diol, could occur as eight isomers. All of these may be separable on paper chromatograms by analogy with catechins, but all would give identical anthocyanidin since it has no asymmetry. The stereochemistry of substances of this class in grapes has not yet been investigated, so it need not be discussed, but it will undoubtedly become a subject for attention in the future, as it has for other plant systems (Drewes and Roux, 1964).

Very old descriptions can be found in the literature which, interpreted with hindsight, show the presence of anthocyanogens in grapes or wine. For example, Mulder (1857) describes Faure's experiments on the conversion of a colorless extract from wine into a red-violet solution. Laborde (1908a,b, 1908-1910, 1909) demonstrated that certain tannic substances, notably those from grape skins or various grapevine parts, produce red pigments very similar to those normally present in red grapes or wines if the material is autoclaved for 30 min in 2% HCl at 120°C. Ripe white grapes and green grapes or young sprouts of both types gave the reaction. The red pigment was concluded to come from nonglycoside tannoids precipitable with mercuric acetate, since no sugar could be demonstrated in hydrolysates. Oak gallotannins did not give the reaction. Malvezin (1908) confirmed Laborde's experiments with "Muscat blanc" grapes and emphasized that oxidation was required to produce a high level of the red pigment.

Rosenheim (1920b) described extracting dried young leaves of "Black Hamburgh" and "Esperione" grape varieties, which are red, in 1% hydrochloric acid with amyl alcohol until all but a trace of red pigment was removed. Boiling the residual solution with 20% HCl produced an intense wine-red pigment which was readily extractable into amyl alcohol. The total recoverable anthocyanidin was 48% from the preformed anthocyanins and 52% from his leucoanthocyanin. He observed the same reaction with unripe purple grapes and with ripe white grapes which, he noted, had a high leucoanthocyanin content. Rosenheim postulated that the colorless substance was a 4-glucoside of the colorless anthocyanidin pseudobase, equivalent to a 3,4-dihydroxyflav-2,3-en. Presumably because he coined the name and postulated a plausible explanation, Rosenheim is often credited with discovering anthocyanogens. In fact, he referred to similar observations by Willstätter and Nolan (1915), but evidently was unaware of Laborde's studies.

Laborde (1908a) had shown that there were two types of anthocyanogen in grapes, one soluble in concentrated ethanol and the other soluble only in more aqueous solutions. Combes (1922) showed that a substance existed in green "Frankenthal" grapes which could be converted into anthocyanidin if it was dissolved in amyl alcohol and heated with acid. If it was heated in acidic aqueous solution, a red phlobaphene formed. The original substance was a condensed tannin by several classic tests, including precipitation with formaldehyde in acid solution, with alkaloids, and with bromine water, and therefore was not a pseudobase.

Robinson and Robinson (1933) followed up Rosenheim's work, and report isolating a white amorphous leucoanthocyanin from muscat grape skins. "Corinth" ("Zante" currant, *Vitis vinifera*) leaves and "Almeria" variety flesh and skins contain anthocyanogens which produced cyanidin, as shown by the series of qualitative extraction tests developed by the Robinsons. The "Almeria" seeds were noted to be especially rich in leucocyanidin.

As a part of a survey of the occurrence of anthocyanogens in plants (Bate-Smith, 1954b; Bate-Smith and Lerner, 1954; Bate-Smith and Metcalfe, 1957; Bate-Smith and Ribéreau-Gayon, 1959) leucocyanidins were reported in skins and seeds of purple grape, and both leucocyanidin and leucodelphinidin in leaves or twigs of *V. vinifera*. *V. voinieriana* seeds were negative, but other *Vitis* were usually positive, with *V. vinifera* seeds giving strong leucocyanidin and leucodelphinidin tests. Masquelier and Point (1956) isolated from berry skins, twigs, stems, and leaves of 5 white *V. vinifera* varieties a light-brown amorphous leucocyanidin. This preparation was soluble in water, alcohol, and acetone, but insoluble in ethyl acetate. Cyanidin, among the degradation products of this preparation in acid, was identified by chromatography. Phloroglucinol and protocatechuic acid were the products of alkaline fusion. The structure 5,7,3′,4′-tetrahydroxyflavan-3,4-diol was proposed.

As part of a study of anthocyanins of grapes and wines (Ribéreau-Gayon, 1957; Ribéreau-Gayon and Ribéreau-Gayon, 1958) leucocyanidins were noted on paper chromatograms. The leucocyanidins were found to be a major portion of the total phenols of red wines but were very low in wines prepared from juice alone. The skins were relatively low in leucoanthocyans, and the seeds were major sources of these substances for red wines. Several other studies have demonstrated and quantitatively estimated the anthocyanogens in grapes and wine without appreciable study of the chemical nature of the substances involved.

Somaatmadja *et al.* (1965) detected and isolated by paper chromatography two anthocyanogens in aqueous alcholic extracts of "Cabernet-Sauvignon" skins and two in similar seed extracts, all of which they believed to be different. One from seeds was too small in amount for appreciable study. Upon conversion the other gave cyanidin. One of the skin anthocyanogen fractions also gave cyanidin but had a broad, weak absorbance in the 258–283 mμ region, compared with the 282-mμ maximum for the seed fraction. The other skin preparation gave delphinidin upon conversion. Infrared spectra also were interpreted as showing differences between the fractions, but physicochemical characterization of the fractions was not carried further.

Ito and Joslyn (1964) studied a purified anthocyanogen preparation from "French Colombard" grape pomace which had been prepared by repeated butanol extraction and precipitation with ether. Treatment of this preparation with heat in more concentrated acid solution gave cyanidin, but boiling for 20 minutes in 0.1 *N* HCl gave spots on paper chromatograms identified as *d*-catechin, *l*-epicatechin, *l*-epicatechin gallate, protocatechuic acid, phloroglucinol, kaempferol, quercetin, and an unknown phenol. With seeds and skins separated from "Pinot blanc" grapes, similar preparations were not reported to yield flavonols (Joslyn and Dittmar, 1967).

Peri and Cantarelli (1963) isolated from seeds of white "Trebbiano" grapes a leucocyanidin stated to be essentially chromatographically homogeneous. It was separated from ether-soluble catechin by solubility in ethyl acetate and insolubility in ether.

In studies of 12 varieties of grapes (10 *V. vinifera*, 2 *V. labrusca*; 5 red, 7 white) the total extractable phenolics were recovered, and representative two-dimensional paper chromatograms were presented of seeds, skin, pulp, and juice (Singleton *et al.*, 1966a). Although qualitative and quantitative differences were found by variety and ripeness, the chromatograms, particularly those from seeds, had a number of phenolic compounds in common in all samples of each tissue. Column-partition chromatographic separation of the seed extracts removed a more soluble catechin fraction and gave an ethyl-acetate-mobile fraction containing six of the substances giving discrete spots on the paper chromatogram of the typical grape seed. This group of substances, isolated as a light-tan amorphous solid insoluble in ether, gave a strong test for acidic conversion to cyanidin which was identified by paper chromatography.

A second, nondialyzable amorphous light-brown material from an alcohol wash of the column was free of these components and soluble in aqueous alcohol but insoluble in ethyl acetate. It gave a very strong conversion to cyanidin, and a portion purified by dialysis had an apparent molecular weight of 5300 by temperature-differential vapor-pressure osmometry (Rossi and Singleton, 1966a,b).

Joslyn and Dittmar (1967) prepared four cyanidin-yielding anthocyanogens, two from seeds and two from skins of "Pinot blanc" grapes. In each case one was extractable with butanol and not with ethyl acetate, and the other was extracted by ethyl acetate. Each of the four had distinguishing properties. Methyl derivatives and crystalline acetates were prepared from the two seed fractions. Analytical and spectral data suggest marked differences between the two preparations. Apparent molecular weights were about 1400 by vapor-pressure osmometry and about 700 by freezing-point depression. The ethyl-acetate-soluble leucocyanidin from the skins gave *d*-catechin, *l*-epicatechin, and additional unidentified substances as well as cyanidin upon reaction in hot acidic butanol. The butanol-soluble leucocyanidin from skins gave cyanidin and 57.5–65.5% glucose, the only sugar. This apparent procyanidin diglucoside gave a molecular weight of about 700 by freezing-point depression. The activation energies of the cyanidin conversion reactions indicated that a grape phlobatannin preparation was a larger molecule than any of these preparations (Joslyn and Goldstein, 1964b), and the skin leucocyanidins were larger than those isolated from seeds. Masquelier *et al.* (1965) fractionated the leucoanthocyanidin from wine by chromatography on cellulose, fractional precipitation with salt, and ethyl acetate extraction. In all cases two fractions were recovered, which were concluded to be a monomeric and a polymeric portion. Ribéreau-Gayon and Stonestreet (1966b) concluded from the low mobility on paper chromatograms of ethyl-acetate-soluble leucoanthocyanidins from red wine that they were small polymers of 2–8 flavonoid units.

It is obvious that the definitive structures for the anthocyanogens of grapes and wine have yet to be determined. It is equally clear that the problem is complex. From the available data it appears probable that there are smaller, but probably dimeric or substituted rather than simple, flavan-3,4-diols present in grapes and in wine. Larger molecules of anthocyanogens are also present, and these also appear to be composed of more than the simple polymeric equivalent of flavan-3,4-diols. The rather sharp apparent differentiation into these two groups, small and large, suggests that there is not a simple random series of all possible degrees of polymerization but, rather, specific anthocyanogen entities of two (or more) types. This in turn suggests enzymatically controlled production and not simple chemical polymerization either in the cell or as artifacts (although artifacts undoubtedly are easily produced and are an unavoidable part of aged wines).

The tannins of grapes or wine have been studied, along with other vegetable tannins, for a very long time. Grape seed extracts contain tannin in the leather-maker's sense. Davy (1819) recommended grape seeds bruised in cold water or pounded oak gallnuts as typical substrates for demonstrating tannin reactions, such as astringent taste, precipitation with gelatin, and so on. He also said that the purest species of tannin is that obtained from the seeds of the grape, presumably because it gave him the whitest precipitate with gelatin. Nachet (1829) and Rabak and Shrader (1921) reported satisfactory use of grape seed tannin in tanneries for making good leather with excellent color, but grapes have not been a major or continuing source of supply for that purpose.

Natural vegetable tannins may be classified as hydrolyzable, with gallotannic acid from oak galls the classic example, or as condensed or phlobatannins. Owing perhaps to the fact that tannic acid from oak galls was so well-known that persons loosely referred to all tannins as tannic acid, or perhaps to the recovery of small amounts of gallic acid, the statement can be found in many references that tannic acid occurs naturally in grapes or wine. This is in spite of the fact that the true tannins of grapes and wine have been known for many years to be condensed tannins (Thudichum and Dupré, 1872; Dekker, 1913). Since oak gallotannic acid may be added to wine under certain processing conditions, it might be found in wine, but tannins of the pentadigalloyl glucose type or any other of the hydrolyzable tannin types do not occur in wine or grapes in appreciable amounts. Durmishidze (1955) reviewed the problem and initiated new studies to answer the question. He found small amounts of bound (only) gallic acid in grapes, as noted previously, but attributed the amount produced to hydrolysis of the catechin gallates present. The total (bound) gallic acid was no more than 5% of the total phenolic substances he found in grape seeds, and considerably less in leaves, skin, or stem tannin. Careful studies of sugar released by hydrolysis led him to the conclusion that grape tannins contain no sugar. Certainly, much too little was present to account for any appreciable galloyl-glucose or ellagitannin level.

Condensed or flavonoid tannins like those of the grape are sometimes called phlobatannins, from their tendency to convert, when heated in dilute aqueous acid, into brown or red-brown insoluble "phlobaphenes." Catechins convert to similar phlobaphenes, and for many years it was believed that the condensed tannins were polymers of catechins. The phlobaphene precipitates from natural condensed tannins have also been called tannin-reds, a revealing name since we now know that flavonoid tannins of plants are very usually if not invariably anthocyanogenic substances, giving the reactions expected of flavan-3,4-diol polymers (Bate-Smith and Swain, 1953; Geissman, 1963). Tannin preparations from grapes are anthocyanogenic but appear to include other moieties in addition to simple cyanidin precursors.

A true vegetable tannin in the leather-tanning sense is any polyphenolic constituent capable of turning hide into leather. For best results this reaction requires a molecular-weight range of about 500–3000 (White, 1957). Whether or not a phenolic substance is a tannin, then, depends on its molecular weight. Since dimeric flavonoids would exceed 500 molecular weight, all these dimeric and larger substances we have been discussing as catechins and anthocyanogens are also condensed, astringent tannins (Joslyn and Goldstein, 1964a; Freudenberg and Weinges, 1962). The same reactions, mainly hydrogen bonding, which enable tannins to convert hide protein into leather cause tannins to precipitate from wines with proteins, polyamides, poly(vinylpyrrolidinone), etc. (Singleton, 1967a,b). This too is related to molecular weight and geometry. The phenolic substances of wine can be modified by partial and, to a degree, selective removal of tannins by such agents (Rossi and Singleton, 1966b). Such reactions have structural implications. For instance, an ethyl-acetate-insoluble condensed tannin from grape seeds with an apparent molecular weight of 5300 was precipitated by the gelatin added in greater proportion than was an ethyl-acetate-soluble leucoanthocyanin which, by analogy with Joslyn and Dittmar's (1967) work, should have a molecular weight of 600–1500. In studies with hydrogen-bonding synthetic polymers in chromatographic columns, however, much more of the smaller molecule was adsorbed. Such work has many implications and applications in commercial wine processing and in separation research, but only limited conclusions can yet be drawn from a structural viewpoint.

Durmishidze (1955) found that the total phenolic fraction from grapes soluble in ethyl acetate had an apparent average molecular weight of 370–550. Subtracting the identified amounts of catechins enabled him to deduce that a considerable fraction of this preparation consisted of dimeric units which he considered paired catechins, but noted their convertibility to cyanidin. He also reported that a considerable portion of tannin was soluble only in alkali, and presumably of larger molecular weight, in grapevine parts other than fruit, but a small percentage of such material was found in the grapes themselves. Wucherpfennig and Franke (1964) separated wine phenols by gel permeation chromatography and found a fast-moving, presumably large-molecular fraction, and a much greater amount of a somewhat smaller-molecular fraction. Somers (1966a) used gel filtration techniques and a specially prepared gel to get quantitative recovery of the wine phenols processed. From consideration of vanillin-HCl versus Folin Denis colorimetry (Goldstein and Swain, 1963) and the exclusion limits of the gel he concluded that most of the tannins of a 1959 wine were in the molecular-weight range 2000–5000, but some went up to 50,000. Ribéreau-Gayon and Stonestreet's (1966a,b) postulation of 2- to 8-unit polymers would give molecular weights of about 600–2300.

Masquelier *et al.* (1965) studied tannic anthocyanogens of wine by three fractionating techniques, cellulose column chromatography, salt precipitation, and solvent extraction and paper chromatography. The data are interpreted as showing the presence of monomeric leucoanthocyanidins, polymers of up to about 5 monomeric units, and larger anthocyanogenic tannins. Diemair and Polster (1967a,b) fractionated the protein-precipitable tannins of red wine by methods including thin-layer and polyamide or gel permeation column chromatography. The polymeric tannins separated from anthocyanins, and monomeric flavonoids were further fractionated into a brown fraction with 5,000–10,000 and a blue fraction with 1000–5000 estimated molecular weight. The condensed, anthocyanogenic, hide-powder-precipitable tannin from *V. rotundifolia* leaves has been studied as an enzyme inhibitor (Porter *et al.*, 1961; Porter and Schwartz, 1962; Bell *et al.*, 1965). It was isolated by caffeine precipitation and had a molecular weight by ultracentrifugation of 17,000-20,000. It appears that there are sizable flavolan polymers in grape tissues and that, owing to relative insolubility, easy precipitability alone or with proteins from disrupted tissue, and inability to dialyze through membranes, these large molecules may be missing from wines or other preparations from grapes.

The occurrence of anthocyanin-like subunits within the polymeric tannins of aged red wine (Somers, 1966a; Diemair and Polster, 1967a,b) has been known indirectly for some time. For example, Ribéreau-Gayon and Peynaud (1935) showed that the red pigment of wine exists in two forms, one which will pass through a dialysis membrane and one which will not. The large, nondialyzable form is precipitable with gelatin, and forms haze in chilled wine. Since the simple anthocyanins of wine gradually convert during aging to a more and more associated or polymeric form (Berg, 1963; Ribéreau-Gayon and Stonestreet, 1965), it had been believed that the anthocyanin-bearing "tannins" of wine were artifacts rather than grape constituents.

This question has been reopened by work of Somers (1967a), who reported the isolation of a red tannin fraction from the skin of the "Shiraz" grape having a molecular weight of 2000 or more. This fraction represented 42% of the weight of phenolic substance recovered from the skin of one berry and 12% of the pigment (547 m$\mu$ absorbance). These results indicate biochemical incorporation of anthocyanidin moieties in red grape skin tannin. However, such units certainly do not appear to be present in tannins from white grapes or anthocyanin-free parts of red grapes (seeds for example). This fact, the observation that the "polymeric" red-brown material at the origin represents a smaller and smaller part of the total pigment of fresh grapes with increasingly mild and rapid paper chromatographic procedures, and the possibilities for change during the processing of the sample, leave room for doubt that this red tannin is a grape metabolic product instead of an interesting further example of the tendency of condensed tannins to react with themselves and other phenols to produce larger molecules. It is noteworthy that Diemair and Polster (1967b) reported no reaction between their red wine tannin fractions and vanillin in hydrochloric acid. It appears that artificially produced polymeric phlobaphenes are unreactive in this test, whereas natural anthocyanogenic tannins do react (Hillis and Urbach, 1958).

Although the presence of anthocyanidin moieties in native grape tannin appears to us to be questionable, the release of bound catechins from grape tannins under mild conditions indicates their presence in these polymers. The detection of *l*-epicatechin gallate, a relatively labile substance, as a degradation product from grape tannin by Ito and Joslyn (1964) has been verified by Su and Singleton (1968, unpublished data). The relative proportions of cyanidin and catechins which are produced by degradation of grape tannins are variable, depending upon the reaction conditions (Ito and Joslyn, 1964; Joslyn and Dittmar, 1967; Peri, 1967a,b). King (1955) has suggested that leucocyanidin may disproportionate into catechin and cyanidin. A relatively large production of cyanidin compared to other products is produced from anthocyanogens under oxidative conditions (Jurd, 1966a; Peri, 1967a,b). In this context the occurrence of *l*-epicatechin gallate bound in grape tannin suggests that this unit was incorporated as such into the tannin because, with one exception of which we are aware (Santhanam *et al.*, 1965), leucocyanidin gallates have not been reported in nature.

### 3. Anthocyanins

The pigments of red, blue, purple, and "black" grapes have received the most study of any class of phenolic substance in grapes, and pigments have perhaps received more study in grapes than in any other plant. Since, in addition, the subject has been summarized thoroughly (Ribéreau-Gayon, 1959, 1964a; Webb, 1964; Harborne, 1967), it is somewhat condensed here.

Chromatographic methods have shown us that the typical grape skin contains several anthocyanidin derivatives and that it is very difficult to achieve chromatographic purity of any one pigment by combinations of methods such as lead precipitation, immiscible solvent extraction from water, or picrate crystallization. Without chromatographic procedures, progress was laborious and limited, but considerable work was done. The properties of these red, water-soluble pigments were studied as a class, and their relation to "tannin" or colorless phenolic substances of grapes and wine was realized. Much practical and observational data were determined, but progress in structure identification was limited owing to the complexity of the pigment mixtures in grapes (Maumené, 1874). This complexity led Gautier (1892a,b, 1911) to conclude that each grape variety had its own characteristic "ampelochroic acid" pigment. He and Carpentieri (1908) suggested that pigments could be used to differentiate species and varieties of grapes. Sostegni (1897) found pyrocatechol and phloroglucinol among the products of alkaline degradation of grape anthocyanins.

Willstatter and co-workers distinguished European grape anthocyanin from other anthocyanins by various reactions, isolated it as a crystalline product, identified it as a monoglucoside of a dimethyl delphinidin derivative, and named it oenin (Willstätter and Everest, 1913; Willstätter and Zollinger, 1915, 1916). The isolation of syringic acid from degradation of oenin (Anderson and Nabenhauer, 1926; Karrer and Widmer, 1927) showed that malvidin was the constituent anthocyanidin and proved its structure to be a 3,4′,5,7-tetrahydroxy-3′,5′-dimethoxyflavylium salt. Oenin (enin) was shown by synthesis to be malvidin-3-$\beta$-glucoside (Levy *et al.*, 1931).

The pigment of several non-European grape varieties was found to be similar. The isolate from "Isabella" variety was malvidin glucoside (Anderson, 1924; Anderson and Nabenhauer, 1926). Those from "Ives," "Seibel," "Clinton," "Norton," and "Concord" varieties were apparently mixtures of malvidin and petunidin glucosides, and in some cases degradation indicated the presence of *p*-hydroxycinnamic acid and diglucosides (Anderson, 1923, 1924; Anderson and Nabenhauer, 1924, 1926; Shriner and Anderson, 1928). Levy *et al.* (1931) noted that the original isolate from "Fogarina" grapes (Parisi and Bruni, 1926) was about 45% malvidin derivative and also contained petunidin and delphinidin derivatives. Cornforth (1939) found malvidin to be the anthocyanidin of an Australian wild grape, *Vitis hypoglauca.* Brown (1940) presented evidence that in *V. rotundifolia*, variety "Hunt," the pigment muscadinin was the 3,5-diglucoside of petunidin (3′-methoxydelphinidin).

Robinson and Robinson (1932) reported, following a study by extraction and color tests, that "Hanefoot" (probably a red variant of "Muscat of Alexandria," usually called "Hanepoot") grapes contained cyanidin-3-glucoside, and "Colmar" grapes contained enin. The anthocyanins found in surveys of plants were summarized (Lawrence *et al.*, 1938, 1939) to show malvidin-3,5-diglucoside in *V. heterophylla* fruit, but malvidin-3-glucoside in all other grapes investigated. Cyanidin glycosides figured prominently in *Vitis* leaves. It was noted that fruit pigments in the Vitaceae belonged in the complex, acylated group and were not single, easily identified anthocyanins (Robinson and Robinson, 1934).

Column chromatography had been used for some time to produce pigment patterns with normal wines, especially as an aid in detecting added dyes or prohibited blending in adulterated wines (Flanzy, 1935b; Mohler and Hämmerle, 1935, 1936; Gentilini, 1941; Ramos and de Oliveira, 1949), and paper chromatography was used later for the same purpose (Ruf, 1952). Sadow (1953) tried various methods for isolating and studying pigments from "Carignane" and "Zinfandel" grapes and wine and preparations from "Norton" made by Anderson (1923). Silicic acid column chromatography and countercurrent distribution were the best of the procedures investigated. Four major anthocyanin fractions were reported, but other, usually fugitive, fractions were noted. It was concluded, based upon comparison of properties with synthetic malvidin, petunidin, and delphinidin, plus the formation of glucosazone, that the major fractions were malvidin, malvidin-3-β-glucoside, delphinidin, and delphinidin-3-β-glucoside. The fraction identified as malvidin increased from 33 to 67 mg/liter of wine during one year of aging in bottles. The total of the four fractions was about 102 mg/kg of fresh weight for "Carignane," and 134 mg/kg of "Zinfandel," with the malvidin-3-β-glucoside fraction representing over half to two-thirds and only 12–20% of each of the other fractions present. Later work in the same laboratory, however, reported 6–12 bands on a slightly longer column with the same "Zinfandel" samples (Stasunas, 1955). Malvidin, delphinidin, and petunidin glucosides were reported present.

Sastry and Tischer (1952a,b) applied paper chromatography to the study of an acidic methanol extract from skins of "Concord" grapes. Three major bands were found, and evidence was presented that they represented malvidin-3-glucoside (enin), malvidin diglucoside (malvin), and free malvidin. Shortly thereafter Ribéreau-Gayon (1953) also applied paper chromatography to the study of grape anthocyanins. This has been the major tool in nearly all research on identification of grape anthocyans since that time.

Ribéreau-Gayon (1953) found in skins of the two *vinifera* varieties, "Cabernet" and "Merlot," ten pigments which gave similar paper chromatographic patterns in these two varieties, but the pattern of ten pigments found in the hybrids "Seibel 7053" and "5455" was different. The indicated possibility of differentiating pure *V. vinifera* varieties from those resulting from crosses with other species was of considerable commercial interest and was actively pursued. Despite reports to the contrary, there is now general agreement that *V. vinifera* does not produce diglucoside pigments whereas most if not all of the other species do, particularly those used in crossbreeding (Ribéreau-Gayon and Sudraud, 1957; Ribéreau-Gayon, 1959, 1963b; Webb, 1964; Deibner and Bourzeix, 1965). The reports that pure *V. vinifera* varieties contained diglucosides are attributed to misinterpretations or contaminations.

Pelargonidin apparently does not occur in *Vitis* species, but cyanidin, peonidin, delphinidin, petunidin, and malvidin occur commonly. No other anthocyanidins have been authenticated in grapes. In common with other natural systems the anthocyanidins do not appear to occur in detectable amounts in the free state, but only as glycosides. Owing to hydrolysis after disruption of the normal berry, anthocyanidins may occur in deteriorating berries or processed samples and have been reported in wines. Glucose is the only sugar which has been found regularly and demonstrated conclusively in grape anthocyanins. It apparently never occurs as other than a monosaccharide in grape anthocyanins. Fructosides and galactosides have been reported in isolated cases, but from available data these identifications are believed to be incorrect (Webb, 1964; Ribéreau-Gayon, 1964a,c). The pigments are frequently acylated. The cinnamic acids *p*-coumaric and caffeic acid have been shown to be the common acylating moieties. Reports of chlorogenic acid, malic acid, and tartaric acid as anthocyanin-acylating agents have not been verified, and are also believed to be incorrect (Webb, 1964; Harborne, 1967).

The paper chromatograms of the skin pigments from grapes or from very young red wine are quite complicated and usually show many anthocyans. The 3-monoglucosides of the 5 anthocyanidins alone would account for 5 spots. If the anthocyanidins were present in the experimental samples this would make 10 substances. Acylation of all of the glucosides with one agent would raise the normal complement to 10 and a second acylator to 15, excluding anthocyanidins. Non-*vinifera* varieties with diglucosides and two acylators could have as many as 30 anthocyanins without consideration of diacylation possibilities, etc. The anthocyan chromatographic pattern obtained with a given grape variety is evidently rather reproducible with careful technique, but there are likely to be considerable differences between varieties. That there are qualitative and relative quantity differences in the anthocyan complex among varieties is obvious from examination of a number of such chromatograms (Rankine *et al.*, 1958; Albach *et al.*, 1959; Ribéreau-Gayon, 1959). The patterns obtained from 167 *V. vinifera* varieties showed 2 to 9 anthocyans, with a rather similar pattern of 7 or 8 pigments being most prevalent. The wild *V. vinifera* fruit was very similar to the commercial varieties, showing a pattern of 7 pigments and lacking an eighth that was missing also in many cultivars. Most other *Vitis* species, as would be predicted from the usual diglucoside presence, have a greater number of different anthocyans. For example, Ribéreau-Gayon (1959) reported 17 for *V. lincecumii*; Rice (1965) finds 17 paper chromatographic spots, four of which give two anthocyanidins upon hydrolysis, in "Concord" (pure?

*V. labrusca*). However, Zamorani and Pifferi (1966) have reported 21 anthocyans in "Merlot" wine, a *V. vinifera* variety.

Examples taken from relatively recent work upon the identity of pigments in grapes may be illustrative of complexities and varietal differences. Akiyoshi *et al.* (1963) and Albach *et al.* (1963, 1965a,b) have extended the survey previously mentioned (Rankine *et al.*, 1958; Albach *et al.*, 1959) to the detailed chemical identification of pigments separated on paper chromatograms. The major (82%) pigment of the "Flame Tokay" variety was isolated by preparative paper chromatography, and shown by sequential partial hydrolysis to be a monoglycoside. The sugar liberated was shown to be glucose by cochromatography on paper with known sugars. The anthocyanidin had the paper chromatographic and spectral properties of cyanidin, and alkaline degradation gave phloroglucinol and protocatechuic acid. The position of glucose linkage to the cyanidin was deduced from spectral data, particularly the lack of fluorescence expected from a 5-glucoside, and from a comparison with known 3- and 5-glucosides produced from partial hydrolysates of cyanidin-3,5-diglucoside isolated from "Chrysler Imperial" roses. The same pigment, cyanidin-3-glucoside, was identified in "Red Malaga" and "Emperor" table grapes.

Albach *et al.* (1963), also by hydrolysis and identification of products, partial hydrolysis, alkaline degradation, and fluorescence interpretation, identified peonidin-3-glucoside in a variety, "Fresia," selected as rich in this component. They (Albach *et al.*, 1965a) then studied three chromatographically separated faster-moving bands from the variety "Tinta Pinheira." This variety was chosen for a relatively high content of these presumably acylated pigments. The fastest band was shown to be interconvertible into the second of the three, and this conversion was pH-dependent. The two forms were suggested to be the violet quinoid anhydrobase and the red oxonium salt. This double-banding appeared to be limited to the *p*-coumaroyl group of derivatives, evidently because of coincidence of a pH gradient on the paper with the region of the *p*-coumaroyl substances near the solvent front. Another possible isomerism, cis-trans of the acyl portion, has been reported by Birkofer *et al.* (1963) to give double-banding in some solvents.

The acyl groups were readily lost in acidic solution or, of course, alkaline, but were more stable in methanol without addition of acid. Partial hydrolysis studies showed that removal of the acyl group released *p*-coumaric acid from the fastest pair of bands of acylated pigments, and caffeic acid from the slowest band. The anthocyanins released were, in decreasing proportion, malvidin-3-glucoside, peonidin-3-glucoside, and small amounts of petunidin-3-glucoside and delphinidin-3-glucoside.

The proportions of the 4 glucosides within the *p*-coumaroyl, caffeoyl, and free glucoside fractions of the grape anthocyanins were about the same. By methylation and gas chromatography of the acyl-sugar it was shown, barring acyl migration, that the *p*-coumaroyl unit was on position 4 of the sugar of the acylated anthocyanidin-3-glucosides (Albach *et al.*, 1965b).Assuming the most probable $\beta$-linkage in the glucoside and the trans configuration in the acyl group, this would be the first completely elucidated structure of an acylated anthocyanin from grapes and one of the few from any source (Birkhofer *et al.*, 1965; Harborne, 1967). The possibility of acyl migration, however, must be considered, as noted by Albach *et al.* (1965b), and insufficient comparative results are available to estimate the biochemically normal situation with respect to site of acylation of grape anthocyanins.

Taken together, these reports from Webb, Kepner, and co-workers show that the common anthocyanin grouping in *V. vinifera* grapes is, in order of increasing mobility on paper in *n*-butanol-acetic acid-water (4:1:5), a small unknown band, delphinidin-3-glucoside, petunidin-3-glucoside, malvidin-3-glucoside and/or cyanidin-3-glucoside, peonidin-3-glucoside, caffeoyl derivatives, and *p*-coumaryl derivatives of these anthocyanins (two bands). Although cyanidin was not identified among the acylated pigments, this may be because the variety most studied lacked, or was low in, this pigment. Since the proportions of the anthocyanidins in the two acyl forms and in the free glucosides were similar, cyanidin glucoside would appear likely to occur in *vinifera* grapes in acylated form as it evidently does in other species, and it has been so reported (e.g., Liuni *et al.*, 1965).

Relative proportions of these pigments are indicated by Rankine *et al.* (1958), and quantitative percentages can be calculated from data presented by Albach *et al.* (1959). The average relative content as percentage of the total densitometric scan of paper chromatograms of the pigments corrected to a common reference standard peak area were, for 117 acyl-pigmented varieties: 0.2% unknown, 3.4% delphinidin glucoside, 5.0% petunidin glucoside, 41.2% malvidin glucoside (plus cyanidin glucoside), 10.1% peonidin glucoside, 13.6% caffeoyl-bearing pigments, and 26.5% *p*-coumaroyl-bearing pigments. The 8 varieties tested lacking acyl pigments were 0.5% unknown, 0.9% delphinidin glucoside, 0.2% petunidin glucoside, 70.8% malvidin (plus cyanidin) glucosides, and 27.6% peonidin glucoside.

In other instructive examples, Smith and Luh (1965) identified the pigments of "Rubired" (a non-*vinifera* cross) grapes by similar techniques, also including chromatographic separation, partial hydrolysis, alkaline degradation, and spectral analysis. They found the full series of

3-glucoside, 3,5-diglucoside, and 3-(*p*-coumaroyl) glucoside for malvidin and peonidin. Delphinidin and petunidin were present as the 3-glucosides and 3,5-diglucosides, and additional anthocyans were present but not identified. Chen and Luh (1967) similarly studied the anthocyanins of "Royalty" grapes, a cross involving *V. rupestris* with *V. vinifera.* On two-dimensional paper chromatograms 18 anthocyan spots were found. In decreasing order of concentration the pigments identified were: malvidin-3,5-diglucoside; *p*-coumaroyl malvidin-3,5-diglucoside; *p*-coumaroyl malvidin-3-glucoside; malvidin-3-glucoside; peonidin-3,5-diglucoside; caffeoyl malvidin-3,5-diglucoside; peonidin-3-glucoside; *p*-coumaroyl peonidin-3-glucoside; and *p*-coumaroyl peonidin-3,5-diglucoside. The presence of cyanidin-3-glucoside, and petunidin-3,5-diglucoside was also indicated.

Somers (1966b) studied the "Shiraz" ("Petite Sirah") variety grown in 3 regions in Australia. In all locations the proportions of the individual pigments as part of the totals were very similar. Cyanidin was not found, nor, as expected for a *vinifera* variety, were there any diglucosides. Peonidin, delphinidin, petunidin, and malvidin were all present, each as a 3-glucoside, two different 3-(*p*-coumaroyl)glucosides, and a 3-(caffeoyl)glucoside. The nature of the difference between the two *p*-coumaroyl derivatives of the same glucoside was not elucidated. One form was more stable than the other, and it seems possible that they were interconvertible forms, as noted by Albach *et al.* (1965a). Since, in the stable case but not in the less stable, hydrogen peroxide oxidation was able to show the acyl group to be on the sugar (Harborne, 1964b) and the molar proportions of acylation were not certain, multiple acylation, acylation on different sugar hydroxyls, or acylation on the anthocyanidin was theoretically possible. Partial hydrolysis, however, in no case gave more than two pigments.

Koeppen and Basson (1966) studied the pigments of "Barlinka" grapes and found them typical of *vinifera* varieties. Peonidin-, delphinidin-, petunidin-, and malvidin-3-glucosides and a mono-*p*-coumaroyl ester of malvidin-3-glucoside were isolated and identified by infrared spectra, alkaline degradation, hydrolysis, spectral properties, etc. In work with "Concord" pigments, Ingalsbe *et al.* (1963) found that blue colloidal precipitates from "Concord" juice consist partly of metal complexes and, most importantly, of *p*-coumaroyl esters of delphinidin- and cyanidin-3-glucosides. Rice (1965), in paper chromatographic studies of "Concord" pigments, found cyanidin, peonidin, delphinidin, petunidin, and malvidin monoglucosides, diglucosides, and in most cases one or more acylated forms for each glucoside. Shewfelt (1966) found 14 paper chromatographic spots of pigments in "Concord" grape juice, and the aglycone proportions were 45.7% delphinidin, 37.1% cyanidin, 8.8% petunidin, and 8.4% malvidin.

The anthocyans of wine would be expected to be the same as those of grapes, and if the wines are young enough this may be true. However, the acyl groups are hydrolyzed very readily and will rapidly disappear from linkage in anthocyanins and appear as free acids in the wine. The glycosides have been reported to hydrolyze to free anthocyanidins, and anthocyanogens have been reported to give cyanidin in wine (Ribéreau-Gayon and Ribéreau-Gayon, 1958; Durmishidze, 1955). Very little anthocyanidin can usually be recovered from wine, however, and the major change appears to be the incorporation of anthocyanins into polymeric red tanninlike molecules.

### 4. Flavonols and Flavones

Flavonols have been known to be in grapes for a long time. Neubauer (1873, 1874) isolated quercetin from grape leaves and proved its identity by elemental analysis and other properties. Traces of quercitrin were also reported on the basis of a quercetin glycoside melting at 170°C and the production of reducing sugar upon hydrolysis. Hilger and Gross (1887) agreed with Neubauer's work, but further details of sugar identification in the glycoside were not discussed. Traces of quercetin and quercitrin were reported in the grape berry. There are a number of other references in the early literature to the presence of quercitrin, but those which we have been able to trace to the original report do not give convincing data that the sugar was specifically identified. Apparently the quercitrin should be interpreted in these cases as quercetin glycoside and not necessarily quercetin-3-rhamnoside. However, it apparently became accepted that quercetin-3-rhamnoside was present in grapes, and it was subsequently so reported without specific study of the glycosidic form. Shunk *et al.* (1894) crystallized from grape leaves a yellow, water-soluble glycoside, and identified the sugar as glucose but did not identify the aglycone.

The identification of quercetin was confirmed and extended to wines, particularly wines fermented before separation of the skins (v. Fellenberg, 1913). All parts of the solids of the grape were found to contain quercetin, largely in glycoside form. Williams and Wender (1952a,b) studied commercial table grapes of the *Vitis vinifera* varieties "Thompson Seedless," "Emperor," and "Tokay" and the *V. labrusca* "Concord." The flavonols were isolated by procedures including hot-water extraction, synthetic resin adsorption, and Magnesol chromatography. They were identified as quercetin and isoquercitrin (quercetin-3-glucoside). The only sugar found when the glycoside was hydrolyzed was glucose, identified by paper chromatography and its osazone derivative. Quercetin was identified by procedures including paper chromatography and the properties of its pentaacetate. The sugar was shown to be attached at position 3 by comparison of the properties of the tetramethyl derivative obtained and those expected if the glucose had been blocking the 3 or the 7 position. A typical yield of isolated crude isoquercitrin was about 9 mg/kg of grapes.

Masquelier and Point (1954) found quercetin-3-glucoside in wines prepared from "Sémillon" and "Sauvignon" grapes. They noted that the flavonol was localized in the skin. It was not detected if juice only was fermented. Dupuy and Puisais (1955) used paper chromatography and reported that quercitrin and probably hesperidin were present in red and white grapes, and free quercetin and kaempferol were found in grape leaves. Hennig and Burkhardt used two-dimensional paper chromatography and a series of colorimetric tests to find three flavonol derivatives which were identified at first (1957b) as quercetin, quercitrin, and rutin but were later considered unidentified (1958, 1960). By paper chromatographic comparison in three or four solvent systems the flavonol aglycones from grapes were identified as quercetin and myricetin. As many as five flavonol glycosides were noted as present in some wines, and two were considered to be quercitrin and myricitrin but no sugar identification procedures were reported.

Nutsubidze and Gulbani (1964) reported from paper chromatographic studies that quercetin, quercitrin, isoquercitrin, rutin, and kaempferol were present at 4–5% in grape leaves of several grape varieties and in other vine parts at much lower levels. Ribéreau-Gayon (1964a,b) studied the skins of three red ("Cabernet-Sauvignon," "Merlot," and the Seyvé Villard cross no. 18,315) and three white ("Sémillon," "Sauvignon," and the Seyvé Villard cross no. 23,410) varieties of grape. The flavonols were separated by extraction with ethyl acetate, and studied by paper and thin-layer chromatography. The free aglycones quercetin, kaempferol, and myricetin were not found in grape skins but were nearly the sole forms present in wines, where they were the products of hydrolysis of the flavonol glycosides of the skins. All three were found in the red grapes, but myricetin was not detected in the three white grape varieties.

The three flavonols were present in the grape skins as four different glycosides. All were found as the 3-glucosides, and quercetin occurred in combination with glucuronic acid also. The glucose and glucuronic acid liberated by hydrolysis of each of the separated flavonoids were identified by paper chromatography. The glucosides were monoglucosides based on their mobility on polyamide thin-layer chromatograms. The substitution was on the alcoholic hydroxyl in position 3 based upon a comparison of ultraviolet spectra (Jurd, 1962). Approximate percentages of the total flavonol content were, for the red grapes, kaempferol glucoside 5%, quercetin glucoside (isoquercitrin) 50%, quercetin glucuronoside 35%, and myricetin glucoside 15%, and for the white grapes respectively 10, 55, 35, and 0%. Bourzeix (1967) found 37–97 mg/liter of total flavonols in wines fermented on the skins with 6 grape varieties, white and red. The analysis involved preliminary thin-layer chromatography and spectroscopy with aluminum chloride.

Pifferi (1966), following a paper-chromatographic study of "Merlot," "Raboso," "Clinton," and "Baco" red wines, reported quercetin, quercitrin, and rutin. Mar Monux (1966), using paper chromatography, found myricetin, quercetin, and kaempferol in white wines plus two other unidentified substances which seemed to be flavonol derivatives. Bourzeix and Baniol (1966), using cellulose thin-layer chromatography, found five flavonols in red wine from the hybrid "Seyvé-Villard 18315." By comparison with a commercial sample, rutin was absent. Quercitrin and isoquercitrin did not separate in this system with commercial samples, but, from double spotting in the same region of the chromatogram, both were considered to be present in the wines. Two flavonol derivatives were not identified, quercetin was identified, and kaempferol was not detected.

From the available data, it appears to us that identification of the rhamnosides rutin, quercitrin, and myricitrin in grapes or wine was incorrect, resulting from misinterpretation of similar glucosides or the glucuronoside reported by Ribéreau-Gayon. This seems particularly plausible in view of the fact that two recent careful studies of the sugars have failed to show rhamnose in grape juice (Esau, 1967) or grapes (Kliewer, 1966). However, Yap and Reichardt (1964) show two-dimensional maps of the phenolics of the leaves of many species of grapes with spots which they identified following hydrolysis studies (details not given) as isoquercitrin, quercitrin, and rutin, all on the same chromatograms. Free rhamnose is found, however, in some wines (Amerine, 1954; Melamed, 1962; Esau and Amerine, 1966; Rice *et al.*, 1968). The suggestion has been discounted that the rhamnosides occur in the vegetative parts of the plant and not in the fruit itself, at least with regard to stems (Masquelier and Point, 1954; Ribéreau-Gayon, 1964a,b). Further study seems necessary for a final conclusion.

Flavones have generally appeared to be absent from grape berries or wine, although Pifferi (1966) suggests that one of the spots found on two-dimensional chromatograms of 4 red wines was a flavone. Dupuy and Puisais (1955) note that they did not find flavones or flavanones in white or red grapes. Nutsubidze and Gulbani (1964) report the presence of luteolin in grape leaves as indicated by paper chromatography. Recently Wagner *et al.*, (1967) have reported flavone C-glycosides in *V. cinera* leaves. They isolated the substances by paper and cellulose column chromatography and identified them by melting point, infrared, nuclear magnetic resonance, and ultraviolet absorption spectra. Vitexin (8-C-β-D-glucopyranosyl-5,7,4′-trihydroxyflavone), isovitexin (6-C-β-D-glucopyranosyl-5,7,4′-trihydroxyflavone), orientin (8-C-β-D-glucopyranosyl-5,7,3′,4′-tetrahydroxyflavone), and isoorientin (6-C-β-D-glucopyranosyl-5,7,3′,4′-tetrahydroxyflavone) were found along with isoquercitrin.

## C. Biochemical, Taxonomic, and Genetic Considerations

### 1. Grapes in Relation to Plants in General

It has been mentioned several times that grape varieties differ considerably in their qualitative and quantitative phenolic composition. It has also been mentioned or implied that the particular pattern of specific phenolic substances normally produced and their relative amounts is primarily a species or varietal characteristic (i.e., genetically controlled), whereas the gross total level of phenols or classes of phenols is more subject to environmental influence. This is true of plants in general and makes phenols particularly interesting from a classification and breeding viewpoint.

The systematic distribution of phenolic substances has been studied for a long time, particularly in enumerating occurrences of readily visible substances such as anthocyanins in various species. More recently the studies have been expanded to include various qualitative chemical tests and paper chromatography. These popular subjects, chemotaxonomy and biochemical genetics of phenols, have evolved a considerable body of information which has been reviewed a number of times in the past few years (e.g., Bate-Smith, 1962a,b; Geissman, 1962; Alston and Turner, 1963; Alston, 1964; Swain, 1963, 1966; Bate-Smith and Swain, 1965; Harborne, 1967).

The phenols are generally synthesized in plants via one or both of two major pathways. The shikimic acid pathway gives rise to the universally essential protein constituents phenylalanine and tyrosine, and therefore must have a very ancient evolutionary origin. The acetate-malonate pathway produces phloroglucinol and most of the other *meta*-oriented polyphenol derivatives. Evolutionarily this also appears to be an old mechanism because lichens, molds, and other plants regarded a primitive are rich in such compounds, but generally lacking in most other phenolic forms of higher plants.

The extension of the shikimic acid pathway to cinnamates and the production of lignin appear to be somewhat later evolutionary developments. Flavonoids are produced via a linkage of both pathways, the shikimic acid route for the B-ring and the acetate-malonate route for the A-ring. Plants more primitive than mosses lack flavonoids, and although flavonoids are found in mosses (Harborne, 1967) they are rare and often of unusual types. As one climbs the phylogenetic tree, more types and often a greater amount of flavonoids are commonly present until, generally, a full complement is present in the angiosperms. For various taxonomic reasons grapes may be considered moderately advanced dicotyledonous angiosperms. For example, evolution is generally (but not universally) believed to have proceeded from nonlignified to lignified, woody, upright plants and then on to less lignified vines, like grapes (Adkinson, 1913; Bate-Smith and Swain, 1965). The reported phenolic composition of grapes is consistent with their being relatively highly evolved. They have a rather full complement of flavonoids, and the anthocyanins especially are relatively advanced, being trihydroxylated in the B-ring, methoxylated, and acylated. Grapes do, however, appear to lack other possible advanced characters in flavonoids, such as 2′, 6, or 8 hydroxylation or aurones (Harborne, 1967).

The phenols found in grapes so far are, with few exceptions, the ones which would be considered the most probable and consistent from a general botanical chemotaxonomic viewpoint. This may be disappointing to a taxonomist wishing for highly distinctive and characteristic compounds, but should be comforting to biochemists wishing to study generally applicable metabolism, and to food scientists interested in common processing problems. There is, of course, a possible danger in finding only what one expects to find, particularly on the ubiquitous paper chromatogram, for such may delay recognition of unusual substances. The grape and wine phenol literature appears to contain examples of incorrect identification not only of a constituent as the form generally expected but also of a chromatographic spot as an uncommon constituent when it could as well be interpreted as a more common one. Fortunately such errors are now being minimized by more careful reevaluation and improved techniques. Since qualitative phenol composition is helpful in classifying plant species, it seems equally valid to consider the reported phenols of grapes against the general background of knowledge of systematic occurrences of phenols of different types.

Since flavonoids are products of both the acetate-malonate and shikimic acid biosynthetic pathways and grapes are rich in flavonoids, it is reasonable to expect the presence in grapes of all the common intermediates and accompanying compounds involved. Shikimic and quinic acid have been found; *p*-coumaric and caffeic acids are present as esters in anthocyanins. Since tartaric acid is relatively rare in the plant kingdom and so important in the acids of the grape, the presence of a caffeoyltartaric acid derivative is perhaps less surprising in grapes than in chicory, and its absence in other plants is no surprise. The presence of quinic acid, however, makes one expect that it can assume the role normal for it in other plants and produce chlorogenic acids in grapes, albeit perhaps in abnormally low amounts. Indications of perhaps a greater role for ferulic acid in grapes than in some other plants is consistent with the generally high level of methylation of phenolic groups found in grapes.

By analogy with other woody dicotyledons, the lignin of grapes would be expected to contain units analogous to syringic and vanillic acids (Brown, 1965). Apparently the ratios have not been studied in grapevine lignin, but about three syringic to one vanillic unit would be expected if grapes are similar to other plants in their group. The methoxyl contents of grape berry lignins reported by Fernández and Sanchez-Gavito (1949) seem low for the expected syringic content. Considering the high prevalence of the gallic type of hydroxylation and the generally high level of methylation in the pigments, a high proportion of sinapyl-derived lignin seems likely in grapes. In view of the rather wide variations in hydroxylation and methylation pattern in other phenolics of grapes of different types, a study of the syringic–vanillic ratio in lignins of selected varieties and species of *Vitis* would be of interest.

The detection of liberated fragments after somewhat drastic "hydrolysis" treatments seems to us to require study of the nature of the "bound" form. For example, the presence of free vanillin or syringaldehyde in lignifying tissue (Ryugo, 1962) might lead to oxidation to the corresponding acids. For another example, partial conversion via the typical alkaline degradation has already been suggested as a possible source of syringic acid from malvidin. The occurrence of 2-hydroxy substances such as salicylic acid in grapes seems very unlikely from a botanical viewpoint, as well as insufficiently substantiated from a chemical viewpoint. The coumarins are of somewhat limited, though scattered, appearance in plants (Dean, 1952, 1963; Brown, 1963; Stanley, 1963; Harborne, 1964a). Despite this statistical likelihood of their absence in grapes, the possibility seems to remain, although Bate-Smith and Lerner (1954) stated that coumarins were virtually confined to those families *without* leucoanthocyanins in their leaves. The coumarins reported in grapes or wine are, however, suspected to be artifacts, as previously discussed, and, if that is true, other 2-hydroxy derivatives related to *o*-coumaric acid would seem unlikely. A particularly difficult compound to explain is gentisic acid. Reports to date (Tomaszewski, 1960; Ribéreau-Gayon, 1963a; Ibrahim and Towers, 1960) suggest very widespread liberation from plants, including grapes, by hydrolysis. Ibrahim *et al.* (1962) note that gentisic acid is more common in acidic than in alkaline hydrolysates of plant material, but occurs in both. Although it has been suggested as a lignin constituent, it has not been shown to be there and the corresponding coumarin or cinnamate is unknown. That some of these liberatable compounds, such as gentisic acid, are degradation products of already known phenols is suggested by the fact that there do not seem to be sufficient unexplainable phenolic substances on general chromatograms of grape extracts to account for the diversity and level of these substances if they are simply depsides or glycosides.

As a woody dicotyledonous plant the grape would be expected to contain leucoanthocyanidins, and it happens to contain an unusually high amount of such anthocyanogenic condensed tannins, particularly for a nontropical plant, and also in the seeds, where appreciable tannin is considered relatively rare (Dekker, 1913; Bate-Smith and Ribéreau-Gayon, 1959; Earle and Jones, 1962; Haslam, 1966). Perhaps initially more surprising is the evidently relatively high level of leucocyanidin compared with leucodelphinidin in the grape, in contrast to the high trihydroxy versus dihydroxylation of the B-ring of the anthocyanins of grapes. Leucodelphinidins, however, are relatively uncommon in other plants compared with leucocyanidins. The apparent absence of the relatively rare leucopelargonidins in grapes coincides with the lack of the pelargonins.

Bate-Smith (1963) has noted the absence of a cinnamic acid derivative analogous to gallic acid in plants, although products such as sinapyl lignin and delphinidin derivatives suggest that it should exist. He has suggested that this role is filled by ellagic acid or its equivalent hexahydroxydiphenic acid. This free acid readily lactonizes to ellagic acid, which, once formed, seems to be a singularly insoluble, stable substance, difficult to separate into its gallic acid fragments. The nonlactonized form appears unlikely to exist very long unless bound in some form like the ellagitannins. The amount of preformed ellagic acid or directly hydrolyzable sugar-bound precursor in grapes, if it exists, must be small. Free gallic acid does not exist in grapes, but readily liberated gallic acid does, and the only identified source is the catechin gallates. We have suggested that this gallic acid is the source of the ellagic acid. It is well-known that gallic acid does not, but that gallate esters do, readily autoxidize to ellagic acid, at least in alkaline solution, even without enzymic participation (Hathway, 1957a,b). Of course, other sources of ellagic acid could be postulated, such as oxidative breakdown of bisepigallocatechins, found in black tea (Roberts, 1962).

This postulation that gallic depsides, catechin gallates in the case of grapes, are the source of small amounts of ellagic acid in samples from grapes has a more important corollary. In line with Bate-Smith's reasoning, this would mean that the catechin (or, more properly, *l*-epicatechin and *l*-epigallocatechin) gallates could serve in place of 3,4,5-trihydroxycinnamate in further biosynthesis. It seems significant to us that *l*-epicatechin gallate is present in high proportion in seeds of green grapes, decreasing or disappearing without the appearance of free gallic acid as the grape matures (Singleton *et al.*, 1966a). This coincides with the buildup of malvidin and lignification of the seed during what must be a period of heavy demand for gallate units. The catechins have been studied little except in tea and, to a lesser extent, forest trees and grapes, but they are very widespread in occurrence. It appears to us, considering their unique occurrence among flavonoids as gallates and without methylation or glycosidation, that they are deserving of much more attention as possible participants in the biosynthesis of other flavonoids and phenols. A relatively high radioactivity appears early in catechins in flavonoid synthesis experiments, as shown by, for example, Hillis and Isoi (1965) and Creasy and Swain (1966).

The known anthocyanins of grape are all either 3-glucosides, the most common, or 3,5-diglucosides, a very common type (Harborne, 1964a, 1967). Although Liuni *et al.* (1965) and Cappelleri *et al.* (1966) have reported the possible presence of a triglucosidic anthocyanin, it is not yet identified. The glycosidation pattern of flavonols, often if not usually, bears no relation to that of the accompanying anthocyanins. Different glycosides of the same flavonol aglycone may occur in the same plant. Pending further evidence to the contrary, identification of the 3-glucosides of the three flavonols and the 3-glucuronoside of quercetin by Ribéreau-Gayon (1964a,b) is considered to exclude the presence of the rhamnose glycosides previously reported. Except that insufficient numbers of flavonol glycosides have been noted on chromatograms from fruit to allow for both sets of glycosides simultaneously, there seems to be no botanical reason to rule out the occurrence of the rhamnose-containing forms as well.

Acylation of the anthocyanins is relatively uncommon in occurrence among plants, although more examples are still being noted. The families that have such structures are rather scattered botanically, and the known distribution is often restricted to one or a few genera of the family. The most common type is acylation of 3-glucosides or 3,5-diglucosides with *p*-coumaric acid, and caffeic acid is also observed fairly frequently. The grape is, again, the "normal" type. Acylation of flavonols is known but not common, and has not been observed in grapes.

The flavonol, catechin, and anthocyanidin hydroxylation patterns reported in grapes are all the most common forms, without unusual methylations, glycosidation (except the glucuronoside of quercetin), or other modifications. Present, aside from the rarer afzelechins and pelargonidin, is a complete list of the most common type of aglycones in these series. The usually high proportion in grapes of trihydroxylated B-ring types (gallocatechin, myricetin, malvidin, etc.) is somewhat unusual among plants, but not unexpected for plants in the Archichlamydae, the less specialized section of the Dicotyledons, which includes grapes. The presence of C-glycosides in the flavone series is actually considered a primitive characteristic (Harborne, 1967) and these compounds are now being identified in more and more plants. The apparent shift in the flavones of grapes toward pelargonidin-type hydroxylation and away from malvidin-type substitution follows the trend set by kaempferol in the flavonols and is the usual finding in plants (Swain, 1963, 1966; Harborne, 1967). In fact, the "normal" flavone analog of delphinidin apparently does not occur in plants, and the malvidin analog tricin is rare (Dean, 1963). The apparent lack of flavanones and relatively low level of flavone derivatives in grapes is consistent with the fact that grapes are woody rather than herbaceous plants (Bate-Smith, 1963). Taxonomic considerations suggest that taxifolin (dihydroquercetin) and other flavanonols should occur in grapevines since, if quercetin occurs in leaves, as it does with grapes, taxifolin can usually be found in the woody tissue (Bate-Smith, 1963). Ampelopsin (dihydromyricetin) is found in a relative of grapes.

### 2. The Grape versus Related Genera

The order Rhamnales includes only the Rhamnaceae and the Vitaceae. Grapes appear to bear no similarity whatsoever in phenolic substances to the Rhamnaceae. Considering those present in Rhamnaceae—cathartics and toxins—this is very fortunate from a food viewpoint (Long, 1917; Salgues, 1962). This is not the place for a detailed summary of the phenolics of the Rhamnaceae, but some of the interesting contrasts can be mentioned.

*Rhamnus purshiana* contains the anthraquinones emodin, aloe emodin, and chrysophanol in the dried bark, which is the drug Cascara Sagrada. The activity of this and similar drugs as purgatives is evidently due to reaction products or glycosides of these anthraquinones, some of which are difficultly hydrolyzable C—C glycosides (Fairbairn, 1959; Thompson, 1957) *Rhamnus cathartica* produces frangulin (emodin-l-rhamnoside), and the name rhamnose harks back to the original isolation of this sugar from *Rhamnus* species (Anon., 1968a; Karrer, 1958). Cascara bitters, containing both flavonols and anthraquinones, have been produced from *R. purshiana*, and a number of the Rhamnaceae have been used for dyeing. In addition to the anthraquinones, which have a very limited but peculiar distribution in plants (including filamentous fungi), the flavonols of Rhamnaceae include some exotic forms not found in grapes or most other plants. Xanthorhamnin, from *R. infectoria*, is the 3-rhamninoside of rhamnetin. Rhamninose is composed of two L-rhamnose and one D-galactose units, and rhamnetin is the unusual 7-methoxy derivative of quercetin. Other similar forms, such as 7-methoxy kaempferol and 7,3′-methoxyquercetin, occur in *Rhamnus* species (Dean, 1963; Harborne, 1964a). Anthocyanogenic tannins are produced by Rhamnaceae, and leucodelphinidin is sometimes found when leucocyanidin is not (Bate-Smith, 1957). In tests by Bate-Smith (1962b), kaempferol seemed to be more prominent in Rhamnaceae than in Vitaceae, and caffeic acid less so. Anthocyanidins of *Rhamnus* species appear to emphasize delphinidin and its methyl derivatives as in grapes (Ribéreau-Gayon, 1963c). Malvidin-3-rhamnoside without acylation has been found in flowers of a member of the Rhamnaceae, *Ceanothus* sp. var. "Autumn Blue" (Harborne, 1963).

If other genera within the Vitaceae are compared with *Vitis*, there is still little agreement with respect to the phenolic substances present, although relatively little work has been done except with *Vitis*. Another problem is that generic classification within the family has not been static or uniform. Virginia creeper is now generally called *Parthenocissus quinquefolia*, but was known as *Ampelopsis quinquefolia*. A number of shifts have been made from one genus, including *Vitis*, to another, particularly to the expansion of *Cissus*. Without attempting to clarify these details, a few comments on the phenols are in order.

Bate-Smith (1962b) found ellagic acid missing in Rhamnaceae and Vitaceae other than grapes which he tested. If our hypothesis is correct on the origin of the ellagic acid in grapes, however, this might indicate greater relative freshness and greater enzymic oxidation in the grape sample, or less change during original drying or sample preparation of the herbarium leaf-twig samples used for much of the study (Bate-Smith, 1957), rather than a truly distinguishing characteristic. Myricetin has been isolated from leaves of *Ampelopsis meliaefolia*, and found along with it, for the first time, was the flavanonol ampelopsin (dihydromyricetin) (Kotake and Kubota, 1940). *Parthenocissus quinquefolia* leaves contain tannin cells (Holm, 1911). They give strong tests for leucocyanidins, and delphinidinogens also are reported (Bate-Smith and Lerner, 1954; Bate-Smith, 1954b, 1962b). This plant has been used in studies on leucoanthocyanin and anthocyanin synthesis in tissue culture (Cronenberger *et al.*, 1955), and cyanidin-3-5-diglucoside is reported to give way to malvidin-3-glucoside and malvidin-3,5-diglucoside as necrosis sets in (Stanko and Bardinskaya, 1962). Dihydrocaffeic acid and pyrocatechol were reported many years ago from leaves of *Ampelopsis hederacea* (Karrer, 1958). Flavones have been indicated in stems and leaves of *Vitis quadrangularis*, which is now classified as *Cissus quadrangularis* (Bailey, 1935; Bhaskara *et al.*, 1939).

### 3. *Vitis* Species and Varieties

The differences in phenolic composition among species of *Vitis*, within varieties of one species, and among intraspecies or intravarietal crosses have been of interest for a long time, but, until recently, study was limited to rather gross, observable differences, such as presence versus absence of anthocyanins in fruit. For example, Hedrick and Anthony (1915) noted that white-skinned fruit was a recessive color in grapes, compared with red or "black" fruit. They found all shades of red to "black" to be possible in seedling vines from crosses between varieties of different color, and that there was no such thing as a simple heritable character for red or "black" fruit. As far as we are aware, all wild species of grapes have colored fruit. Pirovano (1967) found homozygous character for dark pigmentation in "Imperial noir" and "Madelaine royale" varieties as indicated by the fact that 48 seedlings of a cross between them all had violet-black fruit. "Imperial noir" crossed with a white grape variety produced 34 seedlings with fruit colors: 22 violet black, 6 grayish black, 6 red violet, and no white. Other examples were given showing the involvement of more than one gene and complex inheritance patterns in fruit color of *V. vinifera* and hybrid grapes. A frankly red color and a dilute blue color were noted as relatively rare in seedling progeny.

Gladwin (1924) noted that fruit color was correlated with the color of the first new leaves. Except for four varieties in a series mostly of hybrids between *V. vinifera* and other species, white leaf character was found in all those with white-skinned fruit, but with pinkness in the leaf the fruit may be dark- or lightcolored. It was possible for the casual observer to detect all the white-fruited varieties in a mixed vineyard with these 4 exceptions. White leaf character tended to be high in relation to *V. vinifera*, all *V. labrusca* had pink leaves, and crimson leaf character correlated to some extent with *V. rupestris*. Violante (1948) used spectrophotometry and investigated the inheritance and type of pigment between American species, which tend to give bluer grapes and wine, and European grapes, which tend to be redder. Such results contributed to understanding of species differences and inheritance of the ability to produce phenolic substances, but relatively little was learned until the work was coupled with chemical and, particularly, chromatographic differentiation of different phenols.

The first modern chromatographic work on interspecific differences in the phenols of *Vitis* was that by Ribéreau-Gayon (1953), with anthocyanins of the berry skin. He found (Ribéreau-Gayon, 1957, 1959, 1964a) that the anthocyanins of samples of the fruit of 14 different species of *Vitis* were quite different in the relative proportion of different specific pigments, as shown in Table I, abridged from his data. The number of individual anthocyanin pigments separable on two-dimensional paper chromatograms ranged from 5 to 17 in these species. *V. rotundifolia* appears unique in having only diglucosides and lacking acylated pigments in its fruit. This agrees with the rather large morphological differences which cause taxonomists to place this species and *V. munsoniana* in a subgenus *Muscadinia*, separate from *Euvitis*—the true grapes (Winkler, 1962). All the other species listed showed anthocyanidin monoglucosides, and only *V. vinifera* and *V. monticola* samples lacked diglucosides. There appears to be a considerable variation among species as to their normal proportions of cyanidin versus delphinidin derivatives and the proportions of each which are methoxyl rather than only free hydroxyl forms. *V. vinifera* is one of the highest in malvidin percentage and in acylated forms.

Considering the variability now known within varieties of *V. vinifera*, these species differences may not all turn out to be absolute upon further investigation, but they certainly show the existence of important and rather well-defined types of variation in the phenols among *Vitis* species. Ribéreau-Gayon (Table I) found that *V. amurensis*, the only species listed that is of Asiatic origin, lacked unmethylated and acylated pigments in his sample. Cappelleri *et al.* (1966) report delphinidin, cyanidin, and acylated derivatives in 4 different species from Asia as well as results with 5 American *Vitis* species and an *Ampelopsis* species, all with at least some representatives of each of the 5 anthocyanidins, diglucosides, monoglucosides, and acylated forms. Other studies listing qualitative identifications of the fruit anthocyanins of different grape species include those by Reuther (1961), Cappelleri *et al.* (1964), and Liuni *et al.* (1965).

As previously stated, a lack of 3,5-diglucosides among the anthocyanins of *V. vinifera* varieties and their general occurrence in other species commonly used for fruit production or in hybridization are now well

TABLE I
PROPORTIONATE ANTHOCYANIN CONTENT OF FRUIT OF VARIOUS GRAPE SPECIES[a]

| Species | Cyanidin | | Peonidin | | Delphinidin | | Petunidin | | Malvidin | | Total acylated | |
|---|---|---|---|---|---|---|---|---|---|---|---|---|
| | 1[b] | 2[c] | 1 | 2 | 1 | 2 | 1 | 2 | 1 | 2 | 1 | 2 |
| *Vitis rotundifolia* | — | 9 | — | 6 | — | 38 | — | 29 | — | 18 | — | — |
| *Vitis riparia* | 2 | 5 | — | 2 | 14 | 12 | 10 | 17 | 6 | 21 | 4 | 7 |
| *Vitis rupestris* | — | 2 | — | 8 | 9 | 34 | 3 | 22 | 2 | 8 | 3 | 9 |
| *Vitis labrusca* | 5 | — | 10 | 1 | 21 | — | 15 | 1 | 34 | 2 | 11 | — |
| *Vitis arizonica* | 8 | 1 | 14 | 10 | 13 | — | 10 | 1 | 29 | 10 | 4 | — |
| *Vitis berlandieri* | 8 | — | 16 | 2 | 23 | — | 20 | 1 | 25 | 2 | 3 | — |
| *Vitis monticola* | 3 | — | 5 | — | 36 | — | 26 | — | 27 | — | 3 | — |
| *Vitis cordifolia* | 10 | — | 11 | 2 | 15 | — | 18 | 2 | 30 | 5 | 5 | 2 |
| *Vitis rubra* | 20 | 1 | 5 | 1 | 30 | 1 | 20 | 2 | 16 | 2 | 2 | — |
| *Vitis lincecumii* | 29 | 2 | 7 | 3 | 17 | 1 | 8 | 1 | 4 | 1 | 13 | 14 |
| *Vitis aestivalis* | 31 | 3 | 11 | 4 | 31 | — | 10 | — | 6 | 2 | 2 | — |
| *Vitis coriacea* | 58 | 4 | 6 | 4 | 20 | — | 4 | — | — | — | — | 4 |
| *Vitis amurensis* | — | — | 13 | 15 | — | — | 5 | — | 27 | 40 | — | — |
| *Vitis vinifera* | 3 | — | 15 | — | 12 | — | 12 | — | 36 | — | 22 | — |

[a]Adapted from Ribéreau Gayon (1959).
[b]Columns 1 are 3-glucoside relative percentages.
[c]Columns 2 are 3,5-diglucoside relative percentages.

documented. In fact, the absence of diglucosidic anthocyanins is now one well-accepted test contributing to proof that a variety belongs to *V. vinifera* (e.g., Koeppen and Basson, 1966). Although there are recurring reports of positive tests for traces of anthocyanin diglucosides in fruit or wine samples from pure *V. vinifera* varieties, these seem incorrect, a result of such factors as unrecognized crossbreeding in the variety, a "wild" vine or two in the vineyard, transfer via cooperage or other equipment used for both types of wine, or inadequate laboratory techniques (Ribéreau-Gayon, 1960a, 1963b, 1964c, 1965b; Drawert, 1961; Reuther, 1961; Biol and Foulonneau, 1962; Van Wyk and Venter, 1964; Webb, 1964; Cappelleri, 1965; Getov and Petkov, 1966). Appreciable selectivity in removal or destruction of one form of the glycosides over the other in wine does not appear possible (Van Wyk and Venter, 1964; Deibner *et al.*, 1966), although malvone, a stable fluorescent product of oxidation of only malvin (malvidin-3,5-diglucoside), can be an aid to identification of non-*vinifera* wines (Visintini-Romanin, 1967).

The 3,5-diglucosides, particularly malvin, which is likely to be the major one present, are recognized relatively easily since the 5-glucosides have bright brick-red fluorescing spots, in distinction to the much less fluorescent 3-glucosides in the normal anthocyanin series. At least 75 publications list procedures and suggest applications, modifications, and improvements in distinguishing grapes and, particularly, wines, with and without anthocyanidin diglucosides. Detailing these procedures would be tedious here, and reviews and recent papers can be consulted (Drawert, 1961; Deibner and Bourzeix, 1965; Lefèvre, 1966; Ribéreau-Gayon, 1963b, 1965b, 1968; Garoglio, 1962, 1968).

It does seem necessary, however, to discuss briefly the reason for the intense international interest in this subject, especially since a misleading report has appeared in an otherwise valuable book (Harborne, 1967). The quality and flavor of wine depend to a considerable degree upon the variety of grape from which it is made. The varieties which produce the wines with a reputation for premium value today are all or nearly all the old pure *vinifera* stock. In France, wines are produced primarily under three regulated classes: premium wines, with an appellation of origin; wines of superior quality; and wines for current consumption. In the first two categories, the grape varieties which may be used are specified by law and are only the "noble" *V. vinifera* varieties. Hybrid wines are permitted in the third group, but specific varieties may be recommended, authorized, tolerated, or forbidden (Biol and Michel, 1961).

In spite of the laws, however, owing to greater yields, greater pest resistance, etc., hybrid varieties had been encroaching more and more into prohibited territory (Ribéreau-Gayon, 1963c, 1965b). Prior to recognition of the absence of anthocyanin diglucosides in the "noble" varieties there was no good way of detecting the illegal practice, but wine from diglucoside producers can now be detected and must be put into the less expensive wines for current consumption; otherwise, prosecution will result. Even the addition of 1% of wine from most diglucoside-bearing hybrid grapes can be detected. The scrupulous producer and the premium internal and export market can thus be protected from the unscrupulous by finding and eliminating the purveyor of red wine "stretched" or diluted with hybrid wine containing anthocyanin diglucosides. Germany and several other countries besides France have incorporated tests to detect malvin and other diglucosides in red wine in their regulatory procedures for controlling import, export, and internal labeling and fraud with wine. Prohibitions against growing hybrids containing diglucosides in certain areas are based upon logical considerations involving spread of certain pests and quality-economic factors plus, perhaps, illogical political and health-fad opinions. That such regulation was intended to prevent "adulteration" of European wines by California wines (Harborne, 1967) is, however, untrue, even if a significant export of unbottled California wine to Europe occurred, which it does not. Nearly all of the grapes grown in California are pure scions of the same *V. vinifera* varieties grown in Europe. They are free of diglucosides, as their lineage requires. Also, a very large proportion, particularly that which might be available for export in bulk, of California wine comes from white grapes which are *V. vinifera*, but are not subject to this test in any case. Efforts have been made to develop similarly reliable tests to distinguish between wines from white grapes from crossbred grapevines and those from pure *V. vinifera* grapevines, but so far without success (Flanzy and Aubert, 1967).

The *Vitis* species other than *V. vinifera* which are involved most commonly in breeding or direct fruit production are *V. labrusca* in eastern United States and probably *V. rupestris* and *V. riparia* in France and most other *V. vinifera* growing areas (Winkler, 1962; Ribéreau-Gayon, 1960a). In every cross studied so far between these and other species bearing anthocyanin diglucosides and *V. vinifera*, the diglucoside trait is dominant in the cross and seems to be carried equally in the pollen or the female parent. A single gene is involved (Ribéreau-Gayon *et al.*, 1955; Ribéreau-Gayon, 1963c; Durquety, 1958; Boubals *et al.*, 1962) and this appears to coincide with the usual case for the character of 3,5-diglucosides versus 3-monoglucosides of anthocyanidin in plants (Harborne, 1960, 1962; Feenstra, 1962; Alston, 1964). Of course, this means that if a heterozygous cross is self-pollinated, one-fourth or, if back-crossed with a *V. vinifera* variety, half of the seedlings resulting would lack diglucosidic pigments even though not pure *vinifera* in origin. Such ratios are found (Boubals *et al.*, 1962; Ribéreau-Gayon, 1963c), and examples of this—*vinifera* pigment pattern in varieties not pure *vinifera*—exist among commercial varieties (e.g., Biol and Michel, 1961). Therefore, the appearance of anthocyanidin diglucosides in grape skins or in wine is accepted as proof of grape crossing or wine blending with non-*vinifera* species, though their absence does not guarantee that crossing or blending has not occurred.

Interposition of a white-fruited variety in crosses appears to affect fruit pigment in the progeny in complex ways. In general it was not possible to transfer the character for diglucoside production via a white-fruited carrier (Durquety, 1958; Cappelleri *et al.*, 1964). In view of the rather clear evidence for a single dominant gene, complete absence of the diglucoside pigments would seem more probable than production of traces in *V. vinifera* varieties, all of which naturally lack the trait. In any case, differences in anthocyanidin diglucosides sufficient to distinguish *vinifera* varieties from the typical hybrid are proved, and the complete absence of this character in pure *vinifera* seems to be the rule.

Other differences have not been studied so well. It is generally believed that non-*vinifera* wines tend to be bluer, and blue grapes seem more common at least in *V. labrusca* than in *V. vinifera* varieties. A higher relative proportion of delphinidin and less cyanidin or malvidin would be one way of accounting for such bluer color. The data in Table I suggest a trend in that direction. "Concord," a fairly blue *V. labrusca* variety, was reported to have 43% of its pigment (by densitometric scanning of paper chromatograms) as delphinidin monoglucoside, and little malvidin in comparison with 61% malvidin and only 18% delphinidin monoglucosides in a companion sample of "Cabernet-Sauvignon," a "black" *V. vinifera* variety (Robinson *et al.*, 1966b). Delphinidin-3-glucoside was the most predominant pigment in "Concord" samples from 8 sources in studies by Shewfelt (1966), with relative percentages of aglycones of 45.7% delphinidin and 8.4% malvidin. Ingalsbe *et al.* (1963) found delphinidin-3-glucoside and its *p*-coumaroyl ester to be the two most plentiful pigments in "Concord" grapes and hot-pressed juice.

Ribéreau-Gayon (1958a,b, 1960b) reported a generally lower content of monomethylated anthocyanins in species native to America, and a generally low dihydroxy (cyanidin, peonidin) content except in certain species. Schneyder and Epp (1959), on the other hand, did not find a distinguishing spectral difference in borate-complex formation, and therefore concluded that there was no difference between *vinifera* and hybrid wines in level of free *o*-dihydroxylation. Sudario (1953) found that the methoxyl content in grape pigments isolated by picrate precipitation was considerably higher in *vinifera* than in hybrids, but this difference could result from the diglucoside contribution to pigment weight rather than proportionately greater methylation of aglycones.

The genetic aspects of the dihydroxy (cyanidin, peonidin) versus trihydroxy (delphinidin, petunidin, malvidin) pigment inheritance and free hydroxy (cyanidin, delphinidin) versus methylated (peonidin, petunidin, malvidin) character in grapes are apparently complex and not yet clarified. Breeding experiments related to this subject and supplemented by paper chromatographic studies of anthocyanins are reported by Ribéreau-Gayon *et al.* (1955), Durquety (1958), and Reuther (1961). These studies indicate the primary role of inheritance in determining these characters, but the data available appear to be insufficient for useful interpretation. From studies with other plants we would expect that delphinidin would be derived from cyanidin or a dihydroxy-B-ring precursor of both and the methylated pigments from the respective free hydroxyl forms, for example, malvidin from petunidin, and petunidin from delphinidin (Feenstra, 1962; Harborne, 1962, 1964a, 1967). Ripening changes and apparent rates of synthesis in grapes are reported to agree with this expectation (Ribéreau-Gayon, 1958a). The cyanidin-to-delphinidin class change in grapes appears to be too complicated for a single gene (Durquety, 1958; Reuther, 1961).

Subject to the limitations imposed by the species, in many instances the differences are nearly as large between varieties of the species as among different species (Rankine *et al.*, 1958; Durmishidze and Nutsubidze, 1958; Albach *et al.*, 1959; Ribéreau-Gayon, 1957, 1964a,c; Robinson *et al.*, 1966b). Large differences have been reported in the proportion of the total anthocyanin contributed by individual pigments. One individual chromatographic band may represent from zero to nearly 100% of the pigment. The striking fact is that the pigment complex seems to be of relatively constant qualitative and proportionate composition for a given variety during ripening (Ribéreau-Gayon, 1958b; Getov, 1966), regardless of where in the world it is grown (provided the *same* variety or similar clone is involved). We know that considerable color intensity differences are environmentally related in grape berries. Therefore, data to the present indicate, in contrast to suggestions to the contrary, that the pigment complex in the fruit is controlled relatively rigidly by the internal enzymic biochemistry of the individual variety and is not a more or less random environmentally influenced mixture of the possible forms (Harborne, 1967). Data of Shewfelt (1966) on "Concord" grapes from 8 sources in New York and South Carolina suggest that the genetics of the variety control the qualitative composition most rigidly and the relative proportions to a lesser degree, and we know the total amount of pigment to be influenced very much by external conditions.

There are a number of varieties of pure *V. vinifera* which have interesting and important variations in pigment pattern. The acylated pigment forms are absent from the group of varieties related to "Pinot noir" and are characteristic of the Burgundy region of France (Rankine *et al.*, 1958; Albach *et al.*, 1959; Getov, 1966). This difference also has significance in detecting fraud in varietal wines and in the fact that the color of the red wines prepared from these grapes is often low and relatively difficult to standardize and retain, at least according to California experience (Webb, 1964). Ribéreau-Gayon (1964a) reports the same lack in only "Muscat Hamburg" of the series of *V. vinifera* varieties he tested in France; however, Webb (1964) reports a detectable level of acylated pigments in "Muscat Hamburg" in California studies. The genetics of this mutation have not been reported, but self-pollinated "Petite Sirah" progeny lost the ability to produce acylated pigments (Rankine *et al.*, 1958), indicating dominant heterozygous character for acylation in the original variety.

A relatively high level of cyanidin-3-glucoside evidently occurs in "Flame Tokay," "Red Malaga," "Emperor," "Hanepoot" (pink Muscat of Alexandria), and "Muscat Hamburg" (Akiyoshi *et al.*, 1963; Robinson and Robinson, 1932; Ribéreau-Gayon, 1964a; Yakivchuk, 1966). These varieties are all relatively low in total color, and it is attractive to suggest that the loss of the normal ability to synthesize trihydroxy pigments is at least one reason for pink or light-colored grapes. It appears that the grapes with the highest likelihood of very highly colored fruit are those with the complete array of grape capabilities to produce cyanidin, delphinidin, methylated derivatives, diglucosides, and acylated forms. Commercial varieties which lack the ability to produce diglucosides, acylated pigments, or delphinidin derivatives appear to include a higher proportion of lighter-colored fruit or wine, but data to verify this impression statistically are not yet available.

Variations in phenols other than anthocyanins in grapes have been studied much less. Singleton *et al.* (1966a) report considerable qualitative variation in the chromatographically separable phenolics of skins of different varieties of white grapes without identifying the compounds. The phenols of seeds of several varieties were much less variable even when more than one species was involved, but both qualitative occurrence of some constituents and relative proportions did vary among varieties. Henke (1960) found that the flavonoid content per gram of dry leaf was considerably lower for *Vitis cinerea* (4.5–6.5 mg) than for *V.*

*riparia* and *V. vinifera* varieties (15–32 mg). This low content was predominant in crosses of *V. cinerea* with the others. *V. cinerea* produces flavone C-glucosides in the leaves, and these have not yet been identified in other species (Wagner *et al.*, 1967).

Yap and Reichardt (1964), with admirable restraint in naming most of the spots on their complex chromatograms, studied the flavonoids and cinnamic acids in leaves of *Vitis* species, varieties, and crosses. They found some qualitative variation within species, though insufficient for value in classification. Between *Vitis* species, however, the patterns were characteristic enough for taxonomic value. They reported that ecological factors did not influence the qualitative leaf phenol composition, that pest-resistance seemed related to composition, and that characteristic patterns or marked substances could be traced via genetic control in descendants of crosses. Reuther (1961) studied the anthoxanthins as well as anthocyanins of berries in *V. vinifera* and *V. riparia* x *V. vinifera* crosses. Again, pest resistance seemed to be related to phenolic pattern, although this could, of course, be very indirect.

## D. Some Artifacts and Related Changes Affecting Qualitative Composition of Wine or Grapes

The phenolic constituents of the grape may or may not appear or persist without change in the wine. The phenolic complex may be differentially extracted into the fluid for wine. It may be modified by grape enzymes, particularly polyphenoloxidase, as the berry is disrupted. Action by yeasts and other microorganisms may make changes in the phenolic substances present. Specific substances and their relative proportions undergo chemical changes during wine making, processing, and aging. Finally, additional types or additional amounts of phenols may be introduced by various practices, notably storage of wine in wooden containers.

These effects and reactions are often important and often complex. More details and more examples are discussed in other sections, but a short discussion seems desirable here. With reactive compounds it is not unusual for artifacts to arise in the most careful laboratory work, of course. The example of coumarin production on paper chromatograms from cinnamic acids has already been described, and the probability noted of similar reactions in wine. Incorporation of anthocyanins into polymers in wine has been discussed in connection with the qualitative make-up of wine tannins. It appears that similar reactions may involve the other flavonoids and phenols, thus lowering their content and increasing the macromolecular phenolic products. Since anthocyanidins adsorb more tightly on cellulose and are less soluble than their glycosides, as are their polymers, it is easy to confuse the two on paper chromatograms of wine or grape samples and in other tests (Vil'yams and Begunova, 1955). This is believed to be one cause for erroneous reports of the presence of anthocyanidins.

There are many reports of free anthocyanidins in grapes or wine, but an appreciable content is ordinarily impossible to demonstrate. The best evidence in grapes, and in fact in all plants, is that the anthocyanidins do not occur free in nature (Ribéreau-Gayon, 1964a; Webb, 1964; Harborne, 1967). Russian workers particularly (e.g., Nilov and Begunova, 1953; Samvelyan, 1959; Yakivchuk, 1966) have claimed that the fresh grape pigment includes a sizable fraction of anthocyanidins and even that these are converted to glucosides in wine! Although this conclusion seems clearly erroneous, we believe it can be explained by reinterpretation of the experiments. Analysis of anthocyanidins separately from anthocyanins has often depended upon extraction of the putative anthocyanidins by solvents such as amyl alcohol, leaving the more water-soluble glucosides in the aqueous layer. We believe that the acylated pigment glucosides, important in grapes but uncommon in other plants, have caused the confusion by being initially solvent-soluble (interpreted as anthocyanidins in fresh grape extracts) and then hydrolyzing in solution (young wine) to the anthocyanidin glucosides which appear in the aqueous layer.

Acyl groups from anthocyanins seem particularly labile. The caffeic acyl groups appear to be hydrolyzed more readily than the *p*-coumaric in the anthocyanin series, but both disappear rapidly (Webb, 1964). Other phenols which are bound in the grape by esterification are freed by hydrolysis and can be found free in wine even though they generally do not occur free in the grape. Besides the acyl groups from anthocyanins this includes the gallates in the catechin series, the *p*-coumaric, caffeic, or ferulic esters of tartaric acid, and any similar quinic depsides. The hydrolysis rate may be rapid or it may be slow. The complete disappearance of the parent depside may be early or delayed. The chlorogenic-acid-like substances may be detectable for a considerable time in wine.

To return to the question of the occurrence of free anthocyanidins in wine, it appears that they could arise from hydrolysis of their glucosides or from anthocyanogen conversion. The hydrolysis of glucosides in wine seems to be relatively rapid for the flavonol series (Ribéreau-Gayon, 1964a,b), but the hydrolysis rate for anthocyanins is obscured by other reactions such as polymerization. Since Somers (1966a, 1967a) found considerable sugar in the anthocyanin-bearing polymerized tannin fraction, hydrolysis before polymerization does not appear to be required. The natural anthocyanidins with a 3-hydroxyl group are very unstable (Jurd, 1963; Jurd and Geissman, 1963), and release of this group by hydrolysis of 3-glucosides causes rapid conversion to other products, with the result that free anthocyanidins should have only a very transient existence in wine (Yang and Steele, 1958; Berg, 1963; Daravingas and Cain, 1968). The nature of these "other products" is not yet clear, and, as with many other degradation and storage-related reaction products in a complex solution such as wine, they are difficult to study. The 3,5-diglucosidic anthocyanidins seem to be appreciably more stable to decolorization than the 3-glucosides, but appear to brown more readily in wine (Robinson *et al.*, 1966b). The B-ring hydroxylation pattern also affects anthocyanin stability, delphinidin being least stable.

If anthocyanidin is produced in wine by conversion of anthocyanogens, this conversion would be apparently limited to cyanidin and perhaps a small amount of delphinidin. The leucocyanidin content of a wine does not appear to decrease appreciably with time (e.g., Berg, 1963) unless precipitated or removed by adsorption. An increase with age in the relative proportion of cyanidin in the total pigment dissolved in red wine has been interpreted as a result of leucocyanidin conversion (Ribéreau-Gayon and Ribéreau-Gayon, 1958), but we believe, rather, that this results from preferential precipitation of the less polar forms, particularly malvidin derivatives. The same effect is believed to be the cause of a decrease in the methoxy-anthocyanins found in wine with time, and not, as originally reported, actual disruption of the ether linkage (Ribéreau-Gayon, 1963d; Spaeth and Rosenblatt, 1950).

It has been common practice to produce and store wines in wooden containers, and to stopper bottles with cork. Today this is no longer the only alternative, and not all modern wines are ever in wooden cooperage or corked, though existing equipment continues in use and since the older methods have some desired effects, the traditional ways are still quite common, particularly with premium-quality wines. Thus, without specific knowledge to the contrary it must be assumed that commercial wines may have been in contact with wood. The wooden container is somewhat porous, and it is possible, depending on the procedure followed, that diffusion from within the walls can transfer a small portion of the previous contents to the following fill of the container. For that reason it is usual to use a specific wooden container for only one type of wine. Even so, there are reports of many examples of red pigment contamination in white wines, etc., from this cause. As mentioned earlier, transfer of a trace of salicylic acid through this means seems to be possible for some time if a barrel has ever been filled with a wine treated with the once-usual high level of added salicylic acid. Contamination of *vinifera* wines with anthocyanin diglucosides from this source has been mentioned. Loss of alcohol and water through the wood, and transfer of oxygen, also have effects on the phenolic concentration and reaction products in a wine. Only some of the specific substances which wood may contribute to wine will be covered in this section, with later sections giving more amounts and aging relations.

Woods which have been used for wine containers include various species of white oak, the most common, and in specific areas redwood, acacia, pine, chestnut, certain eucalypts and others. A study of heartwood constituents which aqueous ethanol could extract from these woods would reveal several which are low or absent in grapes (Hillis, 1962). Except for oak, relatively little study has been made of the occurrence of wood extractives in beverages. Prillinger (1965) has noted that wines stored in acacia barrels can be identified by yellow dyes with intense fluorescence taken up from the wood. From their properties it appears probable that these substances are phenolic. Redwood and certain other Pinaceae from Pacific coastal areas contain unusual $C_{17}$ phenolic compounds and tannins derived from them (Riffer and Anderson, 1967). Guillemonat *et al.* (1963) presented evidence that cork contains caffeic acid esterified to an unusual $\omega$-hydroxylated $C_{22}$ fatty acid. This compound appeared to be present at 4–20 g/kg of cork.

The oak extractives of interest appear to be primarily the ethanol-soluble lignins, hydrolyzable tannins and their fragments, gallic and ellagic acids, and the aldehydes and acids related to vanillin. Oak barrels could also be a source for rhamnose (Dzhanpoladyan and Dzhanazyan, 1962; Otsuka *et al.*, 1963) in wine, which might help explain the apparently erratic occurrence of this sugar in wine, already mentioned in connection with quercitrin. Scopoletin has been reported as an oak extractant in whiskey (Baldwin *et al.*, 1967). Hennig and Burkhardt (1962) found ellagic acid low, and gallic acid absent, for the first few days of extraction of fresh oak chips, but both became major components with time. Along with these, paper chromatography revealed one unidentified blue-fluorescing substance and eight substances testing for phenolic groups. These eight were postulated, from color reactions, etc., to be ellagitannins. Of these, six gave interesting double-spots in one-phase adsorptive solvents. These components were possible ingredients in wine. Otsuka *et al.* (1965) also found gallic and ellagic acids along with vanillic and syringic acids in distilled beverages aged in oak barrels.

Skurikhin (1959) fractionated the "tannin" extracted by brandy from oak, and concluded from methoxyl determination that 24–45% of the total phenol was lignin. Ethanol-soluble lignin is considered to be an important part of the extractives from wood into beverages, and ethanolysis of lignin is believed to be the origin of at least part of the phenolic aldehydes contributed by oak and other container woods to wine (Skurikhin, 1957, 1959, 1962, 1966; Egorov and Egofarova, 1965). Black *et al.* (1953) identified, by paper chromatography and other tests, vanillin, syringaldehyde, coniferylaldehyde, and sinapaldehyde in alcoholic extracts of oak. In conformity with taxonomic expectations, both coniferylaldehyde and sinapaldehyde were found in several other hardwoods examined, but only coniferylaldehyde in softwoods, except redwood, which contained neither. Others reporting these compounds in distilled beverages stored in oak include Egorov and Borisova (1957), who also reported *p*-hydroxybenzaldehyde, and Martin *et al.* (1965), Baldwin *et al.* (1967), and Skurikhin (1967).

The classic source of quercitrin is the genus *Quercus* (Karrer, 1958), and its presence in wine from oak extraction seems quite possible. Rutin has been reported in aged brandy (Lichev, 1962), but it was not found on paper chromatograms of freshly prepared ethanol extracts of oak shavings. Long contact with wood appears necessary for appreciable extraction.

# III

# Quantitative Phenolic Composition

## A. *Methods of Analysis*

Two broad types of phenol assay are useful—those which attempt to determine all phenols as a group distinct from nonphenols, and those which assay specific individual substances or classes of phenols. Total phenol determinations would appear to be the more basic, since the significance of specific phenols is clearer if viewed in relation to the total phenolic content. The capabilities of the methods used must be remembered in comparing results reported by different workers, particularly in earlier research. The term "tannin" content may mean total phenollike substance in one instance, and true leather-producing tannin or some other fraction in another instance. As an extreme example, all the colloids precipitated by barium hydroxide in a conductimetric titration procedure were called tannin by Dutoit and Duboux (1909). Today we know that pectins and other substances contributed heavily, because the values obtained (2.6–7 g/liter for white wines and 3.5–13 g/liter for red) are almost 10 times the expected values. In fruit and wine studies the comparison standard has traditionally been oak-gall gallotannic acid. Results are often expressed as "tannin" content, although it is well recognized that anthocyanins and other phenols are included unless specific methods are used.

## 1. Total Phenol Content

From the viewpoints of arrangement of hydroxyl groups, carbon skeletons, and other substituents, the natural phenols are rather heterogeneous. The reactions upon which phenol determinations rest are often not unique to phenols and are subject to variability and interference unless great care is taken in their use. In most cases individual phenols differ somewhat in their response in a given analytical procedure. Use of a single phenol as a comparison standard makes the results necessarily somewhat arbitrary. As a consequence, many procedures have been suggested for the estimation of natural phenols, and most of them have been applied to grapes or wine. Of course, even very poor methods may give satisfactory relative values on simultaneously analyzed samples, and much valuable information has been obtained with less than perfect procedures.

The earlier methods of total tannin determination applicable to wine have been reviewed by Joslyn (1950), Amerine (1954), and Durmishidze (1955). Other procedures, not all used in wine analysis, are discussed by King and White (1956) and Pridham (1964). The oldest methods still in common use with wine are those closely related to the method of Löwenthal (1860, 1877) as modified and applied to wine by Neubauer (1871, 1872). Alcohol, which would interfere, is evaporated, and portions are titrated with potassium permanganate to an indigo-carmine end point, before and after treatment with activated carbon (Amerine, 1965). The phenol content is usually expressed in terms of the tannic acid equivalent to the permanganate consumed, but the arbitrary nature of the values is sometimes emphasized by using permanganate index numbers such as ml of 0.01 $N$ $KMnO_4$.

The Löwenthal-Neubauer method depends upon the fact that the natural phenols have a relatively high affinity for carbon and can be readily and apparently completely adsorbed from wine. Furthermore, most potentially interfering substances such as ascorbic acid, sugars, alcohols, lactic acid, and oxalic acid are not readily adsorbed by carbon (Singleton and Draper, 1962; Singleton 1964) and thus are subtracted from the total attributed to phenols. Williams (1952), however, reported that only phenols containing *o*- or *p*-dihydroxybenzene structures in some form reacted in titrations with permanganate-indigo-carmine, whereas phloroglucinol, resorcinol, or tyrosine, for example, did not. Cinnamic acid gave a small consumption of permanganate. Smit *et al.* (1955) also found that Löwenthal methods gave variable values for different phenols. Values for resorcinol and phenol, however, were definite, though depressed. The relative reactivity of pyrogallol was higher than that of catechol in both studies. Nakabayashi (1962) found that the end point in the Löwenthal titration corresponded to the maximum absorbance at 265 mμ of a peak shifted from the normal 275 mμ, but that a competing reaction existed involving loss of the 265-mμ absorbance. The amount of permanganate consumed per unit weight by different catechins and other phenols did not seem related to their phenol content in any simple fashion. Naringin, for example, consumed high amounts of the reagent compared with rutin, and rutin was appreciably higher than most, though not all, of the nonglycosidic phenols. In spite of such problems the method has given many meaningful, reproducible, and characteristic values for the total phenol content of wine. Bukharov and Fertman (1962b) concluded that it was the most suitable procedure for winery use.

Although the Löwenthal-Neubauer procedures are still used in some countries for official legal-regulatory wine analysis, in the United States (Association of Official Agricultural Chemists, 1965) and some other countries they have been supplanted by methods derived from work of Folin and Denis (1912). These methods are based upon the reduction by phenols in alkaline solution of complex phosphotungstic-phosphomolybdic reagents to blue pigments. Swain and Hillis (1959) considered Folin-Denis methods the most convenient for analysis of total phenols in fruit, and their techniques have been widely adopted. The procedure is subject to interference by hydrogen sulfide, sulfurous acid, hydrogen peroxide, ferrous ion, and ascorbic acid (Snell and Snell, 1953; Snell *et al.*, 1961), which may occur in wine. However, unless unusual content was present, only a few mg/liter would be likely to be attributed erroneously to phenol from these sources. Natural ascorbic acid is seldom over 100 mg/kg in ripe grapes, and is ordinarily below 10 mg/liter in wine (Ournac, 1965, 1966; Ournac and Poux, 1966). If necessary, separate analyses can provide corrections for added ascorbic acid, sulfur dioxide, etc.

After a study of the reaction conditions and properties, Singleton and Rossi (1965) recommended several improvements in the procedure, such as use of the reagent of Folin and Ciocalteu (1927), selected alkali-time-temperature conditions, and gallic acid as the standard. Not only does expression of results in gallic acid equivalents avoid the semantic problem with "tannin" content but, fortuitously, the molecular weight per reactive phenolic group is the same as that for pentadigalloylglucose tannic acid and thus gives numerical values directly comparable with those of the previous practice. They found that the molar color yield under the improved conditions was reasonably predictable for the series of phenols they tested. Monophenolic compounds gave similar molar color yields except that phenols with suppressed ionization, such as salicylic acid, gave somewhat lower color yields than normal. The color yield of the polyphenols tested was equal to that predicted from the number of phenolic groups present affected by quinoidal reaction possibilities, i.e., phloroglucinol reacted as a monophenol, catechol and pyrogallol both as diphenols, and *d*-catechin as the sum of catechol plus phloroglucinol, three phenol equivalents. Coumarin and sulfur dioxide interfered only slightly, and quercetin reacted more than catechin, suggesting additional enol contribution by the 3- and 4-position substituents. Ascorbic acid would interfere by producing molar color yield at least equivalent to a monophenol. Similar results, though with somewhat greater variation and less predictability, were found for the older Folin-Denis procedure (Singleton and Rossi, 1965; Swain and Goldstein, 1964). Although further investigation with more model compounds is needed, it would appear to be possible from these data to determine the total phenol content by Folin-Ciocalteu methods and calculate the contribution to that total of individual phenols of known structure estimated by other means.

Since most phenolic substances have intense absorption maxima in the vicinity of 220 and 280 mμ, it might be possible to estimate total phenols by ultraviolet spectrophotometry (Pridham, 1964). In wine, however, the peak at shorter wavelength is subject to great interference by many constituents of wine, and other wine substances such as carbonyl compounds have maxima or appreciable absorbance near 280 mμ. Lipis *et al.* (1964) reported that 280-mμ absorbance of diluted wines, compared with calibration data with tannic acid, gave reproducible estimates for total wine phenols. Tsagareli and Shikhashvili (1959), however, recommended the determination of wine phenol absorbance at 313 mμ to avoid interferences occurring at the shorter wavelengths. Considering that the major phenolic substances of most wines or grape extracts do not have an absorption maximum at 313 mμ, this represents a compromise with good spectrophotometric practice. The standard of comparison was black tea tannin. Nine wines averaged 561 mg tannin/liter by their technique, and 417 or 437 mg/liter in Löwenthal determinations by two laboratories. The 313-mμ absorbance method appeared to give particularly high values for the two sweet wines included, compared with the Löwenthal values. Most other workers who have applied spectral absorbance methods to tannin estimation in wine have combined them with preliminary separations to minimize interference. Differential methods such as spectral shifts in acid and alkaline solution (e.g., Maranville and Goldschmid, 1954; Pridham, 1964) apparently have not been evaluated in wine for phenols as a general class of substances.

Comparisons between the values for total phenols of wine or similar products by different methods show reasonably close agreement considering the degree of reproducibility of the individual methods. Smit *et al.*

(1955) showed considerable disagreement between Löwenthal-Neubauer and Folin-Denis methods on different known phenols, but reasonably close agreement with commercial tannins, and excellent agreement with fruit tannin preparations. Neubauer (1872) reported white musts at about 100 mg phenol/liter, red juice about 175 mg/liter, white wines about 300 mg/liter, and red wines 900–2600 mg/liter. In an extensive collaborative study of the values obtained, Valaer (1941) found that a brandy sample averaged 244 mg/liter total phenol by Löwenthal methods, and 220 mg/liter by the Folin-Denis method, with respective ranges by different analysts of 190–300 and 180–250 mg/liter. Pro (1952, 1955) reported that the Folin-Denis method had the advantages of speed and accuracy over the Löwenthal method, but that in many hundreds of samples of various types of wines and spirits the results were in line with each other. Folin-Denis assays by five analysts gave 250–272 mg/liter tannic acid equivalent for a dry vermouth, 375–385 mg/liter for a sauterne, 1670–1740 mg/liter for a burgundy, and for these and other fruit wines a mean standard deviation of 12.8 mg/liter. Korotkevich (1960), using Löwenthal methods, obtained 210 mg/liter and 143 mg/liter for typical table and fortified white wines, respectively, and 1120 mg/liter for typical table or fortified red wines. Similarly, an ammonium molybdate method he developed gave 147, 200, 1045, and 1200 mg/liter, respectively, for the same wines. Sugar apparently interfered with the ammonium molybdate method. Joslyn (1950) considered 200 to 900 mg/liter typical for the total "tannin" content of white table wines, 100–400 mg/liter for white dessert wines, 1000–4000 mg/liter for red table wines, and 400–1600 mg/liter for red dessert wines. Amerine (1954) summarized much data indicating similar values. Disregarding the origin and wine type, 947 white wines averaged 491 mg/liter, and 405 red wines averaged 1682 mg/liter. Many similar values could be cited, some determined by other methods not widely used, such as bromine absorption (Amerine, 1954).

## 2. Analysis of Fractions of the Total Phenols

Several general methods may be used for estimating or determining specific members or classes of phenol in a mixture of other natural phenols. Often, much can be learned by comparison of the total phenol content of several samples prepared so as to lack certain fractions. For example, Neubauer (1872) concluded that the anthocyanin content of red wines contributed only about 100–200 mg/liter to the total phenol calculated as tannic acid based upon comparisons of wines and musts with different color levels but prepared similarly.

Ultimate identification of specific individual phenolic compounds usually involves preliminary isolation by some countercurrent method, usually chromatographic. Once the compound to be identified is completely separated from all other grape or wine components, its quantitative determination is simple. Isolation losses are often large, however, and, especially if the compound is labile, the amount determined may be well below the true content. To combat this effect the desired compounds are often determined in relation to marker components assumed to be equally labile or as a proportion of an inclusive group of similar compounds determined in total by other means. Quantitations depending upon complete isolation of individual substances, have, for the most part, already been mentioned in connection with qualitative identifications. Since such techniques are somewhat different for each substance and are well understood by food scientists, not much need be said about these techniques in general terms.

Fractional analysis short of complete isolation and determination of each individual substance may make use of specific properties or reactions which are not common to all phenols or may depend on more or less complete separation of groups of substances before analysis by less specific methods. Many procedures that depend on group separations are not very precise, and at best, must be considered estimates. Williams (1952), in work with cider, concluded that the different classes of phenols, such as tannins and nontannins, intergraded so that attempted analysis by fractional precipitation or extraction was generally pointless. Study of data available on wines suggests that this is only slightly overstated.

The several methods based upon heavy-metal salt precipitation seem quite arbitrary and not satisfactorily related to specific fractionation among phenols. If phenol precipitation is complete, as it may be with lead or mercury salts, for example, values are found of the same order of magnitude as with the Löwenthal and Folin-Ciocalteu methods (Laborde, 1908-1910). If precipitation is partial, however, the separations between various types of phenols are seldom sharp. The degree of success has been greatest with separation of anthocyanins and other very soluble phenols from "tannins" in wine. It has been found that the anthocyanidin diglucosides of wines from non-*vinifera* grapes may not be completely precipitated by lead or mercury salts unless special precautions are taken and zinc is more likely to leave anthocyanins in solution (Biol and Foulonneau, 1963; Deibner and Bourzeix, 1963).

Precipitations with zinc salts remove a greater proportion of the total phenols in white than in red wines (Laborde, 1908–1910). Kràmszky (1905) reported that red color was not removed from red wine but other phenols were precipitated by ammoniacal zinc sulfate. A white wine gave a total phenol value of 155 mg/liter by the Löwenthal-Neubauer method, and 152 mg/liter by zinc precipitation. For 14 red wines, however, with a Löwenthal total of 1000–2940 mg/liter, he subtracted the amount precipitated with zinc, and the difference, 200–800 mg/liter, was attributed primarily to anthocyanin. Unfortunately, not only anthocyanins are left in solution, and a Tokay (white) with 1400 mg/liter total gave 500 mg/liter nonprecipitable phenol by his techniques.

Nègre (1939a,b, 1942–1943) adjusted the pH with ammonia so that gallic acid would not precipitate with zinc acetate. He determined the total tannoids as those permanganate-oxidizable substances precipitable by lead acetate, and found, for example, 850 mg/liter in a "Pinot noir" red wine and 370 mg/liter in a "Chardonnay" white wine. Similarly, the zinc-precipitable oxidizable substance, or "tannin," was respectively 330 mg/liter and 40 mg/liter, leaving (by difference) polyphenols not tannin of 520 mg/liter in the red and 330 mg/liter in the white wine. Comparing his data for the same red variety vinified as a white and as a red wine, and subtracting the nontannin phenol of the white from that of the red, suggests about 400 mg/liter in tannin equivalents for the anthocyanin content of the young red wine. Data from a ferric precipitation method (Meyer, 1918) estimated 0–870 mg/liter total "tannin" in white wine, 500–4400 mg/liter total "tannin" in red wine, and 200 mg/liter as the anthocyanin content in red wine.

To determine true tannins in the leather sense, an obvious method is adsorption on hide or other proteins. Stiasny's formaldehyde reagent precipitates condensed tannins but not gallotannins (Dreaper, 1910). Since Ponte and Gualdi (1931) obtained comparable precipitate removal with formaldehyde and with hide powder, they concluded that the true tannins of wine were all of the flavanol-derived condensed type. Manceau (1895) tested several methods for true tannin content in wine and recommended that Girard's catgut adsorption be combined with Löwenthal titration. With this method he found 8–50 mg/liter of protein-adsorbable tannins in champagne. This figure seems in line with the low level of polymeric tannin to be expected in white wine. Fresenius and Grünhut (1921), however, reported that the apparent adsorption of dissolved solids by hide powder was approximately equivalent to their Löwenthal-Neubauer total. Diemair *et al.* (1951a) found that combined precipitation with lead acetate plus gelatin followed by tungsten-blue determination gave values on a wide range of wines and juices which were consistently only 50 to 85% of those with the carbon-adsorptive Löwenthal-Neubauer technique.

Protein and proteinlike precipitants such as polyamide and polyvinylpyrrolidinone (PVP) act as tannin or phenol adsorbents by essentially the same mechanism, primarily hydrogen-bonding (Singleton, 1967a,b). The use of gelatin precipitation to separate total phenols into true protein-precipitable tannins and nonprecipitable polyphenols was used, for example, in several studies of grape juice (Caldwell, 1925; Reynolds and Vaile, 1942; Webster and Cross, 1944). In heated pressed juice of the "Concord" type, the total Löwenthal phenol content was of the order of 1000–4000 mg/liter, and the portion precipitable with gelatin varied but was seldom over 50%. Zitko and Rosik (1962) presented data showing that gelatin removed less of the original total phenol as the dose was increased, but that there was no sharp "end point" beyond which gelatin removed no more phenol. Pokorný *et al.* (1966) recommended polyamide adsorption instead of carbon adsorption in the Löwenthal-Neubauer procedure as a more selective and convenient procedure. Twenty wines ranging from 510–3380 mg/liter by the carbon procedure were 230–870 mg/liter lower by the nylon procedure, but they were statistically linearly correlated. Since white and red wines were included, a linear correlation suggests no valid fractionation of phenol by types. The data might be construed to show that a relatively constant 230–870 mg/liter of the total phenols was not adsorbable on polyamide, and presumably consisted of the smaller molecules. Herrmann (1963b), however, has shown that appreciable portions of relatively small soluble phenols like caffeic acid, catechin, and rutin may be adsorbed by polyamide. Rossi and Singleton (1966b) found that, depending on operating conditions and particle size, smaller phenols may be adsorbed on polyamide or PVP even in preference to larger. Of course it is known, and was verified in the same study, that there is ordinarily a preferential removal of the larger, more astringent phenolic tannins compared with the smaller, more soluble phenols by the soluble proteinaceous fining agents, but this difference is apparently not clear-cut enough for satisfactory analytical use.

Leucoanthocyanidins or anthocyanogens are ordinarily not glycosides, and in wine they are evidently nearly all polymeric to some degree at least. As a result they have a relatively high affinity for polyamides and similar agents, a property that has been used for partial separation of these compounds from other color-forming substances to improve analysis (Masquelier, 1963; Cantarelli, 1963a,b; Berg and Akiyoshi, 1962). Such separation has not usually been attempted, however, and the amount of red color generated by treatment with hot acid, usually hydrochloric, has been used to determine the leucoanthocyanidin content. The true or astringent tannins of wine and grapes are polymeric anthocyanogens; therefore the leucoanthocyanidin content also measures the tannin content. Laborde (1908–1910) estimated that the quantity of pigment produced by treatment with hydrochloric acid at 120°C was essentially equivalent to the weight of tannin in green grapes or wine. Based on the fact that he found a 1-g/liter solution of Bordeaux-red dye equivalent in color intensity to a 1-g/liter solution of isolated grape pigment, he reported that three red wines had 1400–1600 mg/liter of anthocyanin but that the color increased to 3600–5400 mg/liter when the wine was heated with acid. By difference, the tannin content would have been 2200–3800 mg/liter, which correlated quite well with results of his other methods of tannin determination on the same wines.

Masquelier *et al.* (1959) estimated the monomeric leucocyanidin by subtracting from the total leucoanthocyanin the polyleucocyanidin, the portion precipitable from dealcoholized red wine with sodium chloride. The pre-existing anthocyanin color was subtracted, and the leucocyanidin content was determined from the color generated by heating in aqueous hydrochloric acid in comparison with that produced by a given weight of pure leucocyanidin under the same conditions. In 18 red wines of various ages the total anthocyanogen content was 1100–4160 mg/liter, the anthocyanogen precipitable by salt was 17–66% of that value, and that remaining in solution was 34–83%. The highest proportion of the total was insoluble in the youngest wines. The total anthocyanogen content of white wines by the same workers (Masquelier *et al.*, 1959, 1965; Masquelier, 1963) was zero to 93 mg/liter.

The anthocyanidin yield from anthocyanogens is greatly affected by the conditions of the conversion reaction (e.g., Peri, 1967a,b; Joslyn and Goldstein, 1964b). Yield in water solution is ordinarily quite low but can be improved to 30% or so by the use of butanol or similar alcohols and carefully chosen conditions. Jurd (1966a) improved yields to 75–80% by oxidation with benzoquinone to suppress side reactions. It therefore seems important to use an anthocyanogen standard under the same analytical conditions. Cantarelli and Peri (1964a) used a pure cyanidin chloride solution as a secondary standard but converted the absorbance to equivalent isolated grape seed tannin-anthocyanogens, determined separately. The anthocyanogenic tannin content determined in clusters of several grape varieties was 25–99% of the total Folin-Denis phenol content. Ribéreau-Gayon and Stonestreet (1964, 1966a,b) applied anthocyanogen conversion methods in their important studies of grape and wine tannin.

Catechins, leucoanthocyanidins, and other flavonoids with a reactive phloroglucinol moiety may be determined colorimetrically by reaction with vanillin and similar aldehydes. Quercetin and similar 4-keto compounds do not react (Swain and Hillis, 1959; Bell *et al.*, 1965). Because the phenols reacting in wine were heterogeneous (partially soluble in ether; partially insoluble but precipitable with lead), v. Fellenberg

(1912) considered the vanillin color reaction of no quantitative value. A potential value in qualitatively distinguishing pomace wine from juice wine was indicated. The reaction was quantitatively adapted to wine by Burkhardt (1963) and Rebelein (1965a,b). Absorbance at the maximum near 500 $m\mu$ is measured after reaction of the wine with vanillin in strong hydrochloric acid. Results are calculated with reference to a *d*-catechin standard, but leucocyanidins and condensed tannins react also. The "catechin" content of 98 white wines averaged 74 mg/liter (range 20–330 mg/liter), and 8 wines prepared from fermentation of diluted white pomace averaged 930 mg/liter (range 470–1440 mg/liter) (Rebelein, 1965a,b). Values were determined for 39 red wines, but the average of 2150 mg/liter and maximum value of 7900 mg/liter seem too high and suggest interference from the more intense but similar color of anthocyanins, which increases in absorbance at low pH or anthocyanogen conversions. Comparisons with a series of white musts and wines (Burkhardt, 1963) showed values of the order of 200 mg/liter for total Löwenthal phenol, 30 mg/liter leucocyanidin by a nylon-butanol-HCl method, and 180 mg/liter of "catechin." Values with individual samples were not always consistent, however, and the leucocyanidin value might exceed the catechin value or the catechin value exceed the total Löwenthal value.

Goldstein and Swain (1963) have pointed out that the vanillin reaction is suppressed when flavanols polymerize, and, relative to the Folin-Ciocalteu value, this suppression is an indicator of the molecular size within a group of condensed tannin polymers. Rossi and Singleton (1966a,b) presented data suggesting that a relatively high ratio of vanillin to Folin-Ciocalteu value is indicative of a wine with a high catechin content and a high browning tendency. Ribéreau-Gayon and Stonestreet (1964, 1966a,b) also used the ratio of vanillin color to total tannin, but the total tannin measure was that from anthocyanogen conversion. They reported total leucoanthocyanidins of 1500–4400 mg/liter, and vanillin-to-anthocyanogen ratios of 1.43 to 0.43, in wines from the same source for different years. The vanillin-to-anthocyanogen ratio was lower in the older samples, apparently indicating polymerization with time, in contrast to results of Masquelier *et al.* (1959).

The polymeric substances of grapes and wine have been studied directly by gel-filtration methods. Wucherpfennig and Franke (1963, 1964) obtained two or three phenol elution peaks from gel columns. The peak areas reflected wine or must treatment and appeared capable of separate estimation of simple and polymeric phenols. Somers (1966a,b, 1967a,b) developed a gel-filtration technique which separates the anthocyanins of grape skin into an acyl-glucoside fraction, a glucoside fraction, and a red polymeric tannin fraction. In a single berry of the Shiraz variety the pigmented tannin at 1.08 mg represented 12% of the color and 42% of the weight of the pigment substance. Similarly, the acylated fraction was 0.75 mg, 40% of the color and 29% of the weight, and the anthocyanin fraction was 0.75 mg, 48% of the color and 29% of the weight. In red wine the acylation was lost and the proportion of pigmented tannin was higher and increased with time.

Solvent extraction, particularly with ethyl acetate, has been used to fractionate the phenols in aqueous solutions prepared from grapes or wines. Glycosides, chlorogenic acid, and similar highly polar phenols would not be extracted readily, nor would the highly polymeric condensed tannins. In fact, Garino-Canina (1924) considered that tannin was absent from wines because he found gallotannic acid added to wine extractable with ether, but without such addition no tannin was extracted. Colagrande (1961) reported that, by ultraviolet absorbance (275 m$\mu$) calculated with reference to a tea-tannin standard, white wines contained 380–1100 mg/liter of ethyl-acetate-extractable phenol, rosé wines 940–1550 mg/liter, and red wines 1450–2350 mg/liter. The phenol not extracted was not reported. Margheri (1960), with a similar method and tannic acid standard, found phenol extractable with ethyl acetate in 30 wines to be 200–320 mg/liter for whites, 360–730 mg/liter for pinks, and 580–1520 mg/liter for reds. Tarantola (1951) found that the ethyl-acetate-extractable phenols represented an average of 16.4% of the total Löwenthal-Neubauer phenols of 8 red wines, 40.7% in 2 rosé wines, and 39.0% in 6 white wines. Ribéreau-Gayon and Maurié (1942) and Maurié (1942) reported, by the same techniques, an average of 21.4% of the total phenol extracted by ethyl acetate from red wines, and 30.8% from white. They also reported that values and ratios were similar whether ether was the extractant or whether bromine absorption rather than permanganate oxidation was the means of quantitation. They reported that quercitrin was extracted by ethyl acetate under their conditions.

If one assumes the total phenol content of a white wine to be about 400 mg/liter, and that of a red wine to be about 1500 mg/liter, the extractable phenols would, from these data, represent about 280 mg/liter for the red wine and about 140 mg/liter for the white, i.e., much closer to each other than the totals. It would be expected that catechins, caffeic acid, and similar compounds of relatively low polarity in acid solution would make up the ethyl acetate extract, along with more or less of the flavanol oligomers, depending upon the exhaustiveness of the extraction (Ribéreau-Gayon and Stonestreet, 1966a). Pifferi (1966) detected a total of 33 phenolic substances of various classes, including some rather polar ones, on paper chromatograms of ethyl acetate from 6 equal-volume consecutive extracts of 3-fold-concentrated wine at pH2. Therefore, while the ethyl-acetate-extractable total agrees reasonably well with the expected total of phenols in wine other than condensed tannins and anthocyanins, it should be considered for the present to be an estimate, not a definitive value.

Other solvents may be used but are likely to suffer from similarly imprecise simple batch fractionation. Masquelier (1956) found in white dry Bordeaux wine 10–50 mg/liter of leucoanthocyanidins soluble in saturated aqueous NaCl and extractable with butanol but insoluble in a mixture of 4 volumes of ether per volume of butanol. An additonal 50–300 mg/liter of phenolic substance was extractable from the ether-butanol with an alkaline solution. Sweet botrytized white wines had respectively zero and 100–500 mg/liter of the same two fractions.

Nearly all the phenolic substances of wine can be readily and completely extracted by ethanol from wine saturated with ammonium sulfate (Singleton, 1961). This procedure can be useful, for example, in removing phenols from potentially interfering sugars in sweet wines prior to other analytical procedures (Harvalia, 1961; Peri and Cantarelli, 1963). Widely used as a qualitative method since work of Willstätter and Everest (1913) has been extraction of anthocyanidins with amyl alcohol from aqueous solution which retains anthocyanins. Although such separation is not absolute (Rosenheim, 1920a) it has been very useful and has contributed to the generally accepted observation that anthocyanidins do not exist naturally in a free state in plants to a detectable degree. Durmishidze and Khachidze (1952) estimated colorimetrically total wine pigment, the amyl-alcohol-extractable "anthocyanidin" portion, and, by difference, anthocyanins. Nikandrova (1958) improved the procedure. We have already explained that we believe extraction of acylated anthocyanidin glucosides to confound these results on grape skin extracts and very young wines. With older wines, solvent soluble oligomeric anthocyanin-bearing tannins may interfere, as may degradation products with appreciable absorbance near that of anthocyanins in the visible spectrum.

Anthocyanin estimation in wine and grape extracts by means of the group's visible absorption spectrum has proved to be more difficult than might have been expected. With a few exceptions, such as Laborde's (1908–1910) dye-equivalence method, already mentioned, the earlier workers used methods giving relative color values and made little effort to convert them to pigment content. Similar color comparative methods have been used with white wines to estimate yellow or brown color intensity. Many of these procedures are described in reviews by Kielhöfer (1944), Amerine (1954), Amerine *et al.* (1959), and others. For general winery use they may be very convenient and useful, but a chemically more meaningful determination is desirable.

It has been known for a long time that the shade and intensity of color exhibited by anthocyanins in tissues or solution is greatly affected by pH and other characteristics of the medium. It is only in recent years, however, that these effects are becoming understood in sufficient detail to explain the observations. Direct measurement of pigment by color in wine is subject to special problems. Since anthocyanins chelate metals (Ingalsbe *et al.*, 1963; Somaatmadja *et al.*, 1964), the metals present in juice or wine, would affect the color (Szechenyi, 1964), and other phenols may also influence the color at the pH of wine by further complexing with the metal-anthocyanin (Jurd and Asen, 1966). Measurement of pigment by visible absorbance would be influenced in a complex solution like wine. Sulfurous acid is usually added to wine and affects the color displayed by a given amount of original pigment (Timberlake and Bridle, 1966a, 1967a,c). Color exhibited by flavylium salts, including anthocyanins, in relation to pH has recently been shown to be subject to quite complex reactions (Jurd, 1963; Jurd and Geissman, 1963; Timberlake and Bridle, 1966b, 1967b,c; Harper and Chandler, 1967a,b). Anthocyanidins at the pH of wine are decolorized rapidly and irreversibly, and evidently cannot therefore be an important factor in the color of the red wines. However, the anthocyanins of the type in grapes (3-glucosides) are in reversible equilibrium with their colorless carbinol base at pH of wine. As the pH is lowered, a higher proportion of the pigment is in the red flavylium cation form, but it requires a very acid medium, at least as low as pH 1.5 and apparently, preferably equivalent to about pH 0.6, for 100% conversion to this red form, and thus better quantitation (Timberlake and Bridle, 1967b; Pataky, 1965; Ribéreau-Gayon and Stonestreet, 1965).

The visible absorption maximum of red wines or grape extracts, depending on the solvent, is about 520–540 mμ, about that expected from the isolated pigments. Owing to the problems attendant upon isolation and maintaining purity of anthocyanidins, plus the spectral considerations just outlined, the relatively few values in the literature for the molar absorptivity are rather variable, but the highest, and probably best, values appear to be about 36,000 to 48,000 for all the natural grape anthocyanidins and their glycosides (Ribéreau-Gayon, 1959; Ribéreau-Gayon and Stonestreet, 1965; Somers, 1966b; Stevenson, 1965; Fuleki and Francis, 1968). Experimental values ¼ to ½ as high have been used, and could, of course, produce a calculated wine pigment weight content as much as 4 times as high as would be correct for the proper conditions of measurement.

Since the absorption maxima and molar absorptivity are quite similar for the constituent anthocyanins of grapes, quantitative estimation of them as a group seems practical. A problem exists in that the absorption spectrum of the anthocyanins in wine is superimposed upon an overlapping background of generally nondescript yellow-brown absorption of variable intensity. The ratio of absorbance at 420 m$\mu$ and 520 m$\mu$ (or similar comparisons) shows that older wines tend to have more brown and less red components (Sudraud, 1958). To compensate for this non-anthocyanin color and other effects in different samples, methods have been applied involving measurement of color before and after pH shift or selective destruction of anthocyanin.

Ribéreau-Gayon and Stonestreet (1965) used relatively carefully validated techniques to analyze red wines of various vintages from each of two Bordeaux sources. The two wines, less than two years old, analyzed by color increase caused by raising the acidity from pH 3.5 to 2% HCl, gave 305 and 362 mg/liter anthocyanin, in comparison with a crystalline grape anthocyanin preparation treated similarly. The same wines, analyzed by absorbance before and after decolorization with a strong sodium bisulfite solution, in comparison with the appropriate standards, gave 330 and 385 mg/liter anthocyanin. Older wines decreased to an apparent limit of about 16–20 mg/liter with the pH-shift method, and 20–27 mg/liter anthocyanin with the bisulfite method. Ribéreau-Gayon and Nedeltchev (1965) compared these two methods with the total anthocyanin determined by the method of Durmishidze and Khachidze (1952). The younger wines were generally similar with all three methods (Table II). The Russian method was higher with older wines. The amount of unmodified anthocyanin is quite low in comparison with the terra-cotta-colored "tannins" in old red wines, and the apparent limit value may well be an artifact dependent upon the analytical method (Ribéreau-Gayon and Nedeltchev, 1965).

Vil'yams and Taranova (1951) separated red wine pigment into two fractions. One was eluted from a cellulose column with dilute aqueous acid, and the other by a second elution with acidic 50% ethanol. The first was called anthocyanin, and the second anthocyanidin. In the light of work already discussed and studies by Berg and Akiyoshi (1962) and Berg (1963), it appears that a more correct term for the more tightly adsorbed fraction would be "associated pigment," "polymerized pigment," or "pigmented tannin." Much the same idea was expressed by Boutaric *et al.* (1937), who explained the failure of diluted wines to follow the Beer-Lambert law on the basis of disassociation of micelles of pigment upon dilution. The amounts by photometry reported by Vil'yams and Taranova (1950, 1951) were 115 mg/liter of free anthocya-

TABLE II
ANTHOCYANIN CONTENT (mg/liter) OF THE SAME RED WINES BY THREE METHODS[a]

| Vintage | pH shift method | $SO_2$ decolorization method | Anthocyanidin photometry |
|---|---|---|---|
| 1964 | 262 | 257 | 287 |
| 1964 | 320 | 317 | 315 |
| 1963 | 77 | 79 | 92 |
| 1961 | 102 | 102 | 120 |
| 1960 | 42 | 45 | 91 |
| 1955 | 27 | 32 | 69 |
| 1954 | 17 | 22 | 49 |
| 1938 | 17 | 20 | 53 |
| 1929 | 32 | 32 | 77 |

[a]Ribéreau-Gayon and Nedeltchev, 1965.

nin and 50 mg/liter of polymeric anthocyanin in one red wine, 288 and 260 in a second, and 317 and 300 in a third. Disregarding the separation, the total apparent anthocyanin content for the three wines would be 165, 548, and 617 mg/liter. Samvelyan (1959) reported that the total anthocyanin content of 10 wines was 230 to 705 mg/liter, averaging 403.

Clore *et al.*, (1965) determined the Folin-Denis equivalent of isolated grape pigment in terms of visible absorbance. About 1100 mg/kg of anthocyanin as tannin equivalent was indicated in "Concord" grapes. Durmishidze (1948a) developed a method for anthocyanins based on a Zeisel methoxyl determination of lead-precipitable substances. Of course, this assumes only methylated pigments and only pigments as sources of methoxyl groups, which may be approximately correct in most young European grape wines but far from correct in other samples. Durmishidze and Khachidze (1952) compared this method with their visible absorbance method, and Deibner (1952) comparatively reviewed these and other anthocyanin methods. By the methoxyl procedure the total anthocyanin content of untreated wines was, for example, 91 mg/liter for a Crimean port and 617 mg/liter for a fairly young "Saperavi" red table wine. The methoxyl method and the visible absorbance method respectively gave 1280 and 1250 mg/liter for a "Saperavi" variety red wine less than 2 years old, but 545 versus 75 mg/liter for a similar wine about 20 years old. This indicates that methoxyl measurement is a satisfactory measurement of pigment in young wines but that pigments are converted in older wines to other products without much change in methoxyl content.

Keith and Powers (1966) reported on gas chromatographic separation and potential determination of silated anthocyanins and other flavonoids. A "Concord" grape juice and a wine concentrate gave apparently characteristic chromatogram recordings, but quantitation will require further development. A part of a growing number of studies has been monitoring of liquid chromatographic column eluants to detect and quantitate phenols. Several of this type have already been mentioned in the section on qualitative analysis. Early examples of this as related to wine or grapes include studies of Spaeth and Rosenblatt (1950), Smit (1953), and Joslyn and Smit (1954). The methods of Vuataz *et al.* (1959) have been applied, in conjunction with paper chromatography, by Singleton *et al.* (1966a). A number of studies have applied direct densitometry or other methods to the estimation of relative proportions of phenols, particularly anthocyanins, separated on paper or other usually two-dimensional chromatograms. Examples are cited by Ribéreau-Gayon (1964a), Drawert (1961), and Webb (1964), and illustrated in studies by Somers (1966b), Robinson *et al.* (1966b), Rice (1965), Zamorani and Pifferi (1964), Albach *et al.* (1959), Durmishidze and Nutsubidze (1958), and others.

The Russian workers have made use of various methods of estimating the proportion of pyrogallol to catechol nuclei in the phenols of grapes or wine. The most suitable are the colorimetric methods based upon ferrous tartrate differential color production with the two types of polyphenol (Bukharov and Fertman, 1962c). However, on the basis of the known qualitative make-up of wine, the values reported (Durmishidze, 1955) seem too high where the tannins are very predominant leucocyanidins, the gallocatechin or gallate content apparently a small portion of the total catechins, and the predominant pigment, malvidin, methylated. King and White (1956) studied the method with known mixtures and found it somewhat difficult to control, particularly subject to error when pyrogallol was low with respect to catechol. Further work thus appears to be needed before these results can be interpreted satisfactorily.

Polarographic and other electrochemical methods may be applied to phenol determination. Zuman (1953) reported polarographic determination of anthocyanins in fruit juices. Ribéreau-Gayon and Gardrat (1956, 1957) used potentiometric oxidation-reduction titration with titanium chloride and dichloroindophenol in studying grape phenols, particularly anthocyanins. Bukharov and Fertman (1962a) reported polarographic determination of gallic acid in alkaline hydrolysates of tannin sources, including wines. Mareca and de Miguel (1965, 1966) studied polarography of flavonoids of wine and grapes. Keith and Powers (1965) studied malvidin-3-glucoside and other anthocyanin solutions, including strawberry juices, polarographically, and the effects heat produced on them. It appears from all these studies that the complexities encountered in natural mixtures, particularly red wines, are too great at present for meaningful quantitation by direct application of such electrochemical methods. Harper and Chandler (1967a,b) recently elucidated the rather complicated sequence of pH-sensitive polarographic reactions in solutions of certain synthetic analogs of anthocyanidins, and it may become possible to make determinations of important oxidation stages in wine phenolic fractions suitably separated from potentially interfering polarographically active wine constituents such as malic acid.

## B. Amount and Distribution of Phenols in the Grape

### 1. Harvested Ripe Grapes—Total Phenol Distribution

Ordinarily grapes are harvested by cutting the entire cluster from the vine. The harvested weight of fruit includes the cluster stems. In any of the usual modern procedures of preparing grapes for consumption as fruit or as beverages, the stems are essentially completely removed. Mechanical harvesting, which is coming into use for grapes, involves methods of shaking, etc., which remove many of the berries while leaving more or less of the stem (depending on the variety of grape) on the vine. The stems, therefore, are not normally of direct significance to a food scientist.

The stems, however, have a relatively high phenol content and might contribute significantly in this respect even though their contribution to cluster weight is relatively small. Much of the data available upon stem content is summarized in general texts (e.g., von Babo and Mach, 1923; Benvegnin *et al.*, 1947; Carpentieri, 1948; Winkler, 1962). The different varieties of grapes vary considerably in their proportion of stems. The extreme range found in the literature was from 1.3 to 9.5% of the fresh weight, with 2–7% a usual range for ripe grapes of the more common varieties. Clusters with immature berries or berries affected by molds, dehydration, etc., have a relatively high percentage of stems, and those with very plump berries have a relatively low one. The average stem content of ripe wine grapes was 3.64% calculated for 49 varieties in Europe (von Babo and Mach, 1923), 3.71% for 44 wine varieties in South Africa (Frater, 1924), and 3.44% for 7 vinifera varieties in California (Amerine and Bailey, 1959), giving a weighted average of 3.66%. This agrees well with 3.7% cited as typical in Argentina (Garoglio, 1965) and 4.0% in

Europe (Benvegnin *et al.*, 1947). These values represent complete removal by hand and may appreciably exceed stem recovery by commercial equipment, depending on the fragility of the stems and the adjustment of the equipment. Bioletti (1914) found that laboratory tests gave stem content of 3.70–7.04% in a group of varieties, whereas in his trials only 0.81–3.23% was recovered from a then-commercial type of mechanical crusher-stemmer.

The stems are relatively rich in phenolic components. In 7 varieties, Amerine and Bailey (1959) found 2,400 to 21,900 mg phenol, calculated as tannic acid, per kg of fresh weight of main stem, and 9,100 to 43,700 mg/kg in the lateral stems. Frater (1924) reported a range of 700 to 9100 mg/kg and an average of 3700 mg/kg "tannin" by the Löwenthal-Neubauer method in combined alcoholic extracts of fresh whole stems macerated for 6 weeks at 50°C. Brunet (1912) reported tannin of 15,000–30,000 mg/kg of stems, and Garoglio (1965) reported 20,000–35,000 mg/kg. von Babo and Mach (1923) reported stems of 34 wine varieties to average 27,300 mg "tannin" per kg fresh stems, with 83,100 mg/kg on a dry-weight basis.

*Vitis labrusca* varieties tested in these studies were very similar to *V. vinifera* varieties in percent of stems in the cluster and in tannin content of the stems. In fact, the tannin content of the stems seems rather typical of the rest of the green tissues of the vine. Durmishidze (1955) reported "tannin" to be 40,000 mg/kg of the dry weight of grape stems, and 20,000 mg/kg of the leaves. Diemair *et al.* (1951b) reported 1320–3600 mg tannin per kg in fresh leaves, and 1040–2200 mg/kg in their stems. The suggestion has been made that the stems of red grape varieties contain more total tannin than stems of white berries (Ribéreau-Gayon and Peynaud, 1960–1961). In wider studies, however, there is much overlap and no consistent difference. It is considered probable that, if a statistically significant difference exists, it results from the historical tendency to select small-berried varieties for red wine (more skin, more color, and perhaps more flavor), a tendency not so likely with white grapes. Thus, the average red variety would tend to have smaller berries with concomitantly more stems and more laterals compared with the average white grape at a given cluster weight.

If we consider 3.7% stems to be average, the remaining 96.3% represents the typical berry weight, which calculates to 74 lb of stems and 1926 lb of berries per harvested ton. From data by Frater (1924) the whole clusters of 44 varieties averaged 74.34% moisture when ripe, indicating 1487 lb (about 180 gallons) of water per ton. Calculated another way, the total juice represented about 1731 lb per ton and contained about 380 lb of dissolved solids, mostly sugar, for the equivalent of 190 gallons at 22°

Brix. In fact, commercial yields of juice or natural wine are usually in excess of 160 gallons per ton of grapes and approach 190 gallons per ton, depending upon recovery methods and grape variety. If all the "tannin" of the stems was extracted and dissolved in this juice or wine it could contribute, according to these estimates, from as little as 90 mg of phenol, calculated as gallic or tannic acid, per liter, to 1700 mg/liter or more, with an estimated average of 1000 mg/liter. Therefore the stems could be a very significant source of phenols if they were not or were incompletely removed before operations to produce beverages, such as hot pressing or fermentation. The removal of the stems is usual, however, because the extra tannin is not usually desirable and stems yield off-flavor and adsorb anthocyanins (Pacottet, 1904; Laborde, 1908–1910).

The total phenols of the entire seeded berry, determined after exhaustive extraction of the solid parts and expressed as gallic acid, averaged 3770 mg/kg for the ripest sampling (18.7–22.9°Brix) of 12 wine varieties (Singleton, 1966). Seven white *V. vinifera* varieties ranged from 2250–6060 mg/kg, 3 red *V. vinifera* contained 2920–5420 mg/kg, and the *V. labrusca* varieties "Catawba" and "Delaware" respectively contained 4060 and 4360 mg/kg. Estimating as for the stems, if all the berry phenols appeared in the juice or wine, the average content would represent 4580 mg of tannic or gallic acid equivalent per liter, and the range about 2700–7400 mg/liter, not considering any possible contribution from stems. Included in this study (Singleton, 1966) were the varieties "Calzin" and "Petite Sirah," which are capable of producing very astringent tannic wines, so the suggested maximum phenol content is presumably realistic. Although we have analyzed experimental wines as high as 6500 mg/liter and have seen slightly higher values in the literature, this is unusual and such wines are not commercially acceptable. As a rule, it appears that the total phenol consumed in the typical food product would be somewhat less than the total berry content unless the entire berry is eaten, seeds, skins, and all.

In the same varieties (Singleton, 1966, 1968, unpublished data) the total phenol of the berry was distributed, on the average, about 1% in the solid pressed pulp, 5% in the juice, 50% in the skins for red varieties and 25% for white, and the remaining 46–69% in the seeds. In terms of proportion of fresh weight, the seeds were about 6%, skins about 15%, and pulp plus juice about 79%. Frater's (1924) data on 44 ripe wine varieties averaged 6.79% skins, 2.91% seeds, and 90.30% pulp and juice. He found an average of about 18% of the total berry "tannin" in the skin and 82% in the seed. Although his data also indicate a tendency for "black" grapes to have more phenol in the skins, some white varieties were among the highest and some pigmented grapes were as low as the usual white-skinned berry. The average total gallic acid equivalent phenol content calculated from Frater's data was 2040 mg/kg berries, and therefore appears comparable. Aizenberg *et al.* (1966) found total "tannins" including pigments to be 1250–3490 mg/kg in 44 varieties of grapes in Armenia.

Brunet (1912) reported 9–11% skins with 1–3% "tannin" in red and below 1% in white skins, 3–4% seeds with 4–5% "tannin," and 85–89% pulp in the berry with no tannin indicated in it. Work summarized by Carpentieri (1948) indicated skin as 2.8–15.3% of the fresh berry weight, with 0.15–0.56% "tannin" for white and 0.50–4.30% for red skins. Seeds were reported as 1.1–5.9% of the berry weight with 2.1–7.4% "tannin." Benvegnin *et al.* (1947) reported seeds as 5–8% "tannin" and skins 0.5–2.0%, with specific amounts for "Chasselas" berries as 2200 mg/kg, 77% in seeds, and 23% in skins; and for "Pinot noir" 3600 mg/kg, 72% in seeds, and 28% in skins. They noted also that red grapes are not necessarily more tannic than white, and report "tannin" content of 100–300 mg/liter in cold-pressed juice from red or white grapes.

Winkler (1962) lists seeds as up to 10% of the fruit weight and 5–8% tannin, skins 5–12% of the weight and 3–6% tannin, and expressed juice tannin content as 100–300 mg/liter from white grapes and 500–2000 mg/liter from red grapes. The last figure must include teinturier and hot-pressed juices. Data from von Babo and Mach (1923) indicate that skin content ranged from 2.1–24.1% of the fresh berry weight but was more uniform, of course, in proportion to the dry substance, averaging about 3.2% of the berry dry weight. The tannin of the skins of several varieties (including *V. labrusca*) was reported as 2,100–33,000 mg/kg berries. Similarly, the average seed content of 10 varieties was 2.9% of the fresh weight and reportedly contained 1550 mg tannin per kg berries.

Korotkevich *et al.* (1950) also report that ⅔ of the "tannides" of the grape are in the seed, but that total content was 2,500–22,300 mg/kg of berries in 40 varieties, with an average of 13,770 mg/kg for a dry year and 5000 mg/kg for a normal year. Skin weight was 4.4–17.0% and seed weights 2.2–7.9% of berry weight. Skin pigment values reported were 15–450 mg/kg of berries, skin tannin 70–4360 mg/kg, and seed tannin 1,500–17,800 mg/kg. Beridze (1965) reported that a sample of physiologically ripe "Rkatsiteli" grapes had 68,600 mg of water-soluble and 5,000 mg of alkali-soluble tannin in 1 kg dry weight of skins, 102,400 mg water-soluble and 8,800 mg alkali-soluble tannin per kg dry seeds, and 73,600 mg water-soluble tannin and 12,800 mg of alkali-soluble tannin per kg of dry stems. Calculations from these data suggest a total berry tannin content of 5300 mg/kg fresh berries if only the water-soluble tannin is considered, and about 6000 mg/kg if the total recovered by water plus alkali is considered. These values fall within the range reported by others for various grape varieties and also indicate that relatively mild solvents such as water and aqueous alcohol can extract by far the major portion of the total berry phenol as also noted by Durmishidze (1955).

Kondo and Zvezdina (1953) reported, on the basis of analyses of 17 varieties, that 1 kg of grapes gave 50–125 g of peels containing 83–1280 mg of pigment and 444–3080 mg of total tannins, and 23.4–82.0 g of seeds containing 1430–9790 mg of total tannins. Leaving out the varieties too low in color to give a satisfactory red wine, the average pigment content of 12 varieties was 788 mg/kg. The average of all varieties was 9.6% skins and 4.4% seeds. The tannin content, in mg/kg berries in skins and seeds was respectively 1017 and 3478 for the white or pink varieties and 1848 and 4318 for the red varieties. Values for the tannin equivalent of the phenols of cold-pressed "Concord" juice averaged 515 mg/kg in tests by Noyes *et al.* (1922). With *V. vinifera* varieties the values have tended to be lower, e.g., 250 mg/liter (Jacobs, 1944). The pulp and juice have increasing phenolic content in the following order: middle layer, nearest the seeds, and nearest the skin (Garoglio, 1965; Benvegnin and Capt, 1936).

Many other reported values could be cited, but these seem typical and sufficient to allow predictions and estimates for the average variety. For any specific sampling of grapes, actual analysis would be necessary owing to considerable variation by variety, vineyard location, year, and processing technique. Table III presents values considered typical for the amounts of total phenol calculated as gallic or tannic acid in the major fractions of the grape berry obtained by combining data recalculated from the reports above, including values from California, Russia, France, South Africa, and elsewhere. It is interesting that these values seem comparable to those from a number of other fruits (e.g., Lee, 1951; Jacobs, 1944) except for the high contribution from the seeds. The apparent tendency for white grapes to produce less total phenol in all tissues, as shown in Table III, may only reflect the specific varieties analyzed, and there is a great amount of overlap among the two groups. The difference with respect to the skins, however, is believed highly significant. The difference is approximately that of the anthocyanins themselves, suggesting that as the variety loses the ability to synthesize anthocyanins no great amount of phenolic precursor is diverted to other products. Rather, the usual effect appears to be a decrease in total phenol equivalent to the pigment lost.

### 2. Specific Phenols—Amounts and Tissue Distribution

The quantitative analytical data are generally not yet as specific and complete as one would wish for defining the typical amounts of each

TABLE III
TOTAL PHENOL DISTRIBUTION IN *Vitis vinifera* WINE GRAPES[a]

| Portion | Red varieties | White varieties |
|---|---|---|
| Skins | 1859 | 904 |
| Pressed pulp | 41 | 35 |
| Juice | 206 | 176 |
| Seeds | 3525 | 2778 |
| Total | 5631 | 3893 |

[a]In gallic or tannic acid equivalents, mg/kg berries.

phenol in each grape tissue. More data are often available on wines than on the grapes themselves, so that improved estimates can sometimes be made by working back from the wine analyses. Of course, the wine content may represent less than the original grape content, and generally could not represent more. It is clear that the different classes of phenols are distinctly different in their distribution among the tissues of the grape berry and other vine parts. Many of these differences have already been mentioned, and others are discussed in connection with wine composition.

Except for the relatively few teinturier (dyer) varieties, which have anthocyanin in the pulp of the berry, the anthocyanins are confined to the relatively tough skin cells of the grape. It is thus possible to produce colorless juice from most red grapes, and the composition of clear fresh juice seems relatively constant among all varieties. The phenols in such juice from most varieties are predominantly the chlorogenic-acid-like caffeic and *p*-coumaric derivatives, which are highly water soluble and highly polar (Singleton *et al.*, 1966a; Hennig and Burkhardt, 1958). Evidence has already been cited that the total phenol content of clear, pulpfree juice is about 100–300 mg/liter, calculated as gallic acid, for *V. vinifera* and about 500 mg/liter for *V. labrusca* varieties. It is reasonable to assume that this approximates the typical content of caffeoyl tartrate and related phenols in grapes. Of course, the true content of chlorogenic-acid-like substances would be overestimated to the degree that other phenols and ascorbic acid contribute to this apparent total in a given sample. Jurics (1967a) lists 125 mg/kg as the chlorogenic acid content of grapes.

The other phenols are largely localized in (or adsorbed on) the relatively solid parts of the grape. The pulp remaining after peeling, deseeding, and pressing out the juice of the grape usually contains only a small amount of extractable phenols (Singleton *et al.*, 1966a; Singleton, 1966, 1968, unpublished data; Cantarelli and Peri, 1964a). In those instances where the pulp retains an appreciable fraction of the grape's phenol it is commonly in a variety like a "Muscat," which has a firm pulp and is difficult to press free of juice. Paper chromatograms of extract of pulp show, of course, the phenols typical for the juice, but also small amounts of phenols characteristic for the skins and seeds of the variety. This may be caused by the inevitable inclusion of tiny aborted "seeds" and fragments of inner skin detached in peeling. Also, if the skin cells are killed or disrupted, skin phenols will diffuse inward and may appear in larger quantity in the pulp or juice. It appears that if a typical variety produces a juicy, easily pressed berry, and if all possible contamination from cell layers near the seed and skin is eliminated, the content of extractable phenol in the solid part of the pulp is insignificant in terms of the berry total and contains no phenols unique to the pulp. The phenols of pulp solids may be important, however, as contributors to white wine and also may adsorb and include protein-tannin precipitates produced from phenols which would otherwise appear in the juice.

The anthocyanin content can obviously range from zero, in a variety devoid of skin pigment, to a maximum for the darkest-colored teinturier variety. This maximum cannot be defined very well at present, but can be estimated to be at least 2500 mg/kg by comparing data on "Alicante Bouschet," a very dark teinturier (Yang, 1943), with data on other fruits indicating about 4500 mg/kg for black raspberries, compared with about 400 mg/kg for red raspberries and 350 mg/kg for strawberries (Daravingas and Cain, 1965; Sondheimer and Kertesz, 1948). "Saperavi," with anthocyanin at about 1280 mg/kg, and "Concord," at about 1100 mg/kg, have already been cited. "Barlinka," a characteristically "black" table grape of *V. vinifera* lineage, had, by actual isolation and weighing, malvidin-3-glucoside at a level of 514 mg/kg, its mono-*p*-coumaroyl acyl derivative at 27 mg/kg, peonidin-3-glucoside at 23 mg/kg, petunidin-3-glucoside at 2 mg/kg, and delphinidin-3-glucoside at 2 mg/kg, totaling at least 568 mg of anthocyanin per kg of fresh grape skins. Puissant and Léon (1967) report samples with anthocyanin contents as high as 2000 mg/kg berries for "Baco," a dark-colored hybrid; 890 mg/kg for "Cabernet," a good red wine grape; and 450 mg/kg for "Gamay Beaujolais," a fairly light-colored red wine grape. Data previously cited suggest that varieties suitable for red wine production have an average pigment content of about 800 mg/kg.

Subtraction of this pigment value from the total apparent content of phenols in red skins leaves about half, or 900–1100 mg/kg, similar to the original total in white skins. The major fraction of this total appears to be anthocyanogenic tannins. An average of 54% of the total phenol from 3 white grape varieties was determined to be anthocyanogen by Cantarelli and Peri (1964a).

Rakcsányi (1964) found that the phenols of grape skin were about

25% tannins precipitable with hide powder, and 75% nontannins. Clore *et al.* (1965) reported that ¾ to ⅔ of the total phenol content of deseeded berries (i.e., mostly skin phenols) was tannin rather than anthocyanin. Considering these and data cited earlier it appears probable that the polymeric tannins typically account for about half of the grape skin phenol or about 500 mg/kg grapes.

The remaining 500 or so mg/kg of phenol in grape skins evidently includes the catechins, the flavonols, the caffeic-acid- and lignin-related derivatives, etc., which are known to be present but are not yet well quantitated. Durmishidze (1955) reported *d*-catechin as 15% and gallocatechins as 51% of the ethyl-acetate-soluble phenols recovered from "Saperavi" skins. This would appear to account for as much as 150 mg/kg + 510 mg/kg, respectively, or essentially all the remaining phenol. The available data do indicate that the flavonols and other phenols of grape skin are present in relatively low amount, but it also appears that the indicated content of gallocatechins is too high and part of the catechins may be present in oligomers rather than free.

The majority of the seed tannin, 37–98% (Cantarelli and Peri, 1964a), is anthocyanogenic tannin of two size ranges, smaller and larger. The remainder is apparently almost entirely catechins and catechin oligomers (Singleton *et al.*, 1966a; Durmishidze, 1955). In "Saperavi" seeds Durmishidze (1955) finds 13% of the seed phenol fraction is *d*-catechin and about 45% gallocatechins plus catechin gallates. "Rkatsiteli" seeds were reported to have 68% of their extractable tannin as catechins of the several types, with half being gallocatechins (Durmishidze, 1955). Based upon paper chromatographic comparisons, relative ether solubility, etc., it appears that the usual content of all catechins would be less than ⅓ of the extractable phenols of seeds or not over about 1000 mg/kg. Jurics (1967a,b) found a total of only 141 mg/kg for *d*-catechin and *l*-epicatechin, but the manner of extraction may not have recovered all the catechins from the seeds. The anthocyanogenic tannins and related compounds of all sizes would be about 2000 mg or more, by difference, in the seeds of 1 kg of average berries.

It appears from these estimates that the major phenolic fractions of typical grapes in a gross quantitative sense would be about 2500 mg/kg anthocyanogenic tannin, ⅘ in the seeds and ⅕ in the skins, about 800 mg/kg anthocyanins in the skins of red grapes, about 200 mg/kg of phenolic cinnamic acid derivatives in the juice, and relatively small amounts, largely in the skins, of most of the other phenolics except the catechins and related substances. The catechin class total appears to be variable, and at this point is relatively uncertain. It appears to be potentially as high as 1500 mg/kg, with ⅔ in the seeds and ⅓ in the skins. It is obvious that only rough estimates are possible with the approach taken here. A refined "balance-sheet" inventory of phenols of grapes in any one variety or an "average" variety will require improved simultaneous analyses, which can be developed only from new experimental study. It will be necessary to consider such factors as the fact that, on a controlled Folin-Ciocalteu assay basis, one mmole of *d*-catechin (290 mg) and one mmole of quercetin (302 mg) would respectively assay as about 1.5 moles and 2.5 moles, or 255 mg and 425 mg, of gallic acid or tannic acid (Singleton and Rossi, 1965).

## C. Vinification and the Amount of Phenols in Typical Wines

There are innumerable analytical values available for total phenolic content of various wines, several of which have already been cited. A few examples and tabulations can give an idea of the variation and average composition in a total gallic or tannic acid equivalent sense. Neubauer (1872) reported that samples of white juice and must contain 56–141 mg/liter, teinturier juice 141–225 mg/liter, white wines 246–426 mg/liter, and red wines 910–2610 mg/liter. Hilgard (1896) reported Bordeaux clarets had a minimum of 1500, a maximum of 2000, and an average of 1750 mg/liter. Similarly, Jura-type red table wines were 2000–3000, averaging 2500 mg/liter, in phenols, ordinary red wines from the south of France had 1000–2000, averaging 1500 mg/liter, and Italian red wines had 1500–2700, averaging 2200 mg/liter. Laborde (1908–1910) found 2550–4120 mg/liter in 21 red Bordeaux wines, 2000–5360 mg/liter in 12 other red wines, and 250–800 mg/liter in 6 white wines. Madeira (white) wines (da Silva, 1911) had 500–1950 mg/liter, with a tendency for low quality to be associated with high tannin although high tannin was found in some wines of the highest quality.

Amerine and Winkler (1941) found that 25 rosé wines prepared from 4 varieties of grapes averaged 452 mg/liter (range 200–900 mg/liter) while 20 red wines from the same varieties averaged 1335 mg/liter (range 500–2800 mg/liter). In addition, "Alicante Bouschet," a red-juice teinturier, averaged 740 mg/liter in wines prepared from juice and 1633 mg/liter in red wines fermented on the skins for one day or longer. Analyses of wines from several vineyard climatic regions and several years were presented by Amerine and Winkler (1944) for a large number of grape varieties. Averages of their data for white wines of 12 of the better-known white varieties of wine grapes all fell in the relatively narrow range of 300–493 mg/liter. Wines prepared from the varieties

"Trousseau," "Aleatico," and "Mission," which are noted for very low-colored red wines, averaged 626–735 mg/liter. The varieties "Gamay," "Pinot noir," "Grenache," and "Aramon," noted for relatively light-colored and mildly astringent red wines, averaged 1233–1304 mg/liter in phenol content. More robust red table wines are ordinarily produced from "Zinfandel," which averaged 1378 mg/liter, "Cabernet-Sauvignon," 1516; "Petite Sirah," 2120; and "Refosco," 2270.

Red dessert wines averaged 1014 mg/liter from "Tinta Madeira," and 1104 mg/liter from "Muscat Hamburg." The teinturier varieties "Alicante Bouschet," "Salvador," and "Alicante Ganzin" respectively averaged 1972, 2792, and 2840 mg/liter. More recent analyses reported by the same workers (Amerine and Winkler, 1963) cover some of the same and additional varieties. The variety "Teinturier," from which red-juice varieties take their class name of "dyer," averaged in 4 instances only 1100 mg/liter. This shows that the red-juice character is not necessarily associated with high total phenol. Many of the red-juice varieties have *V. rupestris* in their parentage and are very highly colored. "Teinturier" is, however, believed to be a pure *V. vinifera* and has a relatively low anthocyanin content. High anthocyanin content, of course, contributes directly to high total phenol assay.

Blyth and Blyth (1927) report that 46 German white wines had a range of total tannic acid equivalent from 160 to 710 mg/liter (average 356). Amerine (1947) analyzed many commercial wines submitted for judging in California. White table wines (399 samples) ranged from about 200 to 1300 mg/liter (average about 550), sherry and white dessert wines (355) had 100–900 (average 488), pink dessert wines (52) had 200–1100 (average about 600), port wines (81) had 400–1600 (average 800), and 282 red table wines had 1000–3800 (average 2050). Many analyses for appetizer, flavored, and dessert wines were summarized by Joslyn and Amerine (1964). California white port (36 samples) was lowest among the white wines in total phenol content (range 200–700 and average 300 mg/liter), and 10 Italian vermouths were highest (500–1100 and average 800 mg/liter. Papakyriakopoulos and Amerine (1956) found that 3 ruby and 3 vintage Portuguese ports ranged in total phenol from 1080 to 2220 mg/liter, with a tendency for higher quality to be related to higher total "tannin" content.

Fessler (1947) reported that typical phenol contents for American commercial wines were 100 mg/liter for hocks, 120 for sherry, 200 for sauterne, 250 for muscatel, 600 for port, 1500 for claret, and 1800 for burgundy. Diemair *et al.* (1951b) reported values which, recalculated to Löwenthal-Neubauer tannin, were 71–1095 mg/liter (average 377) for 90 European white table wines, 320–1693 (average 1000) for 42 northern European red wines, and 133–1054 (average 488) for 46 European white sweet flavored or fortified wines. Singleton and Ough (1962) reported average total phenol contents of 263 mg/liter for 30 white table wines, 2171 for 24 red table wines, 350 for 4 white sweet wines, and 1588 for 8 ports all made in the same California cellar. Clore *et al.* (1967) reported an average of 722 mg/liter for 18 white wines including 8 non-*vinifera*, and 1913 mg/liter for 23 red wines including 13 non-*vinifera*.

Only cursory study of these and similar data is required to notice that white wines tend to be low in phenol, and red wines high, but with considerable overlap. These differences can be explained on the basis of the amount of extraction of phenols from the berry solids. Although it is usual to make white wines from juice only, extraction of berry solids to some extent may be involved in some instances or for some types of wine. Red wines are ordinarily prepared by fermentation "on the skins", i.e., in the presence of the solids of the entire berry. Since some varieties of white grapes have total amounts of phenol in skins, seeds, and pulp that are similar to that of usual red grapes, it would be expected that such grapes, vinified as for a red wine, would give a "white" wine with total phenol content like that of a red wine.

It is an unusual practice to ferment white grapes for several days before separating the fluid, partly because such wines are apt to be very dark yellow or brown (Moreau and Vinet, 1937) and are generally considered unattractive and lowered in quality. As a result there are not very many examples in the literature of tannin analyses of wines prepared by red-vinification methods from white grapes. Amerine and Winkler (1944) list wines from the white "Feher Szagos" variety which had 1200 mg/liter total phenol when fermented on the skins, and only 300–700 mg/liter when made by usual white-wine methods. They also list 2 "Grignolino" wines with red color too light to measure and total phenol contents of 1500 and 1600 mg/liter. Clore *et al.* (1967) report a "Flora" wine with total tannin of 2500 mg/liter, and a "Muscat Ottonel" with 1400 mg/liter, and note that wines in that set of experiments were left long on the skins. These white wine grapes do not produce abnormally high tannin when only their juice is fermented.

Description of older methods indicates that the whole mass of white grapes was fermented for 1 or 2 days and the fermentation then terminated by heating to give Madeira wines with tannin contents as high as 1950 mg/liter (da Silva, 1911). The Kakhetia wines of Russia have no exact commercial counterparts elsewhere, but they are often made by very long storage, 6 months or so, of the fermented wine in contact with the grape pomace (Beridze, 1950, 1956; Nanitashvili, 1957). They may be made from red or white grapes. Although red wines of this type are generally higher in tannin than the whites, neither is as high as might be predicted from the method of production. Red wines of this type from one source had a maximum of 3570 mg/liter, and the comparable white wines were as high as 1970 mg/liter (Beridze, 1950). "Rkatsiteli," a white grape, gave wines of this type of 1350–2900 mg/liter total phenol, whereas "Saperavi," a red-juice variety, gave 3600–3900 mg/liter.

While considering such wines it may be well to consider the maximum tannin possible with red grapes. Above 2000 mg/liter of total phenol may be considered a fairly high level, and 4000 mg/liter is about the maximum which most of the more phenol-rich varieties produce in any normal vinification process. The highest total phenol contents reported by Amerine and Winkler (1944) were 3900 to 4100 mg/liter, for the teinturiers "Salvador," "Alicante Ganzin," and "Alicante Bouschet," and the next highest was 3300 mg/liter for a "Petite Sirah" red table wine. Amerine (1947) reports commercial red wines having up to 3800 mg/liter. Kakhetia red wines may reach as high as 4780 mg/liter (Beridze, 1950). Clore *et al.* (1967) report a "Calzin" wine at 4100 mg/liter of total phenol, and even "Pinot noir," generally low to moderate in total phenol, at 3600 when fermented too long on the skins. The highest wine we have analyzed was a "Calzin" deliberately prepared to have as high tannin content as possible—and that was about 6500 mg/liter. Values as high or higher may be found in the literature, but they appear to be errors or the result of special operations. For example, the pomace removed from a red wine fermentation may be pressed at very high pressure, and the fluid so received, the press wine, is often double or even higher in tannin concentration than the main body of the wine, the "free-run," obtained by simple drainage. Conceivably, by such manipulation, evaporation, etc., small amounts of wines might be produced with any total phenol content up to and including saturation, so no particular purpose would be served by discussing them except in relation to the significance of processes involved. Giashvili (1966) prepared hot-water extracts of pomace for blending purposes at 4500 to 7000 mg tannin per liter.

If the wine with the most tannin is that having had the greatest contact with (and extraction of grape solids) it follows that the wine with the least total phenol is that produced from rapidly clarified juice. Excepting teinturiers, this means white wines, even from red-skinned grapes. Grapes are prepared for wine making by the operation called "crushing," which is usually combined with destemming in commercial practice. The word "crush" suggests a more vigorous operation than is intended, and the effect desired is equivalent to picking the berries from the stems and squeezing each berry between the fingers so that the skin is torn open and the pulp disrupted without appreciable damage to tougher layers of the skin and no breakage or grinding of the seeds or other solid parts. Depending upon the juiciness of the variety, the effectiveness of the crushing, and other factors, about ⅔ of the juice can usually be drained away as free-run, and the rest will be obtained by pressing, often in several increasingly vigorous stages.

The fresh free-run juice is an opaque, slightly viscous mixture containing appreciable suspended solids, and the pressed juice may contain even more. This mixture may be fermented for white wine, or it may be clarified by various procedures before fermentation. The pulp solids and other fragmentary solids in the juice are ordinarily small in actual weight but relatively rich in phenols and include protein-tannin precipitates. To the degree that these solids are precipitated and removed, the total phenol content will decrease as the must is converted to wine, and to the degree that they are extracted and the phenol solubilized by fermentation, the wine phenol content will be raised over that equivalent to the clear fresh juice.

The most desirable, highest quality white table wines of most types are ordinarily those with the lightest color and the freshest, least oxidized flavor. The techniques recommended to accomplish this condition generally coincide with those which would minimize admixture of phenolic substances from the juice with other phenols and preserve juice phenols with minimal change. Generally recommended for highest quality white table wines are the use of free-run juice only, rapid operation with minimum opportunity for oxidation, juice clarification before fermentation, and maintenance of cool temperatures before and during fermentation. As a consequence the minimum phenolic content of white table wine should tend to a relatively constant value at least for a given grape variety, reflecting the native juice content. This value would be higher than that of the pulp cell's vacuole contents to the extent of unavoidable extraction of phenols from cytoplasm and solids and of yeast synthesis, and lower to the extent of precipitation, consumption, or destruction of phenols during processing.

Light color and associated qualities are particularly emphasized in American white port, white sparkling wines, and German and other light white table wines. These wines show a relatively uniform total phenol content, as do the lowest common values among many wines made from juice of nearly all varieties of grape. These minimum total phenol values appear to be about 50 to 350 mg/liter, depending on the variety, probably averaging about 250 mg/liter. The only exceptions appear to be the teinturiers and a few rare varieties, like "Calzin" and one of its parents, "Refosco," that have appreciable levels of tannin in the juice (Singleton *et al.*, 1966a).

If fresh white must strained free of particulate matter is filtered, its total phenol content drops as the suspended fine solids and precipitates are removed. It might be anticipated that fermentation of unfiltered must would lead to solubilization of most of this phenolic substance in the alcoholic solution and to a content of phenols in the wine the same as the musts. In fact, it appears that the total phenol content of white must is usually lowered by fermentation, but it may be raised or left unchanged. Diemair *et al.* (1951b) found that a white juice with 460 mg/liter of phenols dropped to about 200 mg/liter as fermentation was completed and the wine clarified. Berg (1953b) reported that 34 white musts from Davis, California, had a range of total tannin of 40–470 mg/liter (average 250), and the corresponding wines had 110–420 mg/liter (average 200). Similar (31) musts from Napa, California, had 100–560 mg/liter (average 270), and the wines had 10–510 (average 170).

Baragiola and Godet (1914) generally found total phenol to be about half as much in young white wines as in their musts, but in at least one instance it was higher in the wine than the must. Pallotta *et al.* (1966b) report minimum and maximum values for total phenols to be 256–409 for white musts and 298–944 for wines. The anthocyanogen contents of the same samples, in mg of cyanidin per liter, were 54–185 for musts and 48–696 for wines. Nachkov (1967) found leucoanthocyanidin content of 15.6 mg/liter in an "Aligoté," and 25.4 mg/liter in a "Riesling" wine. The lowest found in a series of white Bulgarian wines was 9.4 mg/liter, and only two were over 100, with about 50 appearing typical.

Burkhardt (1963) and Wucherpfennig *et al.* (1964) report on white wines prepared with many variations and analyzed for total phenol, cyanidin production from anthocyanogens, and vanillin-reactive flavanols calculated as catechins. Centrifugation of one "Muller-Thurgau" must dropped these three respective values from 320, 80, and 410 mg/liter to 300, 40, and 390 mg/liter, while fermentation caused a further drop to 250, 30, and 200 mg/liter. After the yeast lees were removed, the wine had respective values of 210, 30, and 170 mg/liter. Other examples of analyses of several "Riesling" musts, paired with their white wine after first racking, were, for total phenols, 270–230, 260–290, 220–220, 660–440, 260–170, 610–360, and 610–340 mg/liter; for leucocyanidins, 10–0, 10–20, 0–0, 210–90, 20–0, 180–60, and 180–50 mg/liter; and for "catechins," 20–0, 60–120, 0–0, 740–300, 70–0, 560–320, and 400–240 mg/liter.

Considered together, these and related data seem to show that when total phenol content in a white must or wine is low, anthocyanogen and catechin content are usually a trace or slightly more. It has long been known that fresh white grape juice ordinarily has very little true tannin and a relatively low polyphenol content (e.g. Mulder, 1857). At one time it was thought that true tannin or anthocyanogen content would distinguish genuine white wines from reprocessed pomace wines or decolorized red wines. However, it was soon found that some genuine white wines gave enough test to interfere (v. Fellenberg, 1944; Fallot, 1909a; Fresenius and Grünhut, 1899).

When the total phenol is low in a white must, that in the wine is usually not much lower than the must, and may be higher. This increase is thought to result primarily from synthesis of phenolic substances by the yeast, particularly tyrosol. The amounts of such increases seem compatible with the tyrosol levels reported in fermented beverages. When the total phenol content is relatively high for a white must, the extra content of phenol is likely to be about equivalent to the leucocyanidins and "catechins" present. The major part of the leucocyanidins and other vanillin-reactive flavanols, however, is usually adsorbed on suspended solids and yeast cells and precipitates in the lees during and shortly after fermentation.

To the extent that the available data may be generalized, it appears that 250 mg/liter of total phenol calculated as gallic or tannic acid is a realistic estimate for a typical white wine with minimal phenols from sources other than pulp cell vacuole juice. It also appears that this value would be made up of about ½ caffeoyl tartrates and related hydroxycinnamates, perhaps ¼ tyrosol, and relatively small amounts of other phenols. Despite some variation (apparently small) by variety, white wines over this minimal level of phenol would probably contain other phenols from the more solid parts of the grape in proportion to the amount extracted less the amount precipitated. The caffeoyl tartrates and other juice phenols seem sufficiently soluble that little of them is precipitated, at least in the early stages. It appears that the catechins and leucocyanidins are the major fractions readily transferring from solids to white wine, and that the anthocyanogens are somewhat more likely to be precipitated by the yeasts and other proteinaceous solids during winemaking.

From the previous discussion it is obvious that the total phenols of white wines can run the gamut from the minimal 250 mg/liter to the full fermented-on-the-skins level of around 2000 mg/liter. It appears, however, that there is not a simple uniform transition through all possible intermediate levels until there is actual breakdown of the firmer tissues. The incorporation of phenols from the softer tissues would appear to account for phenol contents in the white wine above 250 to perhaps 800 mg/liter, but for a phenol level above this an average variety would require diffusion from relatively resistant skin or seed tissues. These firmer tissues apparently remain phenol-retentive as long as they are alive and for a considerable lag period after simple crushing of the grape berry. As a result, the phenol content of wines prepared from freshly crushed grapes tends to be grouped in three rough classes: near 250 mg/liter for minimal solids contact; and around 500 mg/liter for appreciable soft-tissue solids extraction; and above about 1000 mg/liter as an "avalanche" of phenols enters from skin or seed tissues.

For some types of white wine – for example, wines from "Muscat" varieties – it is a frequent practice to hold the crushed mass of grapes for a few hours before draining off the juice to be fermented. This is done with the intent of raising the level of grape aroma transferred to the juice and also to give a period for autolysis of the pulp tissue with subsequent greater recovery of less viscous juice. Although wines prepared in this way are somewhat increased in content of phenols, the increase is limited as described. Even inoculation with yeast and the onset of fermentation may be practiced without producing a very high phenol content provided that diffusion of phenols from skins and seeds does not proceed too long or too rapidly and provided that the capacity of the proteins and yeast cells to adsorb and precipitate phenols is not exceeded. These effects are illustrated by the analyses reported by Wucherpfennig *et al.* (1964), who found that wine from destemmed white grape must held only 3 days rose to 357 mg/liter, compared with 261 mg/liter phenols for must pressed immediately, whereas wine from must heated to 80°C, cooled, and separated jumped to 515 mg/liter. Gaivoronskaya (1960) found generally similar results in the early stages of fermentation of entire crushed white grapes in that total phenol of the fluid in one experiment rose from 206 to 320 mg/liter in 6 hours but then stayed relatively constant for a considerable time, and in another instance only went from 187 mg/liter to 364 mg/liter in 44 hours.

Even pressing, provided it is not too severe and is applied to unheated, unfermented, freshly crushed sound grapes, has a relatively minor effect on the phenol content of the press juice (Ambrosi *et al.*, 1966; Érczhegyi and Mercz, 1966; Wucherpfennig *et al.*, 1964). Of course, pressure that is severe or suddenly applied and released can disrupt firmer tissues and considerably increase the phenol content of the last press juice. Quintanilla Iniesta *et al.* (1966) reported 25–96 mg/liter of anthocyanogens in low-pressed musts, and 450–850 mg/liter in wines from overpressed musts. Gel'gar (1967) reported that pressure raised rapidly to 5 kg/cm$^2$ and held there for 3 minutes gave more suspended solids and more tannin, 740 mg/liter, than slower pressing, which in 11 minutes gave 380 mg/liter.

Standard processing of white wines sometimes results in less tannin than desired. In such instances, stratagems to raise tannin may be used such as pressing in the presence of the stems or even chopping the stems and adding them to the must for a period before pressing (Humeau, 1966; Escudié, 1939).

Any condition causing death or disruption of the cells of the "solid" tissues of the berry prior to juice separation is very likely to produce increased phenol content in the wine in proportion to the fraction of the skin or seed affected. This is made use of in producing red wines, but is generally undesirable in white wine production. Some of these effects are discussed later, but examples of importance to white wine may include withering, freezing, or molding of the grapes. Moreau and Vinet (1937) note that frosts in Anjou in 1919 during grape harvesting gave white wines that were abnormally high in tannin for the region—at about 570 mg/liter. French Sauternes and other concentrated botrytized wines tend to be slightly higher in total content of phenols than corresponding wines from nonmoldy grapes. Other molds may have similar effects (e.g., Milisavljevic, 1957), but the tannin content, particularly of red wine, from moldy grapes is likely to be lowered by other effects, notably oxidation by grape and mold enzymes of the skin phenols. Gavrish (1966) notes that even withering of the grape by drying at 35°–37°C allows diffusion of phenols from the skin cells so that they appear in the juice.

Grape juice, to the American consumer, is dark purple-red "Concord" juice. Such colored juice is produced by heating the grapes or grape skins to kill the cells and then pressing out the juice. The use of similarly prepared red juice for making red wines has had considerable study because of its potential advantages in processing. To date, fermentation of previously prepared red juice does not account for much of the world's red wine production, but that appears subject to change in view of recent studies of techniques avoiding overheating. Since "Concord" grape juice is an important food product, it is also of interest to compare its phenolic composition with that of wines.

Hartmann and Tolman (1918) discussed the production of commercial "Concord" grape juice by heating the grapes and pressing out the juice. Tannin plus coloring matter calculated as tannic acid was raised to about 2700 mg/liter by heating to near 150°F, and to about 3600 mg/liter by pressing. After storage for 16 months, oxidation and precipitation would cause this value to drop to about 1600 mg/liter. Ingalsbe *et al.* (1963) found that 20–40% of "Concord" pigment would precipitate during cold storage. A series of juices by various commercial producers (Hartmann and Tolman, 1918) analyzed from 700 to 3700 mg total phenol per liter. They noted that more than 150°F was likely to release excessive tannin into the "Concord" juice. Kushida (1960), working with "Koshu" grapes, found that 158°F for 40 minutes gave a bitter, undrinkable product when the juice was converted to wine. Noyes *et al.* (1922) found that "Concord" juices pressed at room temperature averaged 492 mg/liter in 1920, and 539 mg/liter in 1921, whereas corresponding juices produced by 5 minutes of heating at 145°F before pressing averaged 1583 and 2471 mg/liter. "Concord" juices produced on different days in a commercial plant in 1919 ranged in total tannic acid equivalent from 2990 to 4180 mg/liter (average 3310).

Webster and Cross (1935, 1936, 1944) prepared juices by heating to 180°–185°F and pressing in a lard press. Many varieties of American origin were tested. The highest in total phenol was an "America" sample, at 4648 mg/liter, and the lowest was "Mathilda," at 597 mg/liter. "Concord" had 2075 mg/liter, of which 1041 was nontannin on the basis of gelatin precipitation. In general, the lowest total phenol samples had the highest proportion of that total as polyphenols not precipitable with gelatin. In a few of the lowest like the "Mathilda" sample, no gelatin-precipitable phenol was found. Reynolds and Vaile (1942) analyzed grapes and hot-pressed and cold-pressed juices, and studied the reproducibility of the sample. Although replicate samples were reasonably reproducible, different varieties and different years of the same variety were rather variable. Holding fruit before pressing (but evidently after heating to 140° –144°F for 5 min) raised the total phenol of the juice from 1900 mg/liter to 2740 mg/liter at 96 hours, and nearly all of the increase was in material precipitable by gelatin, which rose from 1000 to 1640 mg/liter.

Crowther (1959–1960) reported on the phenolic content and color intensity of wines prepared from "Concord" grapes by different procedures. Fermentation to dryness on the skins gave total phenol values of 1200 mg/liter without heating, whereas fermentation after heating gave about 2400 mg/liter with 145°F and about 3000 mg/liter with 170°F. Fermentation temperature apparently had relatively little effect on phenol and color in the heated samples, and it made little difference whether only juice or both juice and skins were fermented after heating. In the unheated samples, fermentation at 80°F did increase extraction about 10%, as a rule, over fermentation at 50°F. Sastry and Tischer (1952a) found that 250°F caused destruction of anthocyanin if prolonged, and that pigment extraction from "Concord" grapes into juice increased following heating at 210°F for up to 32 minutes.

Heat treatment of European wine grapes is a very old idea for producing red juice, and thence wine, without fermenting on the skins. Red wine from heated grapes was known at least as far back as the year 1400

(Ferré, 1928). The effect of fermentation in producing red wine was explained at one time on the basis that the anthocyanins of the grape were insoluble in juice but soluble in dilute alcohol. Rosenstiehl (1897) appears to have been one of the first to demonstrate that anthocyanins were quite soluble in juice and diffused from the skins if the skin cells were killed by heating to 158°F. Since retention of the pigments, tannins, and other phenols within the cells of skins and seeds depends on maintenance of the life processes of the cells, we now know that pasteurization temperatures, for killing cells and destroying enzymes generally, are the threshold temperature above which these cells rapidly give up their phenols and below which they give them up slowly, if at all. Of course, the minimum killing temperature is time-dependent, not absolute.

Bioletti (1905, 1906) experimented, on both laboratory and commercial scales, and found that high temperature (158°F) for a short time was preferable to a lower temperature for a long time in terms of rapid color release and minimal undesirable flavor. Other data relative to the apparent thermal death point of grape skin cells as indicated by anthocyanin release are reported by Cruess (1920), who found that color "flowed" slowly at 120°F and almost instantly at 160°F or more. Ferré (1928) found that dipping clusters into water at 140°F required 15 min before the color migrated from the skins into the pulp, whereas only a short time was required at 176°F and color bled into the water at 212°F. Joslyn *et al.* (1929) found an inflection point at about 158°F, in that color diffusion from grape skins was slow at 104°–122°F, more rapid at 120°–158°F, very rapid at 158°–194°F, and no more rapid at 194°–212°F. Color levels in juice with the same lot of grapes were approximately equivalent at about 2 min at 176°F, 5 min at 158°F, 18 min at 140°F, 40 min at 122°F, and 60 min at 104°F. Tannin content varied somewhat similarly. Guinot and Menoret (1965) found that heating portions of a certain lot of crushed grapes for 5 minutes gave juice, either free-run or pressed, which appeared to approach a limit of 1400 mg of total phenol per liter. In unheated samples, total phenols were only 35% as high in free-run juice, and 49% as high in pressed juice. The respective values were 42% and 53% at 158°–167°F, 47 and 77% at 167°–176°F, 99 and 94% at 176°–185°F, and 100 and 99% above 185°F. They also found complete destruction of peroxidase and phenolase in the juice above 167°F, and ascorbic acid was increased to 77 mg/liter.

Quickly heating grapes *en masse* to a temperature high enough to kill the cells, followed by rapid cooling so as to minimize flavor modification, has been difficult on the scale required by wineries with the equipment available to them. Many processing schemes and engineering studies have been applied in an attempt to adapt heat treatments for color release to red winemaking. Obviously, the first requirement of such treatment is the production of satisfactory anthocyanin content in the product, which has been the basic criterion of success used in studying such heat treatment. Since grapes differ considerably in pigment content, comparison has usually been against wine from the same grapes vinified in the normal manner and based upon relative visual color rather than in actual quantitative terms. More details of the different procedures used, problems of commercial adoption, and relative merits of different approaches are discussed by Dimotaki-Kourakou (1967), Flanzy (1967), Prehoda (1966), a 1965 colloquium at Montpellier (Anon., 1966a), Valuiko *et al.* (1964), Rankine (1964), and others.

Reports of the superiority of heat treatments over the usual fermentation on the skins must be assessed in the context of the attitudes and problems of the specific region, producer, and consumer. For example, an acceptable red wine is very difficult to make from grapes infected with the "noble mold," *Botrytis cinerea,* which on some white wine grapes is considered desirable. The oxidative enzymes in such red grapes spoil the color of all the grapes and ruin the red wine if allowed to act during vinification. Heat treatment not only liberates the anthocyanins but also destroys these enzymes. In such instances an acceptable wine may be produced whereas otherwise it could not (Anon., 1966a; Biol, 1967; Konlechner and Haushofer, 1958, 1961). This is likely to be important in areas with humid weather at the end of the ripening season, especially if the grapes must be left on the vine as long as possible to encourage adequate ripening before the onset of winter weather.

Another example is furnished by the production of port wines. Fermentation on the skins must be brief so that fermentation can be stopped by ethanol addition, and sugar thus retained. The normal vinification methods may not produce sufficiently rapid color extraction and the wines may be lower in color than the anthocyanin color of the grapes would permit. In such a situation, Berg (1940) produced improved wines by fermenting part of the wine as usual, to carry fruitiness, and part of the wine after heat treatment, to improve color, and blending the two.

Berg and Marsh (1956) noted that the fermentation of heated grapes on the skins or of the press juices from heated grapes gave wines of poor quality whereas the fermented free-run juice from heated grapes gave wines similar in quality to those produced by standard methods. In instances of prolonged contact between juice and pomace after heating, part of the generally negative effect of heat treatment may reflect off-flavors caused by the frequently associated slow cooling. Even if the mass is cooled, however, prolonged contact can lead to excessive tannin extraction and astringent wine of lower quality. Heat treatment, particularly if combined with much mechanical breakdown of the heated skins, often leads to wines that are more difficult to free from suspended solids, which also affects appearance and quality. Since this effect is partly related to destruction of the natural enzymes of the grape, adding pectic enzyme preparations to the crushed grapes before heating beyond the point of tolerance of the enzyme, or after cooling to that point, improves clarification or filtration without affecting the color or tannin very much (Konlechner and Haushofer 1958, 1961; Dimotaki-Kourakou, 1967).

Heat treatment can produce wines which are higher in total phenols than wines from the same grapes produced by fermentation on the skins, just as heating can produce wines of higher anthocyanin content. The significant findings, however, seem to be that red wines from heat-treated musts of sound grapes may have higher than normal total phenol when color is maximum. For at least short periods of holding of hot must, color and total phenol extraction tend to parallel each other, and total phenol is ordinarily lower at the same color level or the same at a higher color level as in the same grapes fermented on the skins. Bioletti (1905, 1906) found that holding grapes at 158°F gave twice the anthocyanin color intensity before tannin reached the level produced by standard fermentation on the skins. Although prolonged holding of the heated mass gave him tannin contents as high as 5520 mg/liter, shorter times gave maximum color with about half as much tannin. Less tannin extraction from the seeds was considered the important factor.

Bujard (1968) found, in young red "Gamay" wines prepared by heat treatment, more than 70% of the color but less than 40% of the total phenol, of wine made in the traditional way from the same grapes. Ferré (1926, 1928) found total phenol in wines from a "Gamay" selection to be 3800 mg/liter for the traditionally fermented sample, 2500 for the sample from heat-treated juice, and 5470 for that fermented on the skins after heat treatment. Ratings for color and softness favored the heated juice sample. Prehoda (1966) found a satisfactory color, a more neutral taste, and economic advantage in wines from red juice made by holding crushed grapes 8 hours at 122°F as compared to usual practice. Fermenting on the skins in all cases, Amerine and Ough (1957) found that "Zinfandel" grapes gave total tannin and relative color of 970 mg/liter and 79% for no heat, 970 and 83% after 110°F heating, and 1230 mg/liter and 87% after heating to 140°F. Marteau and Olivieri (1966) found that wine from red juice prepared by ½ hr at 158°F had 112% of the color of wine from the same grapes vinified traditionally, but very similar or lower content of total phenol, tannin, phenols-not-tannin, and leucoanthocyanidins. Konlechner and Haushofer (1956, 1958, 1961) also generally obtained lower tannin in proportion to color with lower temperature (122°F), but tannin could be increased (although usually to the detriment of flavor) at higher temperature (176°–194°F).

The most promising studies in terms of adapting heat treatment for anthocyanin release to the production of premium-quality red wines have combined the idea of a high temperature, produced and removed in a very short time, with the idea of limiting the treatment to the epidermal cells of the grape. Ferré (1926, 1928) dipped grapes in boiling water for 1–3 minutes and noted that the color then diffused into the berry so that the skin became colorless in 6–12 hours and red juice could be recovered. He also found steam jets to be effective. Amerine and De Mattei (1940) obtained similar results by 1–3 min of dipping into 203°–208°F water, and found that this method would give port wines with 150–200% of the color obtained by traditional methods. Rankine (1964) applied similar methods to laboratory and small commercial lots of grapes and obtained dry red wines of very similar composition and only slightly lower sensory quality than in control wines made as usual.

Neubauer (1944) applied superheated steam at 266°–302°F to whole grapes or freshly pressed pomace for 30 sec to 2 min, and reported that 2 to 4 times the usual color level was obtained. Valuiko and Godin (1961) used 302°–356°F steam for 5–10 sec as the grapes passed jets. Valuiko *et al.* (1964) found that treating grapes for 3 min in juice or water heated to 194°–203°F was sufficient to release color, but 10 min was required for raising the tannin level in proportion. They also tested infrared irradiation, steam at 302°F, and hot air at 338°F on whole grapes. The latter two treatments only required 15–20 sec to plasmolyze the pigment cells, and the temperature just under the skin rose to only 158°F, and inside the berry to 104°F. "Cabernet" grapes fermented on the skins gave wine with anthocyanin content of 306 mg/liter and 1490 mg/liter of total phenol including pigment. The same grapes dipped in water at 194° – 203°F for 3 min gave juice with 393 mg/liter pigment and 1110 mg/liter tannin, which after fermentation had 324 and 470 mg/liter. Similar treatment for 10 min gave 715 mg/liter pigment and 2240 mg/liter total phenol in the juice, which after fermentation had 571 and 1610 mg/liter. These data appear to agree well with our previous hypotheses. If one adds the total phenol of the unheated unfermented juice, 580 mg/liter, to the pigment value, there is relatively little of the total phenol not accounted for in the sample heated for 3 min, and this would appear to be limited to the skin phenols other than pigment. A major portion of these phenols are evidently adsorbed and precipitated during fermentation. The sample heated for 10 min had extra tannin in proportion to the pigment (a ratio of 2.8 mg phenol/mg pigment at 3 min, and 3.2 mg phenol/mg pigment at 10 min), as evidently did the sample fermented on the skins and seeds. Seed participation in tannin contribution in the longer heating and in pomace fermentation is indicated. With the higher contribution of the slower-diffusing tannin (presumably seed tannin but perhaps also skin tannin, owing to molecular size) in the normal fermentation or the 10-min heating, it appears that the ability of the fermenting system to adsorb and precipitate tannin is exceeded and more remains in solution in the wine. The mg total phenol/mg pigment by the data of Valuiko *et al.* (1964) is 4.9 for the young normal wine, 1.4 for the wine from 3-min heating, and 2.8 after 10-min heating. This suggests a much greater proportion of seed tannin in the normal pomace fermentation than in even the sample heated near boiling for 10 min.

Coffelt and Berg (1965) developed and applied equipment capable of heating whole grapes under pressure with steam at various temperatures and at controlled times as short as a few seconds. The pigment absorbance of the juice and its tannin content were linear functions of the temperature in the 230°–290°F range and increased with heating time up to at least 18 sec, but color decreased with 5 min of heating. Their juice analyses with one variety and various holding times up to 5 min after a 4-sec heating at various temperatures suggest that tannin release and pigment release paralleled each other and reached a maximum at about 1200 mg/liter total phenol. One would expect primarily skin tannin release under these conditions, and the 1200 mg/liter presumably represented complete release of the skin's phenolic substances. Dry red table wines prepared by their techniques of very brief heating of grapes, rapid cooling, juice recovery, and then fermentation, were judged by a sensory panel as better in a few cases but on the average slightly inferior to wines fermented normally. Port-type wines, however, were generally rated equal or superior in quality. The different varieties and even the same variety from different origins seemed to have different optimum conditions of treatment.

Leglise *et al.* (1967) heated a thin layer of clusters of grapes with flowing open steam at 212°F, and found that after 3 min the surface of the exposed berries was 212°F, the heart of the cluster near the stem was about 176°F, and the inside of the berry under the skin was about 140°F. After 6 min, all were near 212°F. The treated berries were cooled and condensate was removed by blowing air through them. Approximately 3 min brought the temperature near room temperature after 3 min of steaming. Treatment by variations of this technique gave products with total phenols which appeared maximum at about 3700 mg/liter, about the same as normal commercial vinification of the same grapes, but with as much as 142% of the anthocyanin color.

There are not many data, either quantitative or qualitative, on the modifying effects on phenols of heat alone. It appears that, where there are large effects, oxidation is involved, and heat alone generally has relatively little effect within the range which does not produce major flavor changes. Singleton *et al.* (1964b) found that representative samples of the major classes of wine stored 20 days at 128°F under nitrogen were essentially unchanged in total phenol content. There appeared to be a decrease in astringency, and there was a significant increase in bitterness over the originals in the table wines tested. Ingalsbe *et al.* (1963) found a high content of acylated anthocyanins in commercial "Concord" juice and concentrate prepared by heat treatment even though we know these compounds to be relatively unstable (Webb, 1964). Coffelt and Berg (1965) and others have noted the destruction of grape anthocyanin by too high or too prolonged heating, but the nature of the destruction products does not appear to have been studied. Ponting *et al.* (1960) found that the logarithm of the time for a 10% color loss in grape juice plotted versus temperature was linear; as an example, about 40 min at 176°F produced a 10% decrease in color. The red color appeared to be more stable at low temperature in grape juice than in boysenberry or strawberry juice, but had a higher $Q_{10}$ of destruction – 2.7. The decrease in color was related to loss in red without much production of brown, contrary to the usual aging changes. Destruction of pigment by heat, however, is greatly affected by the characteristics of the solution, notably pH and sugar content (Tinsley and Bockian, 1960). Gerasimov and Smirnova (1964) studying white and red wines heated up to 60 days at temperatures to 158°F, found a number of changes on paper chromatograms. Although there appeared to be formation of protocatechuic acid, pyrocatechol, and phloroglucinol, the catechins, including epicatechin gallate, were still detectable in ethyl acetate extracts after treatment.

We have been discussing the "normal" red wine fermentation of the whole grape after crushing, but this fermentation is subject to considerable variation. The temperature during the fermentation is important. The extraction of phenols is slow at first, but as the cells of the grape tissue die it becomes more rapid and the ethanol produced by the fermentation is an important factor both in killing the grape cells and as a better solvent for the tannins. The skins of crushed grapes rise to the top of the grape juice, and the evolution of carbon dioxide during fermentation accentuates the formation of a "cap" of grape skins which is buoyed up out of contact with the roiling fluid. Unless this cap is mixed or submerged into the solution at least twice a day, extraction of anthocyanins and other phenols from the skin will be poor.

Assuming proper cap management, however, and in view of the fact that dry red wines commonly have similar final alcohol contents and are fermented at similar temperatures, the main variable related to phenol content is total contact time between the pomace solids and the fluid. Reliable statistics are not readily available on commercial practices in use at various times in various places, but an idea of common practices can be obtained from published observations and recommendations. Mulder (1857) suggested that the proper time for drawing the fermenting wine from the pomace was 2–3 days for Burgundy and base wine for Champagne rouge, 6 days in the Médoc, 8 days in the south of France for ordinary wines, and seldom longer than 14 days for the darkest red wines. Ribéreau-Gayon (1966) notes that in 1900 it was not unknown to hold the wine on the pomace for a month, but today the time on the skins is 5–6 days in Burgundy, 8–15 days in Bordeaux, and 2–3 days in warmer regions. Viard (1892) notes that too long fermentation on the skins gave "hard"-flavored wines, with excess tannin and sometimes even less color than shorter pomace fermentations. Generally, 5–8 days was recommended unless the weather was cold and slowed fermentation, in which case 10–12 days might be better. He notes that Burgundies might be only 20–30 hours on the skins, and Jura wines sometimes 3 months, with 4–8 days usual for everyday wine. The analyses he reports appear to cover a normal range of total phenol, but include values as high as 3620 mg/liter, too high from a modern consumer's viewpoint. Rousseaux (1909a) considered complete fermentation on the skins (8–10 days) desirable unless a high-tannin variety was involved, and warned against leaving pomace beyond the point of complete sugar conversion.

Though proof is not easy from such data, if sufficient examples are examined there appears to have been a shift from longer to shorter average time on the skins, which means a shift from more tannic to less tannic red wines as an average. Wait (1889) says that 6–8 days on the skins was usual in California in her day. Cruess (1934) reported that most California wineries drew off red wines at zero Brix, or 5–7 days on the skins. Amerine *et al.* (1967) report 3–4 days usual in California for present-day dry red wines, with over 5 days rare, and a 4–10 Brix value of the fermenting solution at separation. Data from Amerine and Winkler (1963) clearly show the effects of this shift. Average total tannin content for several red wines of each of 27 varieties of grapes produced in the same cellar from grapes from a single climatic region was 1318 mg/liter in 1935–1941, and 1152 mg/liter in 1946–1958. In only 4 of the 27 varieties were the comparisons reversed. Data from Fessler (1947) seem to show the same effect. Total phenol content, in mg/liter, for average commercial California wines in 1935, 1940, and 1946 were: burgundy, 2400, 1800, 1800; Claret, 2000, 1600, 1500; and port, 950, 700, 600.

Even white wines (e.g., sauterne, 300, 250, 200) seem to show the same trend as the postprohibition producer became more expert and more responsive to the public's taste.

A similar trend to lower time on the skins and less tannic red wine appears to be operating in Europe, particularly since World War II. Marteau (1960) reports that 12–24 hours only, is usual for ordinary restaurant wines, and 3–5 days for typical red wines in France, with rosé and white wines having very short or no skin contact after crushing. He notes that these short times result from a demand for less tannic, soft, "supple" wine and questions whether economics today will permit the required longer aging of high-tannin wines to reach the degree of softness desired. Ribéreau-Gayon and Stonestreet (1966b) report analyses of vintages from 1921 to 1962 from the same Médoc vineyard, in the range 2700–4400 mg/liter prior to 1948, and in the range 1500–2500 in 1952 and later. Since one would expect no change or a decreased soluble tannin in bottled red wine with age, rather than an increase, this clearly indicates a major change in effective pomace contact time between 1948 and 1952, at least in that winery. Other reports suggest similar trends, at least in the areas which have traditionally produced the more robust red wines higher in extractives and phenols.

Coudon and Pacottet (1901a) noted that a "Pinot noir" wine had a total phenol content of 749 mg/liter at 7 days on the pomace, 964 at 12 days, and 1116 at 16 days. At crushing and daily intervals thereafter, Nègre (1942–1943) took subsamples of fermentation on the skins of "Aramon," a low-color variety and, in this instance, of final alcohol below 9%, and found total phenol of 240, 600, 640, 660, 700, 1220, and 1490 mg/liter. The fraction of total phenol precipitable with zinc, the "tannin," accounted for most of the increase on the fourth and subsequent days, and the polyphenol-not-tannin was relatively constant at 140, 310, 350, 370, 410, 450, and 540 mg/liter. Relative red color in the same series was 1, 20, 20, 20, 22, 22, and 26 arbitrary intensity units. A second fermentation, also of "Aramon" grapes, gave: at day 1, 470 mg/liter total phenols, 350 mg/liter phenols-not-tannin, and color of a relative value of 1; at day 3, 780, 580, and 6.5; at day 4, 1030, 660, and 8; at day 7, 1610, 800, and 8.5; at day 9, 2150, 950, and 13; and the press wine on day 9, 4360, 2140, and 20.

Berg and Akiyoshi (1956b, 1957, 1958, 1960, 1962) and Berg (1963) have studied many aspects of visible red color level of wines as related to fermentation. They found (1956b) that total tannin level increases with time on the pomace after maximum color has been reached in the typical fermentation. Such results have had a great deal to do with the tendency toward shorter fermentations on the skins in recent years. They found that astringency correlates fairly well with total tannin content. Drawing off the fermenting wine as soon as desired color level has been reached in a filtered sample is recommended to ensure high quality without excessive tannin. Wines drained at maximum color, an average of 84 hours in a series of samples, averaged 1420 mg/liter total phenol content, and those at the longest contact times, an average of 110 hours, had 1710 mg/liter.

The alcohol of the fermentation appears to be the major factor in extraction of phenols, including pigments, from the skins. Berg and Akiyoshi (1957, 1958) divided a must into two parts, and one part was raised in ethanol (by addition at frequent intervals) to a level slightly lower than established by fermentation in the other part. At 60 hours and an added alcohol of 8.3%, 1240 mg/liter total phenol had been extracted, whereas at 60 hours of fermentation and 9.2% alcohol, a similar color but only 1020 mg/liter total phenol was found. This appears also to illustrate selective precipitation of tannin by yeast. By extraction of pomace with alcohol at higher concentration for a long time, they estimate that generally not more than ⅔ of the grape anthocyanin appears in the wine by the usual fermentation. Valuiko (1966) reported 50% of grape anthocyanin extracted into red wine during fermentation. Ough and Amerine (1962) report that only 41% of the anthocyanin color present at the start of fermentation was accounted for at the end, and that retention by the solids was about equal to that extracted into the liquid. Of the total color extracted during fermentation, Berg and Akiyoshi (1957, 1958) estimated that 80% was usually extracted by the time ethanol level reached 3%, and nearly all of the remainder was extracted by the time 6% ethanol was produced, although grape varieties differed in behavior during color and tannin extraction. Deibner and Bourzeix (1966) found that 4 macerations in 0.1 *N* HCl at 20°C for a total time of 20 hours removed all the anthocyanin from "Carignane" skins, whereas "Aramon" skins required several more treatments at 60°C for complete extraction. Harvalia (1965) reported maximum anthocyanin absorbance very early in fermentation—at 1–3% alcohol.

Color is less in wines fermented to a certain alcohol level than in wines fortified to the same level. Berg and Akiyoshi (1960) found this loss of color during fermentation to be fairly proportional to the color originally extracted and to the alcohol level. This was suggested to result from acetaldehyde produced by fermentation reacting with the anthocyanin and removing it, but it could, again, result from yeast adsorption.

Visible red color of wine is affected by the proportion of anthocyanin that is in the form of colored flavylium salt, in contrast to other forms, such as the colorless pseudobase. Berg and Akiyoshi (1962) and Berg

(1963) have estimated the percentage in the colored form by the color change produced by shifting to a very low pH, and have also studied the spectral color and the anthocyanogen content in relation to fermentation and aging. In samples drawn during fermentation on the pomace, anthocyanin content, percentage in the colored form, and absorbance at 525 mμ were all higher at 9% ethanol than at either 6 or 12%. Leucoanthocyanidins were highest at the highest alcohol level tested. In fermentations of red juice from heat-treated grapes, however, all these values except percent of anthocyanin in the colored form decreased with fermentation. Leucoanthocyanidin content was minimal at 3% alcohol, and then rose, indicating competition between precipitation and the solubilizing effect of ethanol. The maximum proportion of the anthocyanin was in the colored form at 6% ethanol.

Amerine (1955) showed that color extraction in red pomace fermentations in the first 48–72 hours was much higher at higher fermentation temperatures. Fermented to approximately the same alcohol content and with other conditions similar, a single lot of crushed "Cabernet-Sauvignon" grapes reached 1300 mg/liter total phenol in 168 hours at 53°F, 1900 in 108 hours at 68°F, and 2500 in 72 hours at 85°F (Ough and Amerine, 1960). Other examples are given by Ough and Amerine (1966). Color obtained in "Pinot noir" fermentations in the presence of 175 ppm of added $SO_2$ reached a maximum in 2 days at 80°F or 70°F, and in 3 days at 53°F, with the higher maximum anthocyanin color level obtained at 80°F (Ough and Amerine, 1961). The same report showed that a log-log plot of wine tannin, 525 mμ absorbance in the wine, and alcohol content versus time on the skins and seeds gave a linear relationship in each instance over the tested range of about 20–150 hours. It thus appears that if tannin is to be kept below a certain level by shorter fermentation on the skins, anthocyanin content must be sacrificed. To investigate maximum color yield and other factors, Ough and Amerine (1962) blended juice from white grapes with "Cabernet-Sauvignon" must before fermentation, and white wine and red wine from the same grapes after fermentation. Rate of extraction during fermentation appeared to be independent of the concentration in the extracted solution, indicating no important solution gradient or apparent saturation effects, and indicating grape cell permeability as the controlling factor.

Data pertinent to phenol extraction from grape berry solids during fermentation have been obtained by many other workers. Ribéreau-Gayon and Sudraud (1958) showed the important effect of alcohol in another way. Addition of sucrose to a red fermentation so that the final alcohol was increased from 10.5 to 12.5% gave about 110% of the total phenol. Coudon and Pacottet (1901a) added extra pomace from white

"Aligote" grapes equivalent to half the volume of a "Gamay" red wine fermentation. The "Gamay" alone gave 900 mg/liter total phenol, and the extra pomace gave 1874 mg/liter. Red wines from *V. vinifera* grapes are commonly prepared so that about the maximum possible anthocyanin will appear in the wine. With very highly colored grapes, particularly of other species, this may not be necessary or desirable (Crowther, 1959–1960) although such wines are often intended for blending to raise the natural level of color of other wines.

There are many types of "early-pressed" wines. Rousseaux (1909b) describes "wines of one night" or rosé wines with pale pink color, having 12–24 (seldom as high as 48) hours of maceration and fermentation on the skins; café or early-pressed red wines intermediate between the rosé and the full red wine; and "grey wines," from white-wine methods applied to red grapes, i.e., immediate juice separation after crushing. Amerine and Winkler (1941) found that grapes which made a red wine with 1900 mg/liter of total phenols in 3–5 days on the skins, made rosés at about 1 day, with about 10% of the red wine's color and 500 mg/liter of total phenols. Laborde (1911) used rosé wine's relatively high ratio of total phenol to red pigment content but low total phenol and low leucoanthocyanidin content, to distinguish those made properly, with good grapes, from poorer wines made by longer fermentations of low-colored grapes or partly exhausted pomace. Nègre (1942-1943) developed similar tests based upon low zinc-precipitable tannin in short-fermentation rosés. An example from his report of rosé and red wines from the same grapes was a "Carignane" pair, the rosé with 100 mg/liter total phenol, all nontannin, and the red with 2220 mg/liter total phenol, including 560 mg/liter not precipitable with zinc.

Another aspect of the preparation of wines by fermentation of the fluid juice on the solid pomace relates to the effect of the different solid parts which may be included. Pacottet (1904) fermented fractions of "Pinot noir" grapes for 8–10 days and analyzed the resultant wines for tannin. The units reported seem unclear, but, in relative terms, juice had only 8 units of tannin whereas juice plus stems had 100, juice plus seeds 130, juice plus skins 210, juice plus seeds and skins 410, and juice plus seeds, skins, and stems 456. He noted that the sum of the separate fractions was larger than the combined sample, indicating precipitation. The stems gave a taste to the wine and adsorbed considerable anthocyanin. Their inclusion lowered the color considerably.

Nègre (1942-1943) conducted similar experiments with low-sugar low-color "Aramon" grapes and analyzed for total phenols and zinc-precipitable "tannin" by his method. Juice fermented alone had 60 mg/liter of nontannins, with no tannin. Juice plus stems had the same nontannin level as juice alone, but total phenols rose to 280 mg/liter. Juice plus seeds gave 340 total and 220 nontannin; juice plus skins gave 480 total and 240 nontannin; juice plus any two of the other fractions gave total phenols of 680–740 mg/liter; and the whole fermented cluster gave 1200 mg/liter total phenol including 450 mg/liter of nontannins. The sum of the nontannin phenols in the wine from the different fractions was a fairly good estimate of that found in the samples from the combined fractions. The tannin fraction, on the other hand, was considerably lower in the wine from the whole cluster than would have been predicted. Stems contributed a strong herbaceous taste to all samples that included them, and stem removal was considered preferable.

Ribéreau-Gayon and Ribéreau-Gayon (1953, 1954) found that stems raised total phenol content in an experimental wine from 3840 to 5840 mg/liter, and in other instances stems increased total phenol to 124–136% of the value without them. A wine prepared from destemmed grapes gave Coudon and Pacottet (1901a) 446 mg/liter total phenols, whereas a wine from the same grapes with stems had 785 mg/liter. Biol and Siegrist (Anon., 1966b) found 1105 mg/liter without stems, 1522 mg/liter with, and higher tannin but less difference in percentage if the pomace was heated before fermentation. Sudraud (1958) found that stems contributed to browning, as indicated by relative absorbance at 420 versus 520 mμ.

Singleton and Draper (1964) studied the extraction of total phenol from grape seeds in percentage of the total recoverable by exhaustive extraction. Extraction at 52, 68, and 86°F into 0–14% aqueous ethanol was determined at times up to 421 hours. Increasing ethanol gave regularly increasing extraction, although the highest ethanol gave a more prolonged lag and lower extraction at first, presumably because of a hardening effect on the fresh seeds dropped into it. At a given temperature the percentage of the seed tannin extracted increased linearly with the logarithm of the time after about a 3-hour lag. Of course, increasing temperature gave increasing extraction. About 50% of the total extractable phenols of the seed would be extracted, according to their data, into 11% ethanol in 1 day at 86°F, 2 days at 68°F, and 4 days at 52°F. They estimated that total extraction of the seed phenols without precipitation, etc., would be equivalent to 2000–4000 mg/liter in wine, and that about half this value should be extracted in the usual red table wine fermentation. Thus, it appears that although much of the extracted seed tannin must be precipitated it can be an important part of the wine tannin.

Cantarelli and Peri (1964a) studied the anthocyanogenic tannin content in skins, seeds, pulp, and stems of 3 white grape varieties before and after fermentation in the juice at the amounts naturally present. Relatively high proportions, more than ½ as an average, of the leucoanthocyanidins could not be accounted for in either the lees or the fluid. A usually large but variable amount of the total remained in the solids plus lees for all the tissues and all varieties. For the three varieties "Muscat," "Regina," and "S. Anna Lipsia" the amount of leucoanthocyanin contributed to the wine was reported to be, for stems, 27, 87, and 410 mg/liter; for skins, 23, 41, and 97 mg/liter; for pulp, 84, 67, and 94 mg/liter; and for seeds, 8, 110, and 306 mg/liter. Ribéreau-Gayon and Stonestreet (1964) found that fermentation of juice plus skins gave 1300 mg/liter leucoanthocyanin, juice plus skins and seeds gave 2200 mg/liter, and the whole clusters including stems gave 3700 mg/liter, with the flavanol proportion increased in the whole-cluster sample.

It is clear from these and related studies that all the solid tissues of the grape cluster may contribute important quantities of phenols to wines fermented on these solids. It is also clear that the cells of these tissues must be made permeable by heat, alcohol, or another agency before the phenols can diffuse rapidly into the fluid. Among the methods of inducing this diffusion are "asphyxiation" and "maceration" processes. When rendered anaerobic, plant cells break down rapidly. Autolysis and release of phenols into the juice is even induced simply by holding grapes or crushed grapes in masses or in containers so that natural respiration consumes the available oxygen (Flanzy, 1935b). The process can be hastened by sweeping out the oxygen with other gases, such as nitrogen or carbon dioxide. This may be one factor in the early stages of alcoholic fermentation, although the alcohol seems much more important, and introducing $N_2$ or $CO_2$ seems to have little effect if yeast fermentation has started (Berg and Akiyoshi, 1957).

Roson (1967), for example, has prepared rosé wines by macerating crushed "Cabernet" grapes at 54°–86°F for periods up to 3½ days before fermentation. There appeared to be an asphyxiation threshold beyond which juice color increased rapidly. For rosés, exceeding this threshold was considered undesirable, whereas for red wine it would be necessary. The maceration temperature was important for the best shade of color at the proper level. An initial yellow color was overcome as the pink color developed, and approximately 24 to 48 hours at 68°F gave the most pleasing color with the "Cabernet franc" grapes tested. That such treatment may involve much activity by pectolytic, phenolase, peroxidase, and other enzymes has been emphasized (Nègre, 1966; Oparin, 1963). Sudraud (Anon., 1966a) found that immediate pressing of red grapes after crushing gave wine with 16 mg/liter anthocyanins and 100 mg/liter anthocyanogens, whereas maceration overnight at ambient temperature gave 26 and 320 mg/liter, respectively.

Amerine and De Mattei (1940) stored whole "Carignane" berries in open jars, closed jars, and various gases. The color of juice at 51 hours was increased by all treatments (except HCN) over that of a freshly picked sample. That anaerobiosis was a factor is suggested by the fact that the open jar had only 17% increase, the closed jar 38%, and a nitrogen-filled jar 65%. Toxicity may be another factor, for a sample in an oxygen-filled jar increased 65% in color and the carbon dioxide sample increased 41%. Flanzy and Poux (1958) noted that flushing with carbon dioxide before and during fermentation in small lots gave more color in the wine. Peynaud and Guimberteau (1962) also reported that whole fruit was macerated and the polyphenols liberated when held in atmospheres of carbon dioxide or nitrogen. Various procedures have been developed for treatment of grapes under carbon dioxide ("carbonic maceration"), and these have been compared with other traditional and new methods of vinification (Flanzy, 1967; Anon., 1966a; Rankine, 1964; Bazilevskii, 1964; Benard and Jouret, 1962).

The advantages claimed for carbonic maceration seem to be accompanied by disadvantages, and the ultimate place of such techniques is unclear. There often seems to be a lower pigment level (Benard and Jouret, 1962), and perhaps a specific flavor is produced (Nègre, 1966). Laszlo *et al.* (1966) produced only 10–50% of the color of the normal vintage, but tannin was also 10–30% less and the wine matured earlier and was more drinkable. Flanzy *et al.* (1967a,b) found that the total shikimic-quinic acid content of grapes decreases at first under carbonic maceration at cool temperatures (59°F) but increases considerably with time or at warmer temperatures (for example, 1400–2600 microequivalents of acid per liter in 31 days at 77°F).

Mechanical treatment, to the extent of grinding the pomace, not only quickly produces tannin excessive for drinkable wine, but also usually makes the wine very difficult to clear. Therefore, excessive stirring, etc., during cap management is often detrimental even though it speeds the extraction of phenols (Sudraud and Cassignard, 1958; Ough and Amerine, 1960). The more vigorously mixed pomace often appears to liberate more tannin in proportion to red color. Roubert (Anon., 1966a) found that violent stirring gave 30% more color at 520 m$\mu$, 56% more tannins absorbing at 280 m$\mu$, 24% more methanol (indicating pectin hydrolysis) but less absorbance at 312 m$\mu$, suggesting either less oxidation or less caffeic acid derivatives. Since red sweet wines like port must be drawn from the skins and fortified while sugar remains unfermented, and the grapes used are often not highly colored, portmakers have difficulty in obtaining adequate color and tannin extraction. An ancient method of "mechanically" aiding the extraction of anthocyanins for port was nearly continuous marching back and forth by lines of men treading through the fermenting grapes until the time to drain and add wine spirits. More economic modern ways have often turned to heat rather than mechanical treatment to solve the problem, although important parts of modern practice are vigorous and frequent or continuous pumping of the fluid onto the cap.

Fermentation of red wines under relatively high carbon dioxide pressure has been recommended, sometimes combined with periodic release (Klenk *et al.*, 1954; Konlechner and Haushofer, 1956; Gadzhiev and Abramov, 1962). Explosive pressure release seems interesting in that it might disrupt skin cells without affecting the harder seed cells, but it would appear difficult to control on a large scale. Puisais and Lacoste (Anon., 1966a) noted how the degree of skin and stem breakage during destemming and crushing affected color and tannin release to the subsequent wine. Manual stem removal gave a wine with 1580 mg/liter total phenol, whereas two different models of equipment gave 1400 and 2300 mg/liter with the same grapes.

Among other treatments commonly used in wine making, the addition of sulfur dioxide probably has the greatest effect on color and phenol release during vinification. $SO_2$ in solution in fermenting grapes has a number of effects. As a good "anesthetic," it causes the grape cells to release anthocyanins and other phenols, but in high doses it decolorizes wine pigments (Sauzeat, 1932; Astruc and Castel, 1934). It helps prevent oxidation of wine phenols, but it becomes bound during fermentation so that the effective level changes. A number of studies, already cited, illustrate the relative effects of different levels of added $SO_2$ on red wine fermentation (Ough and Amerine, 1961; Harvalia, 1965; Valuiko, 1966; and others). Reports by Berg and Akiyoshi (1957, 1958, 1960) discuss particularly the effect on visible anthocyanin color during fermentation. In general, color and total phenol extraction are considerably increased by adding $SO_2$ or raising the levels of $SO_2$. Total phenol in an unfermented wine made by must fortification was 1660 mg/liter without the addition of $SO_2$, and 1800 with 100 ppm $SO_2$. Addition of 100 ppm $SO_2$ before fermentation gave 1200 mg/liter, compared with 1120 mg/liter in a control sample without $SO_2$. Samples of another must fermented with 100 ppm $SO_2$ gave 990 mg/liter of total phenols in the wine, whereas 250 ppm $SO_2$ gave 1200 mg/liter. Effects on color were comparable and sometimes greater. Sudraud (Anon., 1966a) found 100 mg/liter of anthocyanin and 760 mg/liter of anthocyanogens in a wine fermented from juice separated after overnight maceration of the crushed berries with 100 ppm $SO_2$, compared with 26 mg/liter of anthocyanins and 320 mg/liter of anthocyanogens from similar treatment without $SO_2$.

Many other techniques have been proposed for changing the phenol content of wines by affecting the solid tissues and rate of extraction from them. In most instances relatively little has been published upon the actual effect upon phenol content, but it is possible to estimate the effect from the examples given previously. Rankine (1964) found freezing less effective than heating. Use of 5–45 minutes of ultrasound to cause anthocyanins and other phenols to diffuse from red grapes gave an increase from 630–770 mg/liter in total phenols, and a 1–6% increase in juice yield (Berishvili, 1963; Popovskii *et al.*, 1961). The fermenting tank and associated equipment used, and their manner of operation, affect phenol content. This is especially true in some of the newer equipment designed to operate semicontinuously, with crushed grapes entering near the bottom, exhausted skins taken off the top (and perhaps recycled), seeds removed at the bottom, and partly fermented wine drawn off below the cap. Obviously, phenol and pigment content in the resultant wine could vary considerably with the operating conditions. The limits would probably approximate a rosé and a wine fermented to dryness on the skins prepared in the traditional manner as modified by the temperature and stirring applied.

The yeast used for the fermentation may affect the rate and extent of extraction of phenols from grape solids. One effect, of course, is the amount of ethanol and its rate of production. More ethanol would favor extraction, but a slower production would have the effect of prolonging solids contact in usual winery practice. There appear to be other effects, however. Mosiashvili (1958) found that, among a number of yeasts of different genera, *Zigosaccharomyces bailii* gave as low or lower total phenols when white must was being fermented. With red must, however, considerable increases were found: as large as 2100 mg/liter for usual *Saccharomyces* wine yeast alone, and 4700 mg/liter for *Z. bailii.* The difference was attributed to a greater pectin-hydrolyzing ability of *Z. bailii.* Added pectolytic enzyme preparations have different degrees of effectiveness in increasing anthocyanin extraction, depending on their source, associated enzymes of other types, and method of application. Since pectic enzymes are often combined with warming to the extent that the enzyme can act rapidly but not survive long, it is difficult to generalize about their effectiveness as phenol-liberating agents, but they do have such effects (Wucherpfennig *et al.*, 1964; Sone *et al.*, 1964; Brad, 1966).

The content of specific phenols in typical red wines can be estimated from the specific reports already reviewed and by comparing white wine content, rosé wine content, grape content, and variation in red wine according to processing. It is tempting to carry such calculations further than we already have, but this sort of speculation is easily overdone in that much better experimental information will likely be rapidly forthcoming.

What appears to be needed is not only accurate determinations of individual substances on a wide range of samples, but also studies of the "carbon-balance" type. We need to know specifically how much of phenol "X" is in the grape, how much appears in the wine under different conditions, and where the rest goes. Considering the difficulties of such studies in the complex phenolic system of wine, it is perhaps not surprising that relatively few studies have been made which might be considered phenol-balance studies. Neubauer (1872) gave many data which can be applied, including analyses of solid grape parts before and after fermentation. He reported that after fermentation in red wine the skins and stems retained about 85% of their original total tannin content, and the seeds retained 71%. Data of Cantarelli and Peri (1964a) on leucoanthocyanin distribution and fermentation balance have already been described. Korotkevich and Korotkevich (1962) described methods of obtaining good phenol-balance data, but gave no data. An earlier report by Korotkevich *et al.* (1950) does give much applicable data, notably that 2-49% of the total phenols of grapes, including anthocyanins, appear in the wine. The majority of this phenol appears to come from the skin, but wine phenol values as high as 200% of the skin content indicate important contributions from seeds or other parts. Red table wines averaged perhaps 40% of total berry phenol, and red dessert wines averaged about 30%. Wines from juice only had 2–11% of the berry total. Kondo and Zvezdina (1953) also found considerable tannin still in grape skins after passing through fermentation, and estimated that 12–36% of the berry total of tannides appeared in port wines, 15–63% in other red dessert wines, and 16 and 38% in two red table wines. Again, in some instances, the wine phenols totaled more than the skin alone contained.

Durmishidze (1948b) found that the fraction of the phenols of red wine which is soluble in ether, i.e., primarily the catechins, diffused into the wine early in the fermentation and remained fairly constant thereafter, at about 230 mg/liter in a wine which reached 3180 mg/liter in total tannin and 983 mg/liter anthocyanin. Tarantola (1951) found that an average of 16% of the total permanganate-active substance was extractable into ethyl acetate in 8 red wines, 41% in 2 rosés, and 39% in 6 white wines. This indicates that the catechins and small anthocyanogenic polymers are more readily extracted from the solids, as would be expected.

# IV

# Viticultural and Biosynthetic Relationships

## A. *Histology and Biosynthesis*

At cellular and subcellular levels the formation and occurrence of different phenols in plants are not understood as well as might be desired, though grapes have been among the materials studied more frequently. The general observations which apply to the occurrence of the tannins and anthocyanins in plant cells and tissues are discussed in considerable detail by Lloyd (1911, 1922), Onslow (1925), Guilliermond *et al.* (1933), Joslyn and Goldstein (1964a), and Esau (1965).

The anthocyanins and tannins appear to occur commonly dissolved in the cell vacuole, although they may occur as amorphous masses or coarse or fine granules, and may impregnate the cell walls or occur in the cytoplasm. The anthocyanins have been reported to occur as granules or even as crystals within the vacuole. It appears probable that the granules represent anthocyanin adsorbed or combined with other substances. Ginzburg (1967) describes the formation of tannins in cisternae or small membrane-covered bodies apparently related to the endoplasmic reticulum, which inflate, coalesce, and form the central vacuole as the cell matures. Wardrop and Cronshaw (1962) report the formation of phenolic substance as deposits at the periphery of vesicles which appear to arise by dissolution of starch grains. Politis (1921) described *Vitis vinifera* "Corinth" fruit anthocyanin as forming in tiny grains of color which sometimes elongated to filaments or formed spherules. They sometimes fused and sometimes coalesced to dissolve their pigment in the vacuole. Sometimes a large, bright-red, "oily"-appearing body formed, which grew to be larger than the nucleus. The same general description would apply to the formation of chlorogenic acid as observed by Politis (1948).

Guilliermond *et al.* (1933) noted that small "vacuoles" within the cytoplasm may have anthocyanin or tannin in high concentration whereas others nearby in the same cell may have little or none. The cytoplasm and cell wall itself ordinarily show no anthocyanin in the living cell, but as the cell dies these tissues are stained as the anthocyanin diffuses from the central vacuole or cytoplasmic vesicles. Lloyd (1911, 1922) makes a distinction between astringent, reactive, soluble tannin in young fruit and gelatinous insoluble masses in ripe fruit of *Diospyros kaki.* He also notes the necessity in the fruit cell of keeping protein-precipitating tannin separate from enzymes, and pictures the occurrence of separate sap vacuoles free of tannin in the same cell.

Schanderl (1957) described and illustrated many aspects of the cellular distribution of tannins and anthocyanins in grapes. He reported that tannin and anthocyanin in the berry skin occur together in the same cell, often in the same intracellular container, and suggested that this is why tannin and anthocyanin contents in red wines tend to parallel each other. He finds that anthocyanin is in large juice-filled vacuoles in some varieties but in others is in smaller membranous sacs dispersed in the cellular protoplasm, making it necessary to break down these ventricles to make red wine. Winkler (1936) also notes the parallel relation with low tannin in low-colored grape skins.

Molisch (1916) described the conversion of tannin in the natural state in white grape skins – from the natural state of colorless apparently semisolid globules, which would react with vanillin-HCl, to dispersed semifluid browned phlobaphene, which would not. This reaction, produced by intense sunlight, visibly discolored the fruit. Scurti and Pavarino (1934) picture a similar condition. Negrul' and Lyu (1963) related normal color to cytology of the grape skin.

No plant tissue appears to lack tannins, or at least phenols, entirely (Esau, 1965), though the tannins are commonly very concentrated in certain cells and low in all the others. Bate-Smith and Metcalfe (1957) considered that the number of tanniferous cells was proportional to the amount of tannin in a given herbarium sample. If no such cells were found, then true tannin was considered absent. Grapes do have tanniferous cells in all vine parts (Metcalfe and Chalk, 1950; Esau, 1948, 1965; v. Brehmer, 1928; Schanderl, 1957). These tannin-rich cells may be larger than the adjacent "normal" cells, perhaps because of a tendency to imbibe water (Lloyd, 1913). This condition may apply in near relatives of grapes (Holm, 1911), but the typical grape-tissue tanniferous cell does not appear different in size or shape from its associated nontanniferous fellows. The cells bearing high tannin are often scattered, apparently at random, in the tissue, although there is a tendency in grapes, as in other plants, for them to occur in "trains" or "tubes" of adjacent cells and in close association with vascular bundles (Niethammer, 1931). Typically, an isolated tanniferous cell is surrounded by cells from which tannin appears absent. In tissues forming anthocyanins this also appears true, namely, that the formation of a high level of anthocyanin within one cell appears to have no carryover or "gradient" effect regarding the synthesis of these phenols in its neighbors. Of course, a certain tissue may have high levels of phenols in all its cells.

The tannin of grape pulp is so low as to be nearly absent in the ripe grape berry, as previously noted. There do not appear to be any tannin-bearing idioblast cells in ripe flesh except perhaps around the vascular strands. Tannin is, however, fairly high in the unripe pulp, and appears to be metabolized or caused to migrate to seeds or skins rather than to be merely insolubilized in place, as in some other fruits (de Villiers, 1926; Escudié, 1939; Rafei, 1941). The disappearance of tannin appears to progress from inside toward the skin.

The morphology of the berry skin and seed is described by Winton and Winton (1935) and von Babo and Mach (1923). The skin consists of a cuticle layer, epidermis, hypodermis of at least 8 cell layers, and, in *V. labrusca,* a 3–5-cell layer of thin-walled cells which makes the skin "slip" easily from the pulp (Rafei, 1941). The anthocyanins are confined to the outer layers of cells, perhaps only 3 or 4 outer cell layers, but sometimes extend into soft tissue below the tougher skin (Winkler, 1936; Alwood, 1914) and, in teinturiers, of course, into the pulp. The skin of preserved *V. vinifera* grapes is relatively thin, and that of other species is variable but tends to be about twice as thick (Bonnet, 1903). This is probably one reason why other species tend to be darker colored, though no microscopic survey appears to have been made of living colored grape-skin tissue. Of course, the skin of the grape varies not only in thickness but, owing to considerable differences in berry size among varieties, in surface area also. With a given thickness of anthocyanin- and tannin-bearing cells, and with a given color intensity in these cells, the smaller the berry the more anthocyanin would be available for the wine or juice.

The cells of the grape seed are divided grossly into 3 types: a thin soft outer layer rich in tannins, a heavily lignified layer, and a fatty endosperm. The endosperm contains tanniferous cells, and Ekster (1957) reports an inverse relation between fat and tannin content in seeds in winery pomace. This suggests a significant part of the tannin in the endosperm, and also suggests that cells with tannin lack oil. Cracking the seeds to expose the endosperm to extraction adds greatly to red wine tannin. Seeds of red grapes have been reported to yield more oil than seeds of white grapes (dal Piaz, 1892), which suggests a general competition between fat and tannin for carbohydrate diverted to secondary products. Seeds were also reported to have about 30% oil from grapes ripened in hot climates, and 10% oil from those ripened cold climates. It would then be expected that the cold-climate grapes would have more tannin.

The size, weight, appearance, and morphological details of grape seeds are relatively constant for a given variety of grapes, and therefore serve as some of the best characteristics for variety identification and breeding studies (Chauveaud, 1894; Bonnet, 1902; Bioletti, 1938; Olmo, 1942).

The average number of seeds per berry has some relation to variety but varies some with location and season. The full natural complement of seeds in a grape berry, from the viewpoint of botanical structure, is 4, but berries in a given cluster may contain 0 to 4 or, in few cases, even more than 4 seeds. "Thompson Seedless" and other varieties which produce fruit parthenocarpically commonly have zero fully developed seeds per berry, and other varieties may have a high proportion of undeveloped or empty seeds (Olmo, 1934). The berries with empty seeds (not viable, float on water, but about normal size) often have abnormal color (Olmo,1934). Lazenby (1903) reported averages of 1.2–2.7 seeds per berry among 1 wild and 6 cultivated varieties of grapes. The average weight of 1 fresh seed in these 7 selections was 35–63 mg. Interestingly enough, his data indicate that the major compositional difference between the wild and domesticated grapes was a great increase in pulp content in the cultivar. The wild grape had 28.5% skins, 32.1% seeds, and only 39.4% edible juice or pulp, whereas the cultivated grapes averaged 20.1–24.6% skins, only 3.5–5.3% seeds, and 71.5–75.0% edible juice or pulp. It appears clear that the breeding and development of commercial fruiting varieties has mainly modified the ratio of pulp to seed.

The number of seeds in a berry is correlated directly and highly with its size when ripe, and also influences rate of skin coloration and other aspects of ripening (Müller-Thurgau, 1898; Olmo, 1946; Pastena, 1964; Gärtel, 1954, 1966). The seed appears to be a source of natural grape hormones and to affect the response of grape berries to applied gibberellin and other plant-regulating substances (e.g., Weaver *et al.*, 1962;

Weaver and Pool, 1965). Ribéreau-Gayon and Peynaud (1960–1961) report, for "Malbec," no berries had 0 or 5 seeds per berry, 56% had 1 seed, 34% 2, 8% 3, and 2% 4. Gärtel (1954, 1966), in a sizable sampling of "White Riesling" grapes, found that berries without seeds were 5.9% of the total and had an average weight of 0.55 g; one-seeded berries were 39.1% and 0.89 g; two-seeded were 37.7% and 1.15 g; three-seeded were 14.8% and 1.35 g; four-seeded were 2.2% and 1.67 g; and five-seeded were 0.2% and 1.85 g. On the average, a single seed increased berry weight by 0.26 g; thus, tannin content related to seeds should be relatively constant unless a high proportion of seedless berries occurred.

Gärtel (1966) also reported that volume increased and surface-to-volume ratio decreased as average seeds per berry increased, with clusters that averaged 2.0 seeds per berry being very tightly packed by crowding of the berries. Loose clusters averaged only 0.7 seed per berry. These and similar data show that seed content correlates with many different aspects, such as berry color, stem content, and berry surface, which may have a large effect on the phenol content of grapes and wines in addition to the direct effect of the seed tannin itself. It is certainly conceivable that specific phenols in these tissues are active participants in the regulatory mechanisms producing the associated large effects on sugar content, berry size, etc., and are not simply another incidental manifestation. The known relative influences of different phenols present in grapes on regulation of indoleacetic acid content through activation and suppression of its oxidase would be a major possibility of such direct participation (Gortner and Kent, 1958; Galston, 1967).

It appears that the typical small-berried wine grape variety has an average of about 2 seeds per berry, about 50 mg per seed, with a seed phenol content of about 5% on a fresh basis. At an average berry weight of 1.5 g, the berry would be 6.7% seed by weight and seed "tannin" would be 3.3 g/kg berries. These figures compare well with those cited for direct determinations.

## B. Ripening and Biosynthesis

Changes during development and ripening, in fruits in general and grapes in particular, are relatively well known as regards general appearance, sugars, and acids. Changes in phenols, however, have received much less study. Tannins and anthocyanins are barely mentioned by Amerine (1956) in a detailed review of the literature on grape maturation. Analytical values for total phenol content, and apparently also for the content of most specific phenols, are usually quite variable (as well as more tedious to determine) than dissolved solids or titratable acidity. Different varieties and different samples of one variety may appear to have changes in phenol content which are erratic or unpredictable on a ripeness basis. This situation has tended to focus attention on constituents other than phenols during the development and ripening of grapes. If the samples examined are sufficient, however, definite trends in phenol content can be shown during ripening.

The maturation of a grape berry may be considered to comprise two major periods, development and ripening. The transition between the two stages is marked by a number of changes, not necessarily simultaneous but, individually, generally rapid and easily detected. In terms adequate for our purposes, the development period is characterized by a rapid growth primarily by cell division, and the ripening period by a rapid increase in berry size and density by cell enlargement and sugar accumulation. The beginning of grape berry ripening, called "veraison," is indicated by the initiation of anthocyanin formation in red varieties, the initiation of rapid sugar accumulation, a decrease in acidity and chlorophyll, and the onset of berry softening. About 2 months after veraison, depending on weather and variety, the berry should be ripe for wine, at about 23° Brix. Veraison generally follows a brief pause in berry weight increase as a shift is made between the 2 major stages. This transition is believed to coincide with completion of the development of germinable seeds (Amerine, 1956; Winkler and Williams, 1935; Moreau, 1932).

The total phenol content of grapes as related to ripeness has been difficult to interpret from earlier literature owing to sizable extraneous fluctuations (Caldwell, 1925; Reynolds and Vaile, 1942; Noyes *et al.*, 1922). The analytical procedures were somewhat arbitrary and subject to variability, but the sampling of the grapes and preparation of an extract were also difficult to standardize for reliable comparisons. The phenol content of wines, particularly red wines, could be misleading. Red wines prepared from riper grapes generally had more total phenol and higher anthocyanin content than wines from comparable grapes of less ripeness. This, however, was often the result of a higher sugar content in the riper grapes, giving more alcohol and consequently better extraction of phenol into the wine. It could not be safely concluded from such data that, in fact, there were more phenols or more pigment present in the riper grapes (Winkler and Amerine, 1943).

It appears that phenols must accumulate very rapidly in early development, for phenol content is generally high in percentage in the small green berry before veraison. This early period has not been studied as much as the ripening period. Singleton (1966) presented data on approximately weekly harvests from mid-July to early October or until the fruit was over 20° Brix. Total phenol content was determined after extraction by a procedure verified to recover a very high percentage of all extractable phenols present. The results were expressed both as phenols per unit weight of entire berries and as phenols per single whole berry.

Individually, the 12 varieties studied showed considerable variation not only in amount of phenols present but in apparent fluctuation or progression of phenol content during ripening. There was a general tendency for total phenols, in mg/kg berries, to decrease as berries ripened. The most common pattern appeared to be for phenol concentration to be high in the earliest sample, then decrease for a few harvests, rise again to near or beyond the previous high level, and finally decrease. This pattern was by no means uniform, however. Some varieties increased yet again in the final harvests, and the timing of maxima and minima with respect to Brix or pH was not the same for different varieties.

A general tendency has usually been noted for total phenols to become lower per unit fresh or dry weight as grapes ripen, although exceptions may be found depending upon the timing and developmental span involved (e.g., Gaivoronskaya, 1963; Kolesnik and Kremlev, 1957; Durmishidze, 1955; Durmishidze and Tsiskarishvili, 1950; Beridze, 1950; Sisakyan *et al.*, 1947, 1948; Baragiola and Godet, 1914; Kelhofer, 1908). The common pattern of intermediate fluctuations described above seems very similar to that reported for total phenols for grape juice by Copeman (1927), for grape skins by Venezia and Gentilini (1934), and for seeds or whole berries by Diemair *et al.* (1951b). Barnes (1940) noted a high correlation ($r = 0.78$) between total phenol concentration and acid concentration in juice firmly pressed from thawed samples of ripening grapes. Since acid decreases during ripening, this indicates that total phenol concentration in these samples also decreased during ripening. Reynolds and Vaile (1942), however, found no such correlation in their samples.

Total phenol per individual berry, however, increases during much of the ripening period. The apparent decrease in total phenol per kg of berries results from the fact that phenol synthesis, although appreciable, is less than the weight increase caused by increases in berry size and density. The data for 12 varieties were combined (Singleton, 1966) on the basis of days before or after the juice Brix reached 18, regardless of date of sampling. Total extractable phenol before veraison and 85 days before fully ripe averaged about 80% of the amount present when fully ripe. Average phenol content per berry increased from this point to about 105% of the ripe value, at about 35 days before fully ripe. Average berry content then dropped slightly as ripeness approached, with perhaps a slight rise at final harvest ripeness. It is clear that there is considerable net synthesis of phenolic substances over a major part of the ripening of a grape berry. As specific examples, an average whole "Calzin" berry contained 4.1 mg of total phenols calculated as gallic acid at veraison (90% of the berries were green, berry weight was 0.96 g, and Brix was 8.6), while the ripe berry (100% red berries, 1.53 g/berry, Brix 22.5) contained 8.3 mg of phenols. "Pinot blanc" had 7.5 mg total phenols in the green 0.78-g berry, 10.8 mg in the 1.12-g berry, and 8.1 mg in the 1.30-g ripe berry.

During ripening, the skin weight usually doubles although, owing to the greater increase of the juicy pulp, the skins's percentage of the berry weight actually decreases (Amerine, 1956; Brunet, 1912). Even more than the skins, the seeds, having reached full size early, decrease in percentage of berry weight during ripening. Since the skins and seeds are respectively the site of almost one-third and two-thirds of the berry's phenols, they must be very active in phenol metabolism during this period. The juice and pulp generally account for one-tenth or less of the berry's phenols, but, owing to their great increase in amount during ripening, they would contribute to the overall increase in phenols.

The phenol content of grape juice samples obtained by cold pressing (Noyes *et al.*, 1922) fluctuated apparently erratically around a constant level, regardless of ripeness stage as measured by Brix, acid content, harvest date, or berry size. Similar results are reported by Gatet (1939). Juice obtained by heat treatment (Noyes *et al.*, 1922), representing, of course, more skin extraction, was much higher and more erratic in content of total phenols. One season's results indicated a general upward trend in phenol content with increasing ripeness, but another season did not. Webster *et al.* (1934) prepared "Concord" juice by chopping the berries and very vigorously pressing without crushing the seeds. Berries were harvested on the same dates but at different degrees of ripeness as indicated by berry color of green, red, or purple. The berries in any single color grade decreased with later harvest in total phenol, tannins precipitable with gelatin, and nonprecipitable phenols. Sugar content was similar or decreased in the same samples. When berries of different color groups taken on the same date were compared, however, the phenol analysis appeared to increase with ripeness more often than not.

The anthocyanin content of berry skin starts from zero at veraison and continues to increase until sugar level in the juice is relatively high, and then may fall. Golodriga and Suyatinov (1966) and Tsakov (1963) found that pigment content increased with ripeness to 20°–23° Brix.

Winkler and Amerine (1938b, 1943) found that as "Zinfandel" grapes ripened they increased in anthocyanin content at least until 26.2° Brix was reached. Begunova (1953) found "Saperavi," a teinturier, to increase from 111–197 mg anthocyanin per liter, in the first harvest after veraison, to 253–311 mg/liter at maximum. The highest pigment level was before the final harvest in 2 of 3 instances, indicating that a decrease may occur if the harvest is sufficiently delayed. "Cabernet" grapes in the same study followed a similar pattern.

Ribéreau-Gayon (1964a) found with "Merlot" that anthocyanin per berry increased linearly from about 0.2 mg/berry in early September to 1.6 mg/berry in late October, when the berries weighed about 1.3 g. Anthocyanin content decreased in berries warmed to simulate overmature conditions. Gavrish (1966) similarly found that withered grapes were lowered in anthocyanin content but increased in total phenol content in approximate proportion to the water loss. Clore *et al.* (1965) found that anthocyanin color extracted from "Concord" grape skins increased as long as samples were taken in two seasons, but dropped in the final sample in a season when samples were made at later dates. They also found that the total phenol content of juice from grapes hot-pressed after the seeds were removed paralleled the color level very well and increased except at the very late stages, presumably as senescence began. Durmishidze and Tsiskarishvili (1950) note that the leaves and vegetative parts of the vine are also decreasing in tannin at this time, so that translocation of the phenols from the berry is not indicated. Insolubilization of the tannins so that they could not be recovered would be suspected as the explanation. They reported that the ratio of water-soluble tannin to alkali-soluble tannin increased in the solid tissues but decreased in the pulp. If polymerization and insolubilization is the correct explanation for the decrease in extractable phenols in senescent grapes, the polymers must not be dissolved by alkali.

Diemair *et al.* (1951b) studied the tannin content of leaves, stems, seeds, and berries with and without seeds. Results for berries without seeds were fairly uniform, with a slight downward trend in phenols per unit weight. The phenol content of the stems, seeds, and berries with seeds, however, was characterized by periods of very rapid change interposed among periods of slight or gradual change. The periods of rapid change in one tissue seemed to bear no clear relation either to the general ripeness stage of the fruit or to the rapid changes in the other tissues. The metabolic turnover of phenols in the seeds, however, was clearly great and of major importance to total extractable phenol content of the whole berry at any given moment.

Cantarelli and Peri (1964a) analyzed 7 white or very light-colored grape varieties for total phenols and anthocyanogens during ripening. In nearly all instances both were maximal at intermediate harvests in both the whole cluster and the seeds. The ratio of anthocyanogen to total phenol was somewhat erratic and appeared to be highest in the latest harvest for the whole cluster. It was much more uniform in behavior in seeds, where it passed through a maximum near 1.0 and then fell to about 0.75. This seems to indicate polymerization with attendant poorer conversion to the measurable anthocyanidin. Polymerization of phenols during ripening is also indicated by molecular-weight studies by Durmishidze (1955).

Bockian *et al.* (1955) followed the appearance of various anthocyanins on paper chromatograms as the berry developed color. While there appeared to be a definite sequence of development, the sequence appeared to be in reverse order of disappearance upon dilution. That is, it appears that the pigment which manifests itself first is the one which is highest in the ripe fruit, and that the ratio of the different pigments is approximately constant. This also appears true from other reports (Shewfelt, 1966) and is related to the general observation that paper chromatographic maps are similar for a given variety regardless of sample, even if total pigment content is quite different. Ribéreau-Gayon (1958b), however, reported that the relative synthesis of delphinidin was high when synthesis of anthocyanins was high, and that malvidin increased at the expense of delphinidin when synthesis of anthocyanins was retarded.

Durmishidze (1951, 1955) and Durmishidze and Nutsubidze (1954) report that, in comparisons of green grapes in July with ripe grapes in September, the *d*-catechin content, mainly in the seed, dropped to ⅓ of the green value. Gallocatechin about doubled, *l*-epicatechin gallate appeared in skins only at full ripeness, and overall catechin gallate content decreased about 8–27% during ripening. Singleton *et al.* (1966a) followed changes by semiquantitative paper chromatography of separated phenols from seeds during ripening of 12 varieties. The general trends were common to all varieties. Most of the phenols were not specifically identified, but *l*-epicatechin gallate was reported to decrease during the latter stages of ripening without a corresponding increase in *l*-epicatechin or free gallic acid. Other phenols appeared or increased during ripening, and some decreased or passed through a maximum.

Carles and Lamazou-Betbeder (1958) report that quinic acid stays constant but shikimic acid drops during ripening of grapes. Hulme (1958b) suggests that shikimic acid tends to be low in fruit unless the metabolism is slowing down, as in the very ripe or senescent fruit. This seems to correlate with data of Flanzy *et al.* (1967a,b) showing a rapid increase in the sum of these acids as the grape is killed by anaerobic storage.

## C. Environmental, Plant Physiological, and Agronomic Relationships

A host of extrinsic and intrinsic factors affect the phenolic substances of plants (Swain, 1963). The most clearly demonstrated effect of climate on grape phenols is that of mean vineyard temperature on berry color. Providing that the weather is warm enough to mature the grapes satisfactorily, the cooler the region or season the more anthocyanin pigment is produced in the grape skin. Bioletti (1938) reports that relative values for the same variety would be about 100 in the cooler regions of California, 67 in intermediate, and 45 in a hot region. The effect, noted in all red varieties, appears to be greater in some (Winkler, 1936). "Emperor" table grapes, for example, may range from purple-black to nearly colorless in regions of cool to hot average summer temperatures. Tannin parallels color closely in the skin and responds the same to seasonal mean temperature.

Of course, in the clear, dry "Mediterranean" type of climate which prevails in many viticultural areas, including much of California, a high mean summer temperature is related to altitude, air circulation, and sometimes cloudiness though seldom rain. In other northerly or continental areas, however, seasons are likely to be more variable, with rain, cloudiness, and other factors complicating the interpretation of effects of climate and weather on grape phenols. Ribéreau-Gayon (1958b) suggests sunlight as the more significant influence promoting anthocyanin synthesis, and light certainly has an important effect on berry color. The effect of light and temperature would appear, then, to be competitive in the sense that high sunlight usually correlates with high mean temperature, though they affect anthocyanin content oppositely.

Winkler and Amerine (1937, 1938b, 1943) found that tannin of wine from 6 varieties from several California regions averaged 1200 mg/liter in 1936, a notably warm year, compared with 1100 mg/liter for comparable samples in 1935, an average year. The slightly but insignificantly higher tannin can be explained on the basis of higher alcohol, 12.2%, in the warmer year, with associated greater extraction. In spite of this the wines for the cooler year, 1935, were significantly higher in anthocyanin content. Fessler (1939) reports that averaging values for many samples of wine shows that tannin content is lower in the warmer vineyard regions.

Amerine and Winkler (1944, 1963) presented many analyses for wines, including color intensity and total phenol content. The values are grouped by variety and by climatic region based upon mean summer temperature. Fermentation time on the skins is an unknown variable, but the wines were made in a single cellar. If the red wines are matched by alcohol content, low total phenol shows a clear tendency to be associated with the hotter regions, and low anthocyanin an even clearer tendency toward such association. Table IV shows their average data on red wines of the five varieties "Barbera," "Cabernet-Sauvignon," "Carignane," "Grenache," and "Zinfandel." It appears that climatic region I in California may be too cool for maximum color or tannin development in these *V. vinifera* grapes, and climatic regions IV and V may be too warm. This conclusion must remain tentative in view of the alcohol and fermentation variables, but climate seems quite clearly important to color and tannin. Climatic regions, by their designation, are roughly equivalent to mean temperatures on a ten-year average from April through October that are less than 62°F for region I, 62°–64°F for II, 64°–67°F for III, 67°–69°F for IV, and above 69°F for V. Golodriga and Suyatinov (1966) found that the average anthocyanin content of grapes from 22 varieties was 169 mg/kg in 1963 with a mean August-September temperature of 70°F, 286 mg/kg in 1962 with 66°F, and 358 mg/kg in 1964 with 63°F. The "Cabernet-Sauvignon" samples were affected less, but each of the other selections showed the seasonal effect indicated by the average. It was noted (dal Piaz 1885, 1892) that the oil content of grape seeds was higher in hot countries, so tannin, being inverse, should be lower.

The phenol content of grapes appears to respond to weather rapidly and reversibly. Large fluctuations in phenol content have been found in almost every study which has included repeated analyses of grapes from the same population at several dates during a season. These fluctuations have sometimes been attributed, even by the researchers, to analytical and sampling problems, but rapid response to environmental influences

TABLE IV
COMPOSITION OF RED WINES FROM FIVE VARIETIES OF GRAPES GROWN IN COOL (I) THROUGH HOT (V) REGIONS

| | I | II | III | IV | V |
|---|---|---|---|---|---|
| Number of samples | 125 | 126 | 97 | 219 | 67 |
| Alcohol content (%) | 12.2 | 12.4 | 12.2 | 11.9 | 11.6 |
| Anthocyanin (intensity, arbitrary units) | 239 | 297 | 297 | 188 | 158 |
| Total phenols (mg/liter) | 1260 | 1310 | 1450 | 1100 | 1140 |

seems to be the true cause. This is indicated by the fact that when frequent samples are taken the phenol content changes systematically, not randomly, even though the changes may be rapid; also, berries of varieties in different stages of ripeness tend to fluctuate together in phenol content if enough examples are examined (Diemair *et al.*, 1951b; Singleton, 1966; Baragiola and Godet, 1914; Alwood *et al.*, 1916). Total phenol content may double or drop to half in a week's time in the berry, and variations may be even larger in stems or leaves. Diemair *et al.* (1951b) showed that much of the whole-berry variation, at least in white grapes, was due to variation in the seed's phenol content even late in August, when the seed might be assumed to be metabolically inactive.

On some occasions different tissues appear to respond differently to external influences and to interact in ways difficult to interpret with the knowledge presently available. Nevertheless, this appears to be an area worthy of much more study. Diemair *et al.* (1951b) found a large decrease in alcohol-ether-soluble phenols in leaves on June 12, followed on June 19 by a return to normal and a large temporary increase in water-soluble phenol. Berry phenols increased sharply to a maximum on June 12 and decreased on the 19th. Stems increased almost 3-fold in phenol content in one instance from July 10 to 17, while berries increased only slightly. Mildew infection complicated interpretation, however. Plant injury was associated with a decrease in phenols, and recovery or new growth with an increase.

Caldwell (1925) prepared cold but very strongly pressed juices with relatively high total phenol content, and, therefore, skin extract, from the same New Jersey vineyard of a number of American varieties for 5 years. Total phenols and phenols precipitable and nonprecipitable with gelatin were determined. All three of these phenol measures and acidity appeared to respond to seasonal conditions in like manner but in varying degree. Sugar content was not always the same at sampling, so ripeness was also a source of variation. In general, phenol content was highest in 1922, intermediate in 1921, and low in 1923. A high number of days of sunshine appeared to correlate with low tannin, but rain, hail, and temperature differences complicated the picture. Reynolds and Vaile (1942), in similar studies in Arkansas, also found that year-to-year fluctuations in one variety were frequently as large as differences between varieties, but were unable to correlate this with weather or compositional factors.

Noyes *et al.* (1922) found large differences in phenol content within each season for both cold- and hot-pressed "Concord" juice from the same New York source. Between two seasons the cold-pressed was similar, but the hot-pressed juice was much higher in one than the other. In the season giving the higher phenol content the fruit reached ripe sugar levels almost one month earlier, suggesting a sunnier summer. Data by Crowther *et al.* (1964) suggest that variation in "Concord" juice phenol content was greater by year than by vineyard location in Canada. Clore *et al.* (1965) found that Washington "Concord" grapes without seeds had similar anthocyanin levels in 1958 and 1960 but much higher total phenol levels in 1960. Solar radiation was similar in 1959 and 1960, yet 1959 produced both less anthocyanin and less total phenol.

The change in phenol content of grapes in response to environment can evidently be very rapid but has not yet been studied much on a time basis of less than a week. Webster and Cross (1942) reported that tannin content increased steadily from 5 AM to 11 PM in one set of "Concord" berry samples but reached the highest value at 11 AM in another set. "Tokay" grapes may color poorly, and growers have believed that the upper clusters in a vine color better than the lower because they are cooler, and that an overnight color increase can be seen after cooling and other effects produced by a fog or small shower (Bonnet, 1929). The variation in weekly samples previously cited is so profound that appreciable variation must certainly take place within a single day. Since it has been generally found that flavonoids do not translocate as such and are synthesized in the tissue where they are found, rapid decreases in total phenol in grape seeds and skins (mainly flavonoids) indicate efficient systems for degrading as well as synthesizing these compounds.

Webster and Cross (1942), Webster *et al.* (1934), and Rankine (1964) find that irrigation reduces the color level of grapes. Laffer (1936) reports that both irrigation and excessively wet seasons reduce the proportion of tannins in grapes. Kondo and Zvezdina (1953) reported more anthocyanin and tannin in grapes grown on relatively dry soil. Korotkevich *et al.* (1950) reported that the total tannides in 40 red varieties of grape averaged 13,770 mg/kg in a dry summer and 5000 mg/kg in a normal summer. It appears probable that a major factor in causing these effects would be the juiciness of berries in relation to seed and skin content. Dehydration of berries naturally raises the tannin content on a percentage basis (Gavrish, 1966). Other factors may operate, however. Webster and Cross (1942) and Webster *et al.* (1934) report that cooling a vineyard with water raised the anthocyanin of "Concord" samples. Gortner (1963) finds an inverse relation between soil moisture stress and *p*-coumaric and ferulic acid content in the growing point of pineapple, and Crawford (1967) notes a low phenoloxidase content in plants resistant to flooding.

Photosynthesis and carbohydrate products thereof, naturally, have a crucial relationship to most aspects of plant growth and composition.

One aspect, phenols, however, appears to have some special relationships. For example, the earliest-ripening varieties at least of *V. vinifera*, are white. Although other white varieties may be late, white grapes tend to be sweeter than pigmented grapes on a given date, and seedless berries ripen earlier than their seeded fellows (Mulder, 1857; Müller-Thurgau, 1898). "Pinot noir," one of the red varieties with the shortest ripening season among *V. vinifera*, is noted for a rather low pigment and tannin level, even in cool regions, and lacks acylated anthocyanins. Genevois and Peynaud (1947) have noted that high-tannin varieties of peaches are later in maturing, and the same seems true in general with grapes. The effects seem more than simple diversion of carbohydrates to other products, and suggest phenol participation in controlling grape metabolism.

An inverse relation between the content of plastid pigments (including chlorophyll) and anthocyanin content occurs in grape leaves as well as in other plants (Areshidze, 1963). This is at least partly dependent upon carbohydrate balance, and the red autumn-type coloration of grape leaves may be induced by raising the carbohydrate supply to the mature leaves. Premature autumn coloring of grape leaves will be induced by removal of fruit and immature shoots or by girdling of the canes, and will be delayed by an excessive crop of fruit (Amirdzhanov, 1962).

Anthocyanin formation in the fruit is decreased also by defoliation prior to ripening, and by too large an amount of fruit in proportion to the leaf area of the vine. Winkler (1930) reported that thinning clusters just after the set of berries produced more uniform and earlier coloring of red "Tokay" clusters and better amber color in "Malaga." Loomis and Lutz (1937) found that "Concord" grapes had very good color, as a rule, with over 50,000 $cm^2$ of leaf surface per vine, and that no vine with less than 20,000 $cm^2$ of leaves gave fruit rated very good in color. Cross and Webster (1934, 1935) found that 10 leaves or more per cluster were required for normal coloring of "Concord" in Oklahoma, and that low leaf area was related to both low average color and uneven coloring of individual berries within a cluster. Canes with 2 clusters and reduced to 10 leaves gave 65% of the berries colored, 1 cluster and 10 leaves had 71% colored berries, and a single cluster with no leaves removed had 82% colored berries. A high proportion of purple berries was produced by any procedure giving more leaf area per cluster, such as fertilization with nitrogen or phosphorus under their conditions. The percentage of purple berries was also lowered by shading the leaves or failing to protect the fruit from direct hot sun.

Robinson *et al.* (1949) and Shaulis and Robinson (1953) also found that low leaf area per unit of fruit, i.e., overcropping, was detrimental to fruit coloring, although seasonal effects were apparently considerably greater than the effect of the trellising system. Fruit color was particularly high in 1949 in comparison with the same vines in other years. The year 1949 had the hottest June and July of the series, and the coolest September. Trellising and pruning system per se had no significant effect on berry coloring or tannin in studies by Webster and Cross (1936). Beridze (1965) reported that vigorous vines (those with a high weight of prunings) had a higher tannin content per berry. Crowding of vines by close spacing appeared to increase the content of phenols (mg/g dry weight) in juice, skins, and stems, but had the opposite effect in seeds.

The amount of anthocyanin extractable per unit weight of "Red Malaga" table grapes was increased to as high as double the control level by girdling the trunk or the canes below the clusters at veraison (Weaver, 1952). Very definite (though smaller) increases in anthocyanin content of berries resulted from berry thinning or girdling later.

The effect of light on the grape cluster itself may be considerable, over and above its relationships to vine photosynthesis and high carbohydrate status. The exposure of grape berries to direct intense sunlight may lead to scald, sunburn, and the degradation of skin phenols as described by Molisch (1916) and Scurti and Pavarino (1934). Some exposure to light may be necessary, however, to develop anthocyanin color. As with many other fruits, grape clusters can be found in the field which are red only on the side facing the light. A number of experiments have been made on bagging, foil wrapping, cheesecloth shading, etc., of the clusters themselves (Overholser, 1917; le Roux, 1953; Weaver and McCune, 1960; Naito, 1964, 1966; Naito *et al.*, 1965). The majority of grape varieties tested, particularly the normally very dark-colored berries, show no visible color difference when the clusters are matured in the dark. There do not appear to have been quantitative studies to verify the amount of pigment present, and it is conceivable that lowered total pigment in a "black" grape would not be visible.

Grapes which do show no anthocyanin synthesis in the darkened berry include "Tokay," "Sultanina Rose," and "Molinera Gorda," while anthocyanin was reduced but not eliminated in "Barlinka," "Black Prince," and "Emperor." These tend to be lighter-colored grapes and are noted for natural variability in coloring. Other low-anthocyanin varieties, such as "Red Malaga" and "Aramon," are evidently not sensitive to cluster shading. The color develops normally in the berry skin of the affected varieties in the specific part exposed to light, and intermediate low levels of light have intermediate effects on anthocyanin synthesis. The colorless or depressed-color berries produced by bagging often develop approximately normal color in about 2 days when uncovered to indirect sunlight, and respond to light even after picking. Leucoanthocyanins and anthocyanins, in varieties that show any effects of light, are affected differently (Naito, 1966). The rate of color generation after uncovering suggests that pigment precursors are also suppressed by absence of light in some varieties and not in others. Bagging also tends to lower sugar in the cluster by as much as 3° Brix below that in the uncovered neighboring cluster. Other variables in bagging experiments may be humidity and temperature.

Direct competitive effects between sugar accumulation and anthocyanin synthesis are suggested by de Boixo (1950) from studies of the higher sugar level in the center of a berry than in the periphery. This difference was larger and decreased with ripening of red berries, and smaller but increased with ripening of white berries. The differences, about 1% sugar, seem large, but may be accounted for at least partly by active conversion of sugar to anthocyanins.

It is a well-known principle of plant physiology that "high-carbohydrate status" tends to favor the synthesis of products characteristic of mature plants—fruit, in the organ sense, and anthocyanins, as a chemical example. This has already been illustrated by the effects of girdling in trapping photosynthate in the canes and favoring leaf or fruit coloring, etc. The opposing "high-nitrogen status" favors new growth and protein synthesis, and suppresses fruiting and associated changes by diverting photosynthate to further vegetative growth. Williams (1946) found that adding nitrogenous fertilizer to a "Tokay" planting reduced the anthocyanin content of the berries by 23%. With other varieties the effect was not so large, and in two plots berry color was not suppressed. It appears that adding nitrogen to soil that is truly deficient may even enhance berry pigmentation and give increased yield, but nitrogen addition which does not increase yield is very likely to depress anthocyanin in the fruit. In their area, however, Cross and Webster (1934, 1935) increased the percentage of colored "Concord" berries by fertilizing with nitrogen.

Artyunayan *et al.* (1964) decreased grape pigmentation with nitrogen fertilization but increased tannin and pigment by adding potassium, phosphorus, or all three. The aerial parts of grapevines may redden from potassium deficiency (Ravaz *et al.*, 1933) or other deleterious influences. Mondy *et al.* (1967) recently showed that potassium fertilization lowered phenols in potatoes. It would appear desirable to compare different classes of phenols, such as the caffeoyl derivatives versus the flavonoids, to clarify these effects in grapes. Potassium sulfate added to soil harmed grapevines low in tannin but not vines high in tannin, according to Berezenko and Avazkhodzhaev (1965). The effect was attributed to detoxification by tannin of sulfite formed from the sulfate. Decreased transpiration was noted in the low-tannin vine by the second day, and it could also be postulated that the tannin aids in resisting water loss under the changed osmotic gradient at the roots.

A number of trace elements appear to affect phenol content in plants, but not many relevant data have been found on grapes. Boron deficiency produces a number of effects on grapevines, including abnormally short internodes, distorted leaves, corky spots in stems and sometimes berries, white chlorosis of leaves turning red in some varieties, and severe decreases in crop (Cook *et al.*, 1960). A large number of seedless berries will produce low tannin in the crop which does develop (Gärtel, 1954). In other plants, boron deficiency induces a high phenol level in the green area near necrotic spots in leaves (Perkins and Aronoff, 1956). Lee and Aronoff (1967) suggested from experimental evidence that lack of boron shifts plant metabolism to the pentose pathway, leading to the production of phenolic acids at an abnormally high level. The accumulation of these acids may be the cause of the necrosis. This appears to tie in with the participation of phenols in regulation of indoleacetic acid oxidase (Gortner and Kent, 1958; Galston, 1967). The leaf symptoms of boron deficiency do seem to have features suggestive of injury from 2,4-dichlorophenoxyacetic acid and related to indoleacetic acid metabolism.

Plant hormones have not yet been much studied as related to phenols in grapes. The percentage of berries which color properly in "Concord" is reduced by injurious substances like 2,4,5-trichlorophenoxyacetic acid (Overcash, 1955). Low levels of maleic hydrazide increased berry coloring, whereas high levels decreased it. A kinin, benzyladenine, increased the anthocyanin in stems of "Muscat" grapes (Weaver *et al.*, 1962). Other plant hormones and related substances have large indirect effects on berry phenol content by influencing berry size, seeds per berry, etc. The "hesitation" in rate of increase of berry weight just before veraison is related to a rapid drop and subsequent low level of endogenous auxin (Nitsch, 1962). Relationships between berry size and seed count are attributed at least partly to seed-generated gibberellin (Weaver and Pool, 1965).

Further observations with other plants whet our interest in the phenolics of grapes as influences on and substances influenced by hormonal effects. Twelve days after treatment with 2,4-D the leaves of tomatoes have less than ¼ of the level of rutin of untreated plants, and the stems less than 1/10 (van Bragt *et al.*, 1965). Plants resistant to gibberellin tend to be low in quercetin and other flavonoids, and treatment of a responsive plant with gibberellin tends to give a sizable increase in quercetin (Tronchet, 1961a,b). Gibberellin suppresses the synthesis of gentisic and ferulic acids in artichoke, but not of *p*-hydroxybenzoic and caffeic acids

(Paupardin, 1967). Coumarins and tannins often inhibit seed germination, and dormancy-breaking treatments for seeds (including grape seeds) involve time, leaching, or other reactions capable of inactivating or removing these compounds (Flerov and Kovalenko, 1947). This also may be related to gibberellin activity (Knypl, 1967; Monin, 1967). As grape seeds germinate, their water-soluble tannin content drops 60% and the alkali-soluble tannin rises slightly (Durmishidze, 1955).

Grafting of one variety or species to another, in line with the general lack of translocation of most phenols and flavonoids in particular, does not ordinarily cause any apparent difference in the qualitative phenol makeup of the scion or rootstock. For example, two species of *Tropaeolum* with different quercetin glycosides were found not to translocate their normal constituent flavonols past the stock-scion union in either direction (Delaveau, 1964). It is known that the qualitative anthocyanin pattern of grapes remains essentially the same whether the variety is grafted or is on its own roots. The rootstock, however, may have a great role in the general health and vigor of the vine, and thus affect the quantity of phenols (Trieb, 1967; Curtel, 1904).

The lignin content, tannin content, and storage carbohydrate, especially starch, seem to have important relationships to the maturity and hardiness of vine parts. The amount of total phenols in the vine as a whole is large even without including lignin. The total water- and alkali-soluble phenols of one 14-year-old "Saperavi" vine of 3830 g dry weight was 242 g (Durmishidze, 1955). This total was distributed 17 g in the fruit, 58 g in the leaves, 28 g in the canes, 13 g in the trunk, and 126 g in the roots. For comparison, the same vine with fruit contained 455 g of sugars, mostly in the fruit; 269 g of starch, more than half in the roots; and 320 g of crude protein ($N \times 6.25$), almost ⅓ in the roots.

A mature cane is one which is capable of surviving the winter and giving good growth in the spring. If cuttings are to be taken from canes that are pruned off, they must be capable of growing when propagated in the spring. Cane maturity appears to involve tannin content, though starch content seems more critical (Picard, 1924). Frost resistance and cold hardiness seem to depend on the same factors. The critical factors seem to be mechanically stronger and more lignified tissue, lower moisture level, and a stable hydrophilic colloidal system to bind the water (Ryabchun, 1965; Mikhailov *et al.*, 1965; Ouyan, 1963). Tannin and other phenols may serve as part of the colloidal system, but carbohydrates are important. American species and hybrids thereof that are more resistant to cold are reported to have a higher dry-matter content and a generally higher ratio of tannin to nitrogen than *V. vinifera* (Garino-Canina, 1940-1941). Polyphenoloxidase, peroxidase, and, particularly, catalase activity appear to be related to cold resistance (Marutyan and Mantashyan, 1961; Molchanova, 1966a,b). All three enzymic activities increased during wintering, in general more in resistant varieties and more in the phloem. During a cold period the catalase activity dropped, but with more activity remaining in the resistant vine.

Enzymes related to phenols cannot be considered in detail here, and activity measurements of enzymes from tanniferous tissue can be very misleading unless special precautions are taken to prevent tannin precipitation of the enzymes. As suggestions relating to the fundamental importance of phenols in grapes, it can be mentioned that lower polyphenolase and higher peroxidase and catalase appear related to early-ripening grape varieties (Golodriga and Khe, 1963). Dioecious male vines had higher activity of these three enzymes than did female, and bisexual vines had the lowest peroxidase of all (Golodriga and Khe, 1963; Marutyan, 1954). Increased catalase content appeared to correlate with anthocyanin formation in the berry at veraison (Webster *et al.*, 1934).

## D. Plant Pathological Relationships

Phenolic substances in plants are known to have general, and in some instances highly specific, importance in plant response and resistance to pests and pathogens (Cook and Traubenhaus, 1911; Pridham, 1960; Uritani, 1961; Kuć, 1963; Rubin and Artsikhovskaya, 1963; Goodman *et al.*, 1967). There is a considerable but very scattered body of observational information on grape anthocyanin formation, tannin-related browning, etc., as symptoms related to attack by certain insects or other injurious agents. A few of the more important and definite examples are briefly described.

Infection of grapevines with certain viruses leads to altered phenol metabolism. Early reddening or browning of grape leaves in the fall may be caused by a number of conditions including potassium deficiency, infestation with *Tetranychus* mites, and infection with leafroll virus. Considerable confusion existed originally as to the specific agents causing specific symptoms, but the role of leafroll virus is now clear in causing leaf reddening in red varieties, yellowing in white varieties, and reduced anthocyanin in the fruit of certain varieties (Goheen and Cook, 1959). One of the most dramatic effects of this virus is on the fruit of "Emperor," which on infected vines remains nearly white instead of coloring normally. The virus is transmissible to other varieties by grafting, and "Cardinal" and "Red Malaga," for example, will then produce undercolored fruit whereas the darker variety "Ribier" shows little effect on fruit color (Harmon and Snyder, 1946; Harmon, 1956). Analyses of wines made from "Ruby Cabernet" vines selected for slight, intermediate, and severe red-leaf symptoms indicated tannin and color reduced to as low as half in vines that were affected more seriously; however, reduced sugar (alcohol) caused by the virus was responsible for at least part of the difference in the wine (Alley *et al.*, 1963).

The grape root "louse," phylloxera, is a pest native to eastern North America, and wild *Vitis* species there are often tolerant or resistant to it. Dissemination of this pest caused serious injury or death of *V. vinifera* varieties in many areas in the latter part of the nineteenth century, producing economic panic in viticulture. The pest remains a serious economic problem even though its effects are minimized by grafting scions with desirable fruit to resistant rootstocks. The insect is primarily a root feeder, causing galling and necrosis on the roots. In some climates there is also a flying and leaf-gall-forming phase. Galls on tanniferous plants, including grapes, are usually very rich in tannin. Zotov and Sokolovskaya (1959) found polyphenols in root galls to be 2–3 mg/g dry substance, and undetectable in healthy filamentous roots. Water-soluble tannins increased with gall formation from 20 to 41 mg/g dry weight in the susceptible variety, and even more (14 to 52 mg/g) in the resistant. Additional alkali-soluble phenol was unchanged in the susceptible, remaining at 30 mg/g, but in resistant went from 19 to 58 mg/g dry weight with gall formation. Wines from vines with galls were much higher in tannin, yet similar in other analyses (Salgues, 1960). At present the best estimate is that the phenolic role in phylloxera resistance is large but not specific in the single toxic phenol sense. The resistant variety appears to mobilize phenols and other substances, rapidly forming phellogen and walling off the damaged tissue and the insects (Petri, 1911; Zotov *et al.*, 1966). This rapid reaction in the resistant vine involves enzymic differences related to phenol oxidation and auxin control mediated by phenols (Henke, 1963; Denisova, 1965).

The idea of phylloxera resistance being also influenced by specific unusual phenols or combinations thereof in at least some grape species gains support from recent work showing that resistant species do have different phenolic makeup, that these patterns in crosses are inherited apparently similarly to resistance, and that the resistant *V. cinerea* contains C-glycosides of flavones not found so far in other grapes (Wagner *et al.*, 1967; Yap and Reichardt, 1964; Henke, 1960).

Resistance to other agents may be related to the phenolic composition of grapes in general or certain grape species. Resistance to downy mildew, *Plasmopara viticola*, bears some relation to phenol composition

(Reuther, 1961), but this may be only as incidental markers for genetic character. Nematode resistance is appreciably different among grape species. Although this has not yet been related to the pattern of phenols in grapes, work showing nematocidal effects of raspberry tannin in the soil is suggestive (Taylor and Murant, 1966).

Some of the effects of *Botrytis cinerea* infection of grape berries have been mentioned elsewhere in this report. This mold may, under conditions of high humidity, grow through the healthy grape skin, and then, if a period of low humidity follows, the berry is dehydrated. The resulting musts are desired for certain naturally sweet white table wines, but are very undesirable for red wines because the mold produces a very active oxidase which destroys the red color, resulting in brown, cloudy wines. The mold infects different varieties of grapes more or less readily, and phenol content may be one of the factors. Kokurewicz (1962) patented a procedure to protect grapes from *B. cinerea* by spraying them with 1.5% aqueous tannin. Rubin and Artsikhovskaya (1963) report that *Botrytis* is stimulated by anthocyanins and inhibited by anthocyanidins, and use this to explain slower attack of damaged tissue.

Cook and Traubenhaus (1911) found that tannin was usually more inhibitory to parasitic fungi than saprophytic, with 0.1–0.4% generally strongly inhibitory. Ferulic acid, pyrogallol, pyrocatechol, and vanillin inhibit *B. cinerea*, while chlorogenic and caffeic acids stimulate, according to Ivanova *et al.* (1965). The ability to produce a high level of oxidized phenolic products and thereby inhibit hydrolytic enzymes appears to be a resistance factor to *Botrytis* in the castor bean (Thomas and Orellana, 1963). Contrary to simple desiccation, infection of grapes with *Botrytis* gives a reduction of the tannin that is great in red wines but, in spite of skin tissue breakdown, not much higher than normal in white wines (Nègre, 1942–1943). This is evidently the result of the production by *Botrytis* of an active phenolase which oxidizes and precipitates the phenols. Poux (1966; Anon., 1966a) finds phenolase to be four times as great in moldy berries as in adjacent sound berries.

# V

# Quality and Processing Relationships

There are many aspects to the usefulness of any one lot of grapes for any one purpose, which may be summed up as "quality." Opinions and attitudes of the processor and consumer, as well as the use and the situation, affect the quality considered acceptable or optimum in any instance. If considered only as a little bag of tart sugar water, the grape or a beverage made from it would not be very attractive. Addition to the grape of all constituents other than phenols would greatly improve it, particularly in odor, but it would still not be satisfactory as a fruit or a beverage. The red, yellow, and amber colors would be lacking. The berry structure would no doubt be affected, so that it would be deficient in its plump, crisp, chewiness. Certain character and body would be gone from its taste, and many of the reactions of processing would be affected.

Grape varieties may have important differences in sugar content, acid content, and other factors, but, within limits, adjustment of vineyard location, vine management, harvesting, and beverage processing can produce acceptable levels from any variety. This is generally not true of phenolic content. The factors which make a given variety more desirable include its color, its flavor, its viticultural habits, and its response to processing of various types. All of these do or may involve the phenols. It seems that the phenol content and its distribution should be given more consideration than has generally been the case in deciding the best variety or the best way to use a given variety in a given situation (Gaivoronskaya, 1959). Of course, visible color has always been an important attribute in variety selection, but we should give increased consideration to the amounts of specific pigments and other phenols and their relation to taste, storage stability, and other qualities.

Myers and Caldwell (1939) point out that astringency and its ratio with sugar and acid are important and useful in improving grape juice quality by blending, especially of the juices of different varictics. Relative astringency has been considered in evaluating varieties (e.g., Dennison and Stover, 1962), though often somewhat incidentally. Singleton *et al.* (1966a) noted high tannin in the juice of 2 of 34 varieties tested. These 2 varieties cannot therefore be adjusted in wine tannin satisfactorily by the usual expedient of shortening the time of fermentation on the skins, and early recognition of such facts could save expensive plantings and winery trials or hasten the satisfactory vintage of unusual varieties by special fining, etc. It is believed that we are on the threshold of much greater understanding and application of such knowledge to the technology of phenols as important food constituents, particularly in fruit and wine, as newer techniques of phenol study are improved and more widely applied.

## A. Phenols as Quality and Utility Factors

### 1. Flavor Effects

With some notable exceptions, such as the study of astringency by Joslyn and Goldstein (1964a) and the bitter or sweet citrus flavanones, chalcones, and related substances by Horowitz and co-workers (Harbone, 1964a), specific information on the flavor effects of phenolic substances is difficult to locate. Beythien (1949) tabulates tannins, depsides, and catechins under the class of harsh or rough flavors, and a number of phenolic glycosides including arbutin, syringin, and esculin under bitter flavors. Although many descriptions are available on the sensory effects of phenolic compounds, most of them are based upon tasting by few people and have uncertain applicability in wine or grape flavor interpretation. Among relatively few sensory thresholds determined for phenolic compounds in solution are values of about 10 mg/liter for vanillin, 3200 mg/liter for esculin, and 3200 mg/liter for arbutin (Amerine *et al.*, 1965). "Cream soda" has a vanillin level about 3 times this threshold level (Jacobs, 1959). Standard vanilla flavoring extract is about equivalent in flavoring strength to 0.7% vanillin solution, and coumarin is about 3 times as intense in flavoring strength as vanillin (Merory, 1960). Wasserman (1966) finds respective taste thresholds for the smoke constituents guaiacol, 4-methylguaiacol, and 2,6-dimethoxyphenol in water to be 0.013, 0.065, and 1.65 mg/liter.

West *et al.* (1963) reported that a number of beers tasted normal which averaged 7.1 and 12.5 parts per billion of chlorophenol and phenol, but that beer flavor was abnormal when these respective values were 13.9 and 27.1. Yamamoto *et al.* (1961) and Yamamoto (1961) reported that vanillin and vanillic acid appeared to be formed from ferulic acid during sake fermentation. Vanillin at 1 mg/liter contributed to the "brew" aroma of synthetic sake, and quality was improved by adding 0.05–0.5 mg/liter of vanillin with benzaldehyde and *p*-hydroxybenzaldehyde, which had also been found in sake. Drews *et al.* (1965) report that tyrosol has a strong bitter taste and a phenol-like smell. They found 2–9 mg/liter in usual beer, and up to 28 mg/liter in beer produced by more intensive fermentation. Aso *et al.* (1953) found 42–94 mg/liter tyrosol in fermenting sake mash, and 13–51 mg/liter in commercial sake. Tyrosol was bitter at 14 mg/liter in 15% aqueous ethanol, but not as bitter in sake at 10 mg/liter. Addition of 40 mg/liter of tyrosol to synthetic sake was considered to improve quality. Not all the bitterness of sake could be attributed to tyrosol.

If these data are considered in relation to the composition of wine, it appears that tyrosol is very likely to have significance in the taste of wine, particularly white wine. Vanillin may be important in the taste and odor of wines, particularly those stored in oak containers, and phenolic substances which may enter wine from such sources as charred wood need be present at only very small levels to affect flavor. It seems significant that Underwood and Filipic (1964) find that oxidative and hydrolytic reactions on ligneous material can account for considerable increases of vanillin and syringaldehyde in the conversion of relatively flavorless maple sap to flavorful sirup. The levels of volatile phenols like phenol itself or cresol, which have been reported in some wines, appear too small to be significant, but they might produce subtle differences in flavor.

Ferulic acid and its ethyl ester appear to be significant flavor constituents in sake (Yamamoto *et al.*, 1961; Yamamoto, 1961), producing a pungent taste at 2–3 mg/liter. Esters of gallic acid are bitter, and even more so with longer-chain alcohols (Chiancone, 1949). Fermentation of red wines with the stems produces a "stemmy" flavor which can be described as peppery, harsh, and herbaceous. Although it is not proven to result from the tannins of stems, their involvement is suspected (Nègre, 1942–1943; Laborde, 1908–1910; Pacottet, 1904). The varieties derived from *Vitis riparia* have a grassy "hybrid" flavor which appears to be associated with phenolic substances, although its retention of odor after treatment with diazomethane and lack of precipitation with lead acetate make it doubtful it is a phenol (Bayer, 1958).

Astringency and bitterness are the major taste effects assigned to the phenols. Small polymeric yet soluble tannins, and in foods the leucoanthocyanins, are responsible for the astringency (Bate-Smith and Swain, 1953). Bitterness is a property of many substances in addition to some phenols. Amerine *et al.* (1959) note that red wines containing more than 2500 mg/liter of total phenol will usually taste bitter, but also note that wines of the same phenol content may differ in bitterness. There has been a tendency to equate bitterness with astringency and attribute both to tannin in wine. This seems to be an oversimplification which may obscure the facts, although there is some concordance between the two sensations in wine if only because both depend upon the extraction of grape solids to reach appreciable levels. White table wines of the usual type are never appreciably astringent, for reasons which are obvious from the previous discussion. White wines may be excessively bitter, but a slight bitterness is believed to be a desirable feature. Red wines may be excessively astringent, too bitter, or both. Winemakers speak of "soft" tannin and "hard"-flavored tannin, and old wines may analyze rich in tannin and yet be neither very astringent nor bitter (Peynaud, 1965).

It is interesting that most of the foods for which sensory expertise has been developed in the "professional taster" sense are beverages or beverage-related and tend to be those with a high phenol component in the form of astringency and bitterness in their flavor, such as wine, tea, beer, and cocoa. The color, strength, quality, and "briskness" noted by the tea taster seem to correlate well with phenolic composition (Wood and Roberts, 1964; Geissman, 1962; Stahl, 1962). The black teas contain oxidation products of flavonoids called theaflavins (bright yellow-orange) and thearubigins (muddy brown). The amounts and ratio between these substances have a great influence on sensory features other than the odor of brewed tea. These compounds are evidently purpurogallinlike tropolone derivatives. In combination with caffein, theaflavins contribute "briskness." This is described as a liveliness or cleanness on the palate, suggesting a combination of slight astringency and bitterness. So far, compounds of these types have not been found in wines, but it seems that they could occur there when a comparison is made of the catechin composition and processing of tea and some types of wine. Bitterness is not considered a common attribute of flavonoids, but is of tropolones (Korte *et al.*, 1959).

Terms that the wine taster uses to describe flavors related to tannin include tannic, bitter, astringent, rough, and hard. "Rough" implies astringent harshness, usually without implying bitterness unless so stated. "Smooth" and "mellow" are antonyms of "rough," but tend to imply more than the loss of roughness. "Hard" indicates acerbity, tartness plus astringence and perhaps bitterness, as in unripe fruit. "Supple" or "soft" may be considered the opposite (Marteau, 1960). It is generally agreed that some level of astringency and bitterness is desirable in red wines, but it is not very clear what that level should be or how it can be defined. Tannin acceptability of a series of 13 "Cabernet" wines ranging in total phenol from 1540 to 4620 mg/liter and 17 "Riesling" wines of 90 to 430 mg/liter was scored by 5 expert tasters (Baker and Amerine, 1953). On the basis of acceptability rather than intensity, the tannin levels were not rated significantly different in either group and the color only in the white wines. Average tannin ratings, however, were 6.2 for the "Cabernet" highest in phenol, and 7.2 for the lowest (where higher number indicates preferred level). This was a considerable degree of difference and would surely have been significant if further testing followed the pattern. It appears the tasters could tell the difference, but were not agreed upon the desirable level of tannin. Studies with another series of dry red and dry white table wines (Baker *et al.*, 1966) indicated a rather small negative correlation between tannin and overall red wine quality, and a slightly larger positive correlation for anthocyanin color. Since these factors tend to change together in wine but have opposite effect on quality, difficulty in interpreting results is understandable.

Grünhut (1898) noted that a tannin decrease in wine from 2310 to 2090 mg/liter made a noticeable difference in bitterness. Turbovsky *et al.* (1934) added various tannin preparations to wine, and found that sumac and tannic acid gave less agreeable tastes than grapeseed, oak, or quebracho samples. At added levels of 500, 1000, and 2000 mg/liter, all were considered too astringent in red wines (with unstated original levels of tannin), though the lowest level appeared to produce a desirable improvement in white wine flavor and odor. Cruess (1935) agrees that tannin added at 500 mg/liter improves white wines, and considers 2000–2500 mg/liter the preferable level for red wines, with 3000 mg/liter too astringent for most tasters. In line with our previous discussion of changing practices and preferences in America, this appears perhaps high today. However, a number of indications may be found in the literature that certain wines, particularly dessert wines such as port and madeira-types, are better with more tannin (e.g., Papakyriakopoulos and Amerine, 1956; da Silva, 1911).

Berg *et al.* (1955a,b) reported the threshold in water for a commercial grape seed tannin preparation assaying equivalent to 67.5% tannic acid was 200 mg/liter in two different panel experiments carried to a high level of statistical significance. Experienced wine tasters were better at evaluating tannin astringency than were inexperienced tasters. Tannin raised the sugar level necessary for the detection of difference in sweetness. Tannin in the absence of sugar had a considerable effect in interfering with the recognition of acid, but sugar minimized the effect of the tannin (Hinreiner *et al.*, 1955a,b). Complex interactions of tannin astringency and bitterness with other primary tastes appear to be the rule. In a dry white wine with an original phenol content of 200 mg/liter the minimum detectable concentration of added commercial grape seed tannin was 1000 mg/liter. For a red wine with an original total phenol content of 2000 mg/liter, 1500 mg/liter needed to be added to detect the difference. Even after correcting to the analytical equivalent of pure tannic acid, these values appear somewhat high, so it is believed that the commercial grape seed tannin sample had polymerized to the point of decreased astringency. The astringency of tannin is proportional to its protein-binding capacity, which may change with storage. Joslyn and Goldstein (1964a) indicate for tannin threshold values of about 200 mg/liter and minimum detectable differences of about 400 mg/liter.

Bokuchava and Novozhilov (1946) found different effects on the flavor of tea from different phenolic fractions. Most important was the fraction not precipitated by salt and extractable by ether, probably including polyphenols and catechins, which appeared to be involved in a "fruitiness" taste. Next in importance were the ethyl-acetate-extractables, and the tannins precipitable by ammonium sulfate gave some bitterness. Rossi and Singleton (1966b) separated fresh grape seed extract into chromatographic fractions without cross contamination. The catechin fraction, ether-soluble components, had an approximate detection threshold in water of 20 mg/liter. In white wine with an initial total phenol content of 170 mg/liter the catechin fraction gave no recognizable astringency at the levels tested, although it gave recognizably increased bitterness at 200 mg/liter, and a lowered ability to recognize acidity at 160 mg/liter. A small-molecule fraction soluble in ethyl acetate and including leucocyanidins had in water a threshold of about 20 mg/liter, and in white wine a minimum detectable difference of 120 mg/liter for either astringency or bitterness. At 320 mg/liter it lowered the ability to recognize acid. Most potent was the larger nondialyzable condensed tannin fraction, apparent molecular weight of 5300, with a threshold in water of about 3.5 mg/liter, and a minimum detectable difference in white wine of 20 mg/liter for astringency and 12 mg/liter for bitterness or effect on acidity.

Unfractionated seed tannin extract produced a threshold in water of about 25 mg/liter, and in white wine at 80 mg/liter affected astringency, bitterness, and acidity judgments. It appears from these and previously discussed data that these qualities of white wine are likely to be affected by the levels of these phenols in the wine. Because catechins and smaller leucoanthocyanidins are extracted more rapidly, it appears that the effect on white wines may tend to be more on bitterness than on astringency, though both may be affected. The content of these seed components, disregarding augmentation from skins, is estimated in red wines to be 10 to 100 times the minimum detectable levels in white wine, and therefore certainly significant in their flavor. It would be anticipated, of course, that the minimum detectable difference would be considerably larger when original levels are higher, as in red wines.

We are aware of no detailed studies of the effect of anthocyanins on flavor, or indeed whether they have much flavor. Garino-Canina (1924) states that anthocyanins have much less astringency than tannins, which would certainly be expected from the protein-binding viewpoint. Extremely highly colored must or wine from teinturier varieties tends to have a harsher, coarser, or more bitter flavor than normal wines. This has always been assumed to be from the non-*vinifera* parentage common in these varieties, but it is conceivable that the pigments participate. Obviously, experimental study is needed.

Grape phenols may be modified in wine to produce changes in flavor. The oxidized color and different flavors of sherry, madeira-type, and certain other wines appear to result in major part from phenol reactions. The production of volatile bouquet fractions in these wines and in champagnes has been attributed to reactions between phenols or quinones and amino acids to give aldehydes derived from the amino acids by Strecker degradation (Hodge, 1967). These reactions are discussed in Chapter V,B.

It is generally recognized that the loss of astringency during fruit ripening is related to increased polymerization and decreased solubility of the tannin (Joslyn and Goldstein, 1964a,b). Loss of astringency as polymerization occurs leads to tastelessness rather than bitterness. Depolymerization such as conversion of a polymeric anthocyanogen to anthocyanidins may lead to loss of astringency. These observations are consistent with astringency being dependent upon protein-binding capacity, because the simple flavonoids bind protein poorly and the largest condensed tannins do also in proportion of protein bound or precipitated per unit weight of phenolic substance. It appears, however, that astringency can be produced by polymerization of small phenolic molecules to larger or the breaking of very large to smaller, as well as being lost by these mechanisms. Maier *et al.* (1964) mentioned that heating dates to 140°F gave increased astringency. This was postulated as a partial breakdown, for solubilized leucoanthocyanidin and some anthocyanidin were noted in the dates. The reaction can be visualized as the liberation of one anthocyanidin moiety in the middle of a 10,000-mole-weight low-astringency polymer, giving two high-astringency 5000-mole-weight molecules. On the other hand, polymerization by oxidative or other mechanisms might produce increased astringency as molecular weight increased, and then decreased astringency as the optimally astringent size was exceeded. Reactions of this type in wines appear to deserve study from a flavor viewpoint. There have been reports that limited oxidative reactions can give increased tannin astringency in beverages with high proportions of smaller polyphenols like white wines (e.g., Monties and Jacquin, 1966).

Other examples of modification of phenols which may lead to bitterness in wine include the formation of esculetin and similar products, as already discussed. Esculetin and its relatives are reported to be bitter (certainly, the glycosides are), whereas the precursors, such as caffeic acid, are not notably bitter. Tannin appears to be involved in a bitterness in wine that is apparently of microbiological origin. Voisenet (1929a,b) found that acrolein formation from glycerol by bacteria caused a bitterness in wine, and related this to resin formation. Acrolein alone has an irritating odor, and in brandy has a horseradish-like taste (Holz and Wilharm, 1950). The bitterness appears to arise by reaction of the acrolein with tannins (Rentschler and Tanner, 1951; Herrmann, 1963a), for tannin-free wines did not become bitter from added acrolein whereas wines with tannin did.

## 2. Color

Appearance is an important feature of wine, and a wine's color, along with clarity, is a good indicator of its past history, condition, and probable present and future quality. The colors of most wines are almost wholly the result of their content of various phenolic substances. This was not always so. The extent and the irresponsibility of practices of adulteration which marked food and beverage processing before general enactment of pure food laws are difficult to imagine today. With wine, the most common effort to defraud the customer was by adding dyes to mask defects or imitate better products. More than 30 reviews and books, and at least an equal number of scientific papers published before 1920, pertain directly to artificial coloring of wines. Although various natural dyes were used, some innocuous enough even though dishonest, the synthetic coal tar dyes were used widely when they became available. While the disreputable were busy doctoring their products (artfully and profitably), the early food scientists were busy learning to detect such frauds.

This report is concerned, of course, with the legitimate natural phenolic pigments, but a good deal of the impetus for study and early knowledge about food phenolics grew out of efforts to detect and prohibit extraneous colors in wine, etc. Some feeling for this early era can be drawn from the titles of two books: "A treatise on the adulteration of foods and culinary poisons; exhibiting the fraudulent sophistications of bread, beer, wine, spiritous liquors, tea, coffee, cheese, pepper, mustard, etc., etc., and methods of detecting them" (Accum, 1820); and "Treatise on the manufacture, imitation, adulteration and reduction of foreign wines, brandies, gins, rums . . . based on the French system" (Stephen, 1860). Blyth and Blyth (1927) record some of the abuses and note that a German wine purveyor died when forced to drink his own suspect product, which seems less fair when the amount is noted: 6 quarts. Thudichum and Dupré (1872) describe other more systematic yet drastic efforts to control fraud. A Portuguese producer convicted of adulterating port even by coloring with elderberries was subject to "transportation" for life. A vineyardist with an elder-bush on any part of his land was subject to confiscation of his property and "transportation" for life. This law was in effect in the whole port district and the region bordering it for 77 years (up to 1833).

None of these drastic punishments were very effective until the chemistry of authentic wines and the adulterants became well enough understood to enable enforcement of reasonable laws (Wiley, 1929). It is perhaps comforting that, at least in the matter of wine, America appears more sinned against than sinning—94 red wines from California by early methods of analysis were free of added dyes (Bigelow, 1895).

Methods of detection of aniline dyes in wine depended on fixation of the color on wool under conditions which would destroy or change grape pigments, similar treatment with cotton, failure of the synthetic to precipitate with lead salts, differences in response to pH, formation of different lakes with metal salts, solubility in ether for some of the synthetics, etc. Some of these methods are still useful when coupled with modern spectrophotometry, paper chromatography, etc. (Stanley and Kirk, 1963). The earliest applications of chromatography to wine appear to be those concerned with added color detection (Mohler and Hämmerle, 1935, 1936; Gentilini, 1939, 1941; Ramos and de Oliveira, 1949; Ruf, 1952; Ramos and Pereira, 1954). In fact, Kocsis *et al.* (1941) applied what was very near to paper chromatography in detecting caramel and safflower color in white wine before partition paper chromatography was discovered.

The "natural" color of white wines may include browns of the caramel type if the wine was made from raisins or heated while sweet, as might be true for California baked sherry or Madeira wines. The presence of caramel can be detected by various procedures mostly depending upon solvent partition differences and fluorescence of raisin or other caramel-type colors as compared with other sources of yellow or brown colors in wine (Gentilini, 1966; Valaer, 1947; Mallory and Love, 1945). Most white wines do not appear to contain carbohydrate-derived brown pigments, and those which do are apt to have been heated and be quite brown.

White wines may have a trace of green color, evidently from chlorophyll derivatives, but usually range from nearly colorless through yellow, gold, and amber. Spectrally, these colors all represent relatively nondescript absorbance in the visible region caused by die-away extension of absorbance from maxima in the ultraviolet on into the region of wavelengths longer than 400 mμ. The intensity of the color would depend on absorbance as the visible region is entered, and the shade would depend on the slope and extent of the incursion into the visible region as absorbance is plotted versus wavelength. However, all end-absorbance-type absorption spectra tend to be similar, which means that intensity and hue are related. At low intensity of the brown pigments the wine would look yellow, and at higher intensity it would look increasingly golden to amber.

The substances responsible for the yellow colors of white wine were thought for many years to be flavonols, particularly quercetin, based largely on the fact that quercetin was the only known wine-soluble yellow substance identified in grapes (Genevois, 1951). v. Fellenberg (1913), however, showed that there was appreciable quercetin in wine only if it had been fermented on pomace. Ribéreau-Gayon (1964b) finds only about 15 mg/liter of flavonols in red wines, and none in white wines. Peyrot (1934) concluded that the absorption spectra of white wines and quercetin or quercitrin were not similar. Berg (1953a,b) examined spectra of white wines from 19 varieties of *V. vinifera* and found them quite different in relative absorbance at about 265 mμ, which correlated well with total phenol, and at 325 mμ, which would correlate with caffeic acid and its derivatives. All of the wines decreased rapidly in absorbance at wavelengths longer than about 300 mμ. All wines with a low total phenol content were resistant to browning, whereas those with higher phenol content varied in visible color although they tended to brown less when the chlorogenic-acid-type of absorbance was relatively high.

The best estimate as to the source of the normal degree of yellow color as well as more advanced color in white wines appears to be phenol oxidation products, on the basis of various studies including adsorption of precursors with polyamides, etc. (e.g., Ribéreau-Gayon, 1965d). It is clear that not all phenolic substances of wine contribute equally during oxidation to brown color. Rossi and Singleton (1966a) found that the catechins browned very rapidly when treated with grape polyphenoloxidase. Addition of chlorogenic acid had little effect on this browning, and other grape seed fractions, including small and large leucoanthocyanidins, browned quite slowly. Under comparable nonenzymic conditions the catechins also gave more increase in absorbance at 440 mμ than did the other fractions. The pigment formed by catechin oxidation was not distinguishable, either visually or spectrally, from that of a "normal" white table wine and, during a prolonged period of oxidation, passed eventually, as an aging wine would, to an appearance characteristic of darker wines, such as sherry. They believe that catechin oxidation products are primary factors in white wine color. Oxidation and browning of wine are discussed further in later sections.

As far as color is concerned, browning of white wines is usually held to the minimum consistent with wine type and the processing necessary to produce it. There appears to be increasing demand for lighter color in almost all types of white wine today. This trend appears desirable so long as the desired color can be produced without damaging quality otherwise. To the extent that it might encourage such practices as harvesting unripe grapes or aging wine insufficiently it could be unfavorable to overall wine quality. Crawford *et al.* (1958) discussed U.S. federal standards for minimum color, measurement, and standardization of white wine color from visual and tristimulus viewpoints calculated from absorbance at several wavelengths.

The color of wines has been studied much more with red wines than with white. The color of red wines may vary visibly in intensity from very pale rosé to so dark as to appear nearly black. Wines may have shades of red from nearly orange to bluish or purplish red. This hue depends upon factors such as pH, co-pigments, and metallic lakes affecting the anthocyans themselves, and upon the amount of brown color coexisting in the wine. Since this yellow or brown color appears to be the same type as that present in white wine, the important colors of wines appear to be phenolic substances or phenol reaction products.

A great problem has been specifying the hue and intensity of the color of a wine so as to standardize the color and match it year after year by blending. A number of methods have depended upon matching the wine's color with a series of arbitrary standards which can be specified and presumably reproduced in other laboratories or at another time. These methods have been very useful in studies of the effects of winery operations on color, particularly before spectrophotometry was generally available.

Winkler and Amerine (1938a) have described a number of these procedures, and some more recently proposed techniques include those described by Schneyder (1962), Hidalgo (1961), and Wobisch and

Schneyder (1955). Mackinney and Chichester (1954) reviewed the application of the more definitive tristimulus methods to foods, including wine. Amerine *et al.* (1959) applied the C.I.E. tristimulus method to wines and compared it with simpler but less powerful methods of specifying visual color characteristics. Amerine *et al.* (1965) also discuss the C.I.E. tristimulus and other modern methods in relation to sensory perception of color. Defining a color depends on the three components: dominant wavelength or apparent "average" color; the degree of associated grayness, purity, or saturation; and the brightness or intensity of the color. Tristimulus methods are applied increasingly to specification of visible wine color (Pires Martins, 1963; Jaulmes, 1965; Alexiu, 1965; Anon., 1967b; Ciusa and Barbiroli, 1967).

Among the points noted in those reports is the importance of examining wine color without dilution if it is to relate directly to what the consumer sees. Since tristimulus values are frequently calculated from absorption spectra and spectrophotometers cannot operate if absorbance is too high, darker wines must be examined in thin layers if light-transmission measurements are to be meaningful. Robinson *et al.* (1966a) used a commercial instrument designed to give the trichromatic values directly. They found that dilution not only decreased the intensity of the red color but shifted it in dominant wavelength, usually toward brown though not always. Of course, any dilution of wine would be likely to affect the complex equilibria involving pH, metal ions, tannins, etc., and change the relationship to the color originally present even when buffers were used. Since the parameters of the color perceived by the eye interact, however, decrease of the thickness of the wine observed has an effect on apparent shade of the color as well as intensity (Amerine *et al.*, 1959; Robinson *et al.*, 1966a). This effect makes wine often appear browner in the thin layer near the meniscus in the glass than in the center, which appears to account for some of the "fiery" appearance of wine (Robinson *et al.*, 1966a). Such effects as this make it difficult for spectrophotometry for chemical analysis of the amount of phenols present to be combined satisfactorily and simply with studies of visual effect. Compromise solutions by colorimetry at several wavelengths are not always easy to interpret, but may be useful in visual color estimation. For example, Villforth (1958) used blue, green, yellow, and red filters to characterize red wines colorimetrically with respect to type and vintage. Kushida and Ito (1963) called total color the sum of absorbance at 8 wavelengths between 370 and 610 m$\mu$, and used this in studying both white and red wines. Sudraud (1958) measured the intensity of red wine color by the sum of absorbance at 420 and 520 m$\mu$, and the tint by the ratio of the two absorbances.

Ough *et al.* (1962) found that the sum of absorbance of 420 and 520 mμ will estimate the tristimulus intensity value of a red or rosé wine within 3%. A plot of the absorbance ratio of 420 and 520 mμ versus dominant wavelength could be used to estimate dominant wavelength within ±3 mμ. A trial blend was necessary, however, if wines with very different dominant wavelengths were to be blended to a standard. It appears that, with wine, the color continuum is small and discrete enough, and the trichromatic parameters and observer reactions such, that purity can be ignored and color defined by brightness and hue (Berg *et al.*, 1964). Graphs were constructed by those workers from which the differences necessary for observer discrimination in dominant wavelength or intensity can be read directly for red wines. Panelists tend to prefer a blue rather than an orange color in light red or rosé wines (Ough and Berg, 1959) and can detect very small differences. An experienced panel and an inexperienced panel had similar color preferences for rosé wine, and the results suggested that rosés should have a dominant wavelength of 595–620 mμ, 35–50% brightness, and 25–40% purity to be most pleasing in color (Ough and Amerine, 1967).

Joslyn and Little (1967) studied rosé wine color by C.I.E. tristimulus and by a method they developed involving measurement of red, green, and blue reflectance of 4-mm layers both of wine with a white or a black background and of wine in infinitely thick layers. They found that the thin-layer white-background method was able to show differences in browning and anthocyanin degradation within and between wines which were difficult to establish from spectrophotometric data. Separate analyses were made for total phenol, catechins (vanillin-reactive flavanols), anthocyanins, and anthocyanogens. A red table wine had, per liter, 223 mg of anthocyanin, 1560 mg gallic-acid-equivalent total phenol, and 810 mg flavanols. After heating for 2 months at 50°C. to induce browning there was a small drop in total phenols and a small rise in anthocyanogens, but the flavanols dropped to 500 mg/liter, and anthocyanins to 70 mg/liter. A rosé had 76 mg/liter of anthocyanin, 500 mg/liter total phenol on a gallic-acid basis, and 155 mg/liter flavanol. These were respectively converted by the heat treatment to 29, 490, and 120 mg/liter. Those workers concluded that catechins are more reactive than other flavonoids and serve as a better index of storage changes. Anthocyanin degradation and browning were proportionally different in the three different red and pink wines studied.

The typical bright young red wine has a maximum absorbance near 520 mμ, a minimum near 420 mμ, and a much more intense absorbance extending into the ultraviolet. The spectral differences between red wines and juices of different grapes and other fruit are usually insufficient to distinguish one fruit from another, much less determine proportions in blends. When such differences have been proposed they usually appear to depend on original pH, pH-response differences, and other phenomena (Beyer, 1945; Rice, 1964) whose operation may depend on features other than the specific phenolic components themselves. With chromatographic techniques, of course, anthocyanin patterns obtained on sufficiently young wines can enable identification of mixtures of two different fruits or even different varieties of grapes. The detection of non-*vinifera* species by anthocyanin diglucoside content, and of "Pinot noir" relatives by their lack of acyl pigments, has already been discussed. Successful efforts to detect anthocyanin colors added to "Concord" juice from other sources, including *vinifera* skin extracts, have been described (Rice, 1965; Mattick *et al.*, 1967; Fitelson, 1967). Since elderberries contain a high proportion of cyanidin derivatives, which will complex with borate through the vicinal B-ring hydroxyls, and grapes contain considerable malvidin, which will not complex, it is possible to detect elderberry addition (or other borate-complexing anthocyanidin derivatives) in grape products by retention of red color in borate buffer compared with blue in glycine buffer, both at pH 9.23 (Wobisch and Schneyder, 1952; Saller and de Stefani, 1960).

It has long been observed that as a red wine ages it turns from purplish red to tile-red. As a red wine oxidizes or ages it loses absorbance at 520 mμ and increases relatively in absorbance at 420 mμ. After prolonged aging or oxidative treatment, red and white wines approach the same color, which is one reason that red grapes not suitable for table wines or port can be converted to sherry.

There have been many studies of aging as related to the color of red wines, and of color as an indicator of their age. Erdmann (1879) studied the extraction of the same type of red wines of various ages with amyl alcohol. A color change was noticeable within one year, changing the amyl alcohol layer from violet to yellow-red and the acidified aqueous layer from cherry-red to yellow-red. After 4 years, neither solution had appreciable true red and no blue or green color was generated by alkali. Many examples of later work could be cited, including work on wine spectra. One of the earliest modern studies was by Boutaric *et al.* (1937), who characterized red wine color by the ratio of absorbance at 480/640 mμ and recognized that a maximum existed at 520 mμ.

Sudraud (1958) determined the absorption spectra of wines, and found age to be related to absorbance ratio at 420/520 mμ, with young wines having about 0.7 and a wine more than 50 years old having 1.7. Grape variety, method of vinification, etc., had some effect, but the approximate degree of aging could be fixed by this color ratio of brown to red. Somewhat similar studies were reported by DeGori and Grandi (1955), Stella (1962), and Valuiko and Ivanyutina (1967). Mareca and co-workers studied spectral changes during red wine aging, and correlated the findings with chromatograms of the pigments (e.g., Mareca and del Amo, 1959; Mareca and Gonzalez, 1964, 1965). They found in red wine a yellow class of pigments, a red-brown class, and a red-violet class, which appear to correlate with white-wine-type yellow-brown, anthocyanic tannins (Somers, 1966a), and grape anthocyanins, respectively.

Studies of the sum and ratio of 420 and 520 m$\mu$ absorbance in a series of aged wines from the same sources, and also analyses for anthocyanogens and anthocyanins, were evidently complicated by the longer time of the older vintages on the skins (Ribéreau-Gayon, 1965d). Color intensity increased about as fast in the brown region as it decreased in the red region, giving about the same visual intensity though the color shift toward brown was evident. Findings of Erdmann (1879) are called to mind by the fact that the true anthocyanins essentially disappeared after 4–7 years.

Heat treatment appears to have effects on color similar to those from aging except that the browning increase tends to be less in proportion to the anthocyanin loss, especially in musts and juices (e.g., Ponting *et al.*, 1960; Aso *et al.*, 1956a,b).

Robinson *et al.* (1966b) found the anthocyanin diglucosides more stable to decolorization than the monoglucosides, but with more tendency to brown when added individually to white wine.

Many other factors also influence visible color. Sulfur dioxide tends to prevent the formation of brown, and bleaches not only some of the brown color but anthocyanins as well. Jurd (1964d) has outlined reactions involved in the decolorization and discoloration of anthocyanins. A number of these are considered in other sections. There are a number of reports that the presence of tannin intensifies the red color of wine or juice and shifts it slightly toward blue (e.g., Winkler and Amerine, 1938b; Willstätter and Zollinger, 1916). Gallic acid has been reported to have a similar but weaker effect; however, Jurd and Asen (1966) found no effect of quercetin, chlorogenic acid, or methyl gallate on the spectra of cyanidin-3-glucoside. In the presence of metal ions such as aluminum and iron, complexes between other phenols, the metal, and anthocyanins modify the color exhibited, often in the blue direction. Such complexes exist in nature (Asen and Jurd, 1967) and might be involved in some of the blue pigments of American species of grape. In fact, Jurd and Asen (1966) argue that such complexes are necessary for plants to exhibit colored anthocyanins in the sap pH range of 4–7, because at those pH's the colorless carbinol base is the form of the anthocyanin which should exist otherwise. The metals alone may form complexes with anthocyanins, and these "lakes" have modified colors of importance in food and in "Concord" grape juice (Bate-Smith, 1958; Ingalsbe *et al.*, 1963).

The formation of these complexes depends not only on pH but also on competitive complexing agents such as citric acid (Jurd and Asen, 1966). Szechenyi (1964) added a number of metal ions to cherry juices, and produced color shifts, generally toward blue, with 10 mg/liter or less of copper, iron, tin, or aluminum. Somaatmadja *et al.* (1964) studied anthocyanin chelation with cupric ion, and found that color shifted toward yellow and that molar-complexing ratios were pH-dependent.

The major effect of pH on anthocyanin color and structural form was clarified by Jurd (1963) and Jurd and Geissman (1963), with further important studies by Timberlake and Bridle (1966b, 1967a,c) and Harper and Chandler (1967a,b). In the pH range of interest to us, pH 2–6, the flavylium salts evidently undergo the following transformations:

7,4′-Dihydroxyflavylium ion (yellow) ⇌ ($+H^+$) Anhydrobase (orange)

7,4′-Dihydroxyflavylium ion ⇌ ($+H_2O$) Carbinol base (colorless)

Anhydrobase ⇌ ($+H_2O$) Carbinol base

Chalcone (brownish gold) ⇌ Carbinol base (colorless)

The substitutions involved in the natural anthocyanins play a large role in these equilibria, and the detailed interaction has not yet been studied for the various specific natural anthocyanins in model solutions, much less determined in juices or wines. From the work with the synthetic substances the flavylium ion is clearly the stable form in acid solution. At about pH 3, the anhydrobase and the carbinol base begin to appear. The 50–50 point between anhydrobase and flavylium salt appears to be at about pH 3.8, and the carbinol base seems maximum at about pH 4. The chalcone appears more stable than the two intermediate forms and increases in percentage toward neutrality. The anhydrobase tends to be a low and fairly constant fraction, it seems, and the pseudobase or carbinol base appears quite unstable, with oxidation of anthocyanins in food apparently proportional to the fraction of the pigment in the carbinol base form (Tinsley and Bockian, 1960; Jurd, 1964d).

An OH at position 3, as in the natural pigments, makes the transformation to the colorless carbinol very rapid, and subsequent oxidative degradation facile. Light also affects these reactions. The fact that only hydrogen substituents at position 3 confer stability has been capitalized on by Jurd (1964b) in preparing dyes for food use. Protection of the 3-OH by glycosidation also confers stability, as in grape pigments. Anthocyanin-3-glucosides exhibit more color at juice or wine pH (about 3–4.5) than do 3,5-diglucosides, owing to differences in the proportion of colorless carbinol base relative to red flavylium salt (Timberlake and Bridle, 1967b). Berg and Akiyoshi (1962) and Berg (1963) estimated free anthocyanins based upon their responsiveness to pH compared with nonresponsive yet "red-associated" or polymeric pigments. By gel filtration chromatography Somers (1966a, 1967a) divided the red colors of grapes or wine into a polymeric red tannin fraction and an acyl or a free anthocyanin fraction. From a color viewpoint it is significant that the polymeric forms of wine pigment appear brick-red rather than purple-red.

### 3. By-Products

From a phenolic viewpoint the only by-products from grapes of any general significance are tannin and anthocyanin preparations, although the recovery of bioflavonoids has been considered. Tannin could be recovered from stems or other vine parts, but when grape tannin has been recovered at all it has usually been from the seeds. The seeds have been the most frequent source of grape tannin, partly because they are rich in tannin and can be conveniently screened from relatively dry winery pomace, but primarily because tannin recovery may be combined with recovery of the edible oil from the seeds. The oil is usually the primary recovery sought, and the tannin may not be considered worthy of recovery even when the oil is (Agostini, 1964; Halperin and Goldman, 1950; Frolov-Bagreev, 1948). Although grape-seed tannin is a very good condensed tannin it has a limited specialty market, and competitive products are available inexpensively and in greater supply for general leather tanning, well-drilling "muds," or other uses.

Grape-seed tannin is a good tanning agent for leather and has been so used under locally favorable or emergency conditions (Davy, 1819; Rabak, 1913; Rabak and Shrader, 1921). Rabak used grape seeds from deseeded "Muscat" raisins, and also stems, which gave a poorer product. After the oil was removed the seed hulls were macerated 12 hours in water at 40°C, the macerate was percolated with more water, and the extract was vacuum-concentrated to a paste. The paste was about half solids, which were about 16% tannin and 30% nontannin phenols. It was calculated that 22,000 tons of seeded grapes would give about 1,100 tons of seeds, which would give about 48 tons of tannin extract and 132 tons of oil.

Recommended additives for wine in many old texts were grape-seed tannin and, in fact, crushed grape seeds or winery-prepared extract, particularly as an alternative if tannic acid could not be purchased (Boireau, 1888). Addition of tannin to wines is usually for the purpose of causing satisfactory precipitation and clarification when fining with proteins such as gelatin, although adjustment of flavor and astringency might be practiced on either white or red wines. Grape-seed tannin, as a natural constituent in wine, might seem ideal for these purposes, but, in fact, its addition to wine is not permitted by U.S. regulations. Only drug-quality tannic acid is approved. At least one reason for this is that seed tannin preparations tend to be dark-colored and those formerly imported appear to have been quite crude.

By-products including tannin have been recovered at various times in California (Serre, 1919; Halperin and Goldman, 1950) and in other countries (dal Piaz, 1885; Vermorel and Dantony, 1910; Rasuvaev, 1965). In all cases the successful operation appeared to depend not only upon a particular need for the tannin in wine clarification and other uses, but also upon an integrated operation involving a centralized source of waste to be processed along with recovery of several by-products from the wastes. Nearly all recent reports have been from Communist countries, where the use of grape-seed tannin in wine is evidently widely practiced (Smul'skaya, 1961). The procedures used to recover tannin from grape solids all appear similar. The usual extractant is water or aqueous ethanol, and extraction, concentration, and drying should be done so as to minimize oxidation and modification or degradation by overheating. Some processes call for boiling in water, which gives rapid extraction but seems likely, in the presence of grape acids, to give phlobaphenes and isomerizations with the catechins, and anthocyanidin generation with the leucocyanidin tannins present. This is probably one source of the relatively dark color of the commercial grape tannin we have examined.

A more interesting by-product—one which seems to be enjoying increasing use—is a grape-skin anthocyanin preparation usually called enocyanin. Although not manufactured in this country, it is being imported from Italy and perhaps other countries for use as a food colorant permitted without certification for carbonated and other beverages (Anon., 1966b). It is described as an aqueous-steeped extract of fresh deseeded pomace remaining after pressing grapes for juice or for wine. The preparation of enocyanin was discussed by dal Piaz (1885), and commercial production in Italy was noted by Coste-Floret (1901). The early preparations were evidently made with whole pomace, which probably included too much seed tannin. The fact that highly colored extracts can be prepared from grape skins after they have passed through a red wine fermentation emphasizes the incompleteness of normal red wine extraction.

Commercial production in Italy and legislation pertaining to use of the product are described by Dieci (1967). Fresh seed-free wine pomace is extracted with a sulfurous acid solution, and the extract is concentrated under very low pressure and relatively low temperature. It may be prepared as a liquid, paste, or powder, and is used in Europe in cosmetics, dentifrices, wine, and vinegar, and in the United States especially in carbonated soft drinks. If added to "Concord" juice it can be distinguished by its high malvidin glucoside content, in contrast to the high delphinidin glucoside content of "Concord" juice (Mattick *et al.*, 1967).

Rudnev and Leonov (1961) and Maiorov *et al.* (1963) described extraction of anthocyanins from grape skins with 1% HCl, and the use of these extracts to color fruit and berry wines. Andreev (1965) found that extract prepared with 0.2% $SO_2$ had many advantages over HCl, and recommended a countercurrent extraction at room temperature although 2 replacements in a batch extraction gave 90% extraction. Grebenyuk *et al.* (1966) described equipment to extract whole pomace with hot water so that very little water was required. Yankov and Balkandzhiev (1965) recovered about half as much pigment from fermented pomace as from fresh skins. Extraction was greater with ethanolic than with aqueous acidic solutions. The concentrate would produce red wine color at 2000–5000 mg/liter. Maruyama and Kushida (1965) found that higher alcohol in extracting "Muscat Bailey A" skins gave lower brown color in the red pigment extract and increased stability of the solution color at room temperature. Such natural pigment extracts can be very attractive and useful for coloring food products. Both in the concentrated form and in the colored product they are, of course, subject to the deterioration which is typical for anthocyanins. The relatively high content of malvidin and other methylated derivatives in *V. vinifera* extracts not only gives a redder, less blue color than would the delphinidin nonmethylated series, but also would tend to be more stable to oxidation than the anthocyanin mixture from most other fruits, which tend to have high proportions of less stable pigments.

### 4. Nutritional and Related Aspects

Wine and grapes have had a long career and earned an illustrious reputation not only as pleasant and useful parts of the human diet but also, under some circumstances, as medically useful or especially healthful ones (Lucia, 1954, 1963; Masquelier, 1961a). In some instances the nutritional or medical effect of wine or grapes is the result of their phenol content. The bioflavonoids furnish one example.

Vitamin P was the name given by Rusznyak and Szent-Györgyi (1936) to a flavonol fraction from Hungarian red pepper which helped clear up bleeding of capillary blood vessels in a condition associated with scurvy. There has been considerable controversy over such effects of flavonoids, and, in restrospect, the choice of the name vitamin P (for permeability of blood vessels) appears unfortunate. The name bioflavonoids has been recommended instead, but has not been universally adopted. A number of reviews and summaries of the physiological activity of flavonoids have appeared (e.g., Orzechowski, 1962; Vogin, 1960; Böhm, 1959-1960; Anon., 1959; Griffith *et al.*, 1955; Martin, 1955). From these reports it seems clear that: (1) flavonoids may have a number of physiological effects on animals; (2) flavonoids are not equally physiologically active; (3) flavonoids in effective doses generally lack any deleterious effects; and (4) the value of general human use of bioflavonoids is not clear.

The reported effects of consumption of bioflavonoids include synergistic effects with vitamin C and minimizing such conditions as vacuum-induced purpura, capillary bleeding in diabetes, edema, and inflammation from injury, radiation damage, and atherosclerosis. Some phenolic substances (esculetin, for example) that are not flavonoids have been reported to have "vitamin P" activity (Anon. 1959). Plant phenols including flavonoids have been reported to have many effects not clearly related to capillary fragility, such as bile-flow promotion, diuretic action, sweat promotion, spasmolytic activity, hypocholesterolemic activity, etc. (Böhm, 1959–1960; Mancini *et al.*, 1960; Hörhammer and Wagner, 1963; Schmersahl, 1964). It is obviously impossible to discuss these reports in detail in relation to wine phenols, even though many of the compounds tested are wine or grape constituents.

Some direct studies of wine or grape phenols should be mentioned, however. Lavollay and Sevestre (1944) found that capillary resistance as indicated by the vacuum necessary to produce skin purpura in guinea pigs was increased from 9 cm to 27 cm in about 1½ hours by injecting 1 cc of dealcoholized wine or substances extractable from wine by ethyl acetate. Ingestion of wine itself had a similar effect, though more slowly. The effect was attributed to epicatechin, but leucoanthocyanins appeared to be involved. Scarborough (1945) found the grape to be the richest natural source of bioflavonoids in the series of foodstuffs tested, and found equal amounts in whole black or white grapes. DeEds (1949) reviewed the bioflavonoid story in relation to wine, and noted that part of the effect of grape bioflavonoids in prolonging the life of the scorbutic guinea pig may be an antioxidant effect, helping conserve minimum levels of vitamin C. A fraction from red wine containing anthocyanins eluted from charcoal with acetone and freed of ethyl acetate solubles is an active bioflavonoid, and leucocyanidins from white or red wines are very active (Masquelier and Jensen, 1953a; Masquelier, 1956, 1961a). The amounts necessary to lower sensitivity to vacuum-induced purpura in man appear to be well below the 10–50 mg/liter of leucocyanidins they found in white wines except those which had been botrytised. A major part of the much higher activity of red wine is attributed to the leucocyanidin content because older wines that are low in native anthocyanins are still active.

The flavonol content of red wines would contribute to the bioflavonoid activity (Bourzeix and Baniol, 1966), but the catechins and leucoanthocyanidins seem much more important. For this reason the flavonoids measurable by reaction with vanillin have been considered to be the bioflavonoid fraction of wine. Heintze (1965) found 320–500 mg/kg in onions and peppers, the highest among the fruits and vegetables he tested (grapes not included). Content in whole grapes or red wines would obviously compare very favorably. The bioflavonoid content of grape juice or wine, of course, depends upon maceration or fermentation to increase extraction from skins or seeds (Ournac, 1953).

Russian workers have been very active in investigation of the bioflavonoids of the grape. The Georgian Kakhetia wines are particularly rich in catechins and tannins, owing to their method of vinification, and they are noted in Russia for especially healthful properties (Durmishidze, 1955). The effect of these wines on scorbutic guinea pigs decreased with wine age, as the catechins were oxidized. Phenolic isolates from grapes were compared as to ability to enable guinea pigs on minimal dietary vitamin C to retain this vitamin in their tissues, particularly in the spleen and adrenals. Promotion of vitamin C content was greatest with *l*-epigallocatechin, next with *d*-catechin, and next with seed tannin extracts, in order of their assayed pyrogallol content; malvidin had questionable effect. These data certainly suggest an antioxidant or vitamin C protective effect, perhaps in the dietary mixture itself, since the effectiveness of the phenols appears to decrease in the order of their decreasing ease of oxidation. Tea catechins are quite active bioflavonoids, and their activity is greater in green tea than in black tea, where they are partially oxidized, particularly the pyrogallol derivatives (Anon., 1959; Das, 1963). The catechins and anthocyanogenic tannins of wine have bioflavonoid activity, and the activity seems greater with the more soluble fractions, presumably the less polymeric (Durmishidze, 1958; Durmishidze *et al.*, 1961). Vine sprouts and leaves are rich sources of bioflavonoids, and grape wastes (along with those from other fruits) may be processed commercially for recovery of bioflavonoids (Radoev and Mladenov, 1959; Shnaidman *et al.*, 1965).

Horwitt (1933) isolated "Concord" anthocyans and fed them to rabbits in 500-mg doses. The anthocyans were poorly absorbed; feces were highly colored whereas the urine contained only 1–2% of the detectable pigment. With 100-mg oral doses none was detected in the urine, but when 50 mg was injected subcutaneously the urine became blue-black (red when acidified with HCl). Dogs with gastric fistula injected with the anthocyans showed pigment in the bile and urine within 20 minutes, and in the lymph within 50 minutes. Pourrat *et al.* (1967), using preparations from other fruits, found that anthocyanins were not toxic to rats or mice when administered orally in single doses equivalent to about a quarter of a pound in an adult man. No abnormalities were observed in rats after 3 months of ingesting 6 g/kg/day. Injecting the anthocyanins intravenously produced moderate toxicity ($LD_{50}$ 240 mg/kg for rats), whereas intraperitoneal injection was only about 1/10 as toxic and at low levels of injection capillary resistance was enhanced without protection from vitamin C deficiency. Improvement in visual acuity and darkness adaptation has been reported in humans for a short time after receiving oral doses of about 700 mg of anthocyanins (Alfieri and Sole, 1966; Jayle and Aubert, 1964).

In general, other reports on specific phenols in grapes are similar: little or no demonstrable negative effect, and occasionally a positive one, from oral doses up to much larger than might be expected naturally. As noted by Bate-Smith (Fairbairn, 1959), it would be surprising only if animals were not well able to cope with such frequent and considerable constituents in their diet. Chlorogenic acid by subcutaneous injection in mice in doses up to 200 mg/day showed no toxicity and a slight anabolizing action on the growing animal (Merli, 1965). The catechin complex from tea administered to rats in dietary doses about 10 times the recommended intake for humans (equivalent to more than 5000 mg/70-kg man/day) caused no observable developmental or morphological changes, according to Koroza (Anon. 1959).

Of course, alcoholic beverages can produce intoxication and, with pathological consumption, alcoholism and associated effects. Fabre *et al.* (1950) found that red wine was less intoxicating to experimental animals than was distilled spirit, and believed the wine's tannin to be one of the antidotes to alcohol. Over a period of 2 months Flanzy and Causeret (1952) administered wine, alcohol, and 2 samples of brandy at isoalcoholic levels in comparison with water by stomach tube to lots of 10 rats, each in amounts equivalent to 2 liter/day for a man. Growth was the same as with water with a red wine prepared from a mixture of free run and press wine after fermentation with pomace and stems. Growth was depressed with alcohol and slightly less so with brandy. In further experiments (Flanzy *et al.*, 1955, 1965), wines were tested with grossly different total phenol levels made by fermentation of juice only, rosé fermentation, 4 days on the skins, and very tannic wines made by fermenting stems also or fermenting for 28 days on the skins. The rats drinking wine consumed less fluid than rats drinking water, 94% as much for the ordinary red, 81% for the rosé, and 68% as much for the tannic red wine. The rats were healthy and grew well on the wine, but the rosé and juice gave slightly less growth, presumably because of higher sulfur dioxide content. Two red wines gave the same growth as water; the more tannic wine produced with the stems gave essentially the same growth as water; and the long-fermented wine gave slightly lower growth. Location of the vineyard made little difference on a 15-week regimen, although a "Noah" wine gave a slightly lower growth rate than did a *vinifera* wine or water. A number of other reports (e.g., Jovanovic *et al.*, 1963; Kliewe and Anabtawi, 1964; Breider, 1964a,b; Breider *et al.*, 1965) have considered the relative nutritional effects of wines from *V. vinifera* versus wines from non-*vinifera* varieties. Most of the experiments appear to be subject to serious criticism on grounds, among others, of too few wines of a nature and composition insufficiently defined to justify general conclusions. The most serious effects were reported for chickens. Chickens forced to drink wine or partly diluted wine produced thin-shelled eggs and had poor calcification of leg bones, and the chicks had symptoms resembling perosis. Cannibalism was high in the pens affected most. The differences usually appeared to be worse in chickens receiving non-*vinifera* wine, and juice did not seem to produce the effect that wine produced. The conditions suggest nutritional deficiency, probably choline or methyl-group compounds, the lack of which is known to produce perosis, slipped tendons, and other similar effects in chickens to those reported.

Chickens may be particularly sensitive to choline or other deficiency when forced to drink wine. Kliewe and Anabtawi (1964) tested the effect of various wines on the liver of rats which had been treated with allyl alcohol to initiate liver damage. Wines from non-*vinifera* grapes showed no difference from *vinifera* wines in effects on this condition, and malvidin specifically had no toxic action.

Chickens appear to tolerate red wines less than white wines, and Bosticco (1966) has reported that grape skin anthocyanins lowered hens' growth rate and food conversion efficiency. Deyoe *et al.* (1962) found that 2.5% citrus bioflavonoids had no effect on broilers whereas 5% reduced growth. Patay *et al.* (1961) have reported that forced consumption of a condensed anthocyanogenic tannin from grape stems at 1000 mg/liter in the drinking water led to the death of 3 of 11 rats in 6 months, and at 3000 mg/liter killed 5 of 30 rats in 82 days. At autopsy, liver damage was found. Fuller *et al.* (1967) found that the depressed growth of chickens on a 0.5% tannic acid diet could be completely remitted if methionine, choline, and arginine were added to the diet. Gallic acid is converted in major part to 4-*O*-methylgallic acid in rats and rabbits (Booth *et al.*, 1959). Nonhydrolyzable condensed tannins, as in grapes, are much less toxic when applied to open wounds, such as burns, than are hydrolyzable tannins, such as tannic acid (Wilson and Gisvold, 1956). Handler and Baker (1944) found that tannic acid is not absorbed from the gastrointestinal tract of rats, for no liver damage resulted from feeding tannic acid but did result from injecting it. They also found that growth rates are the same when rats fed normal feed are restricted to the same food intake as rats fed tannic acid in the feed. The restricted growth on tannic acid diets appears to result from aversion to the taste. Condensed tannins should be much more difficult to break down and absorb than hydrolyzable tannins. Korpassy *et al.* (1951) detected a few micrograms of phenol per cc of blood after administering 50–80 cc of Burgundy to 2-kg rabbits by stomach tube.

Herbivores, because of their diet, might be assumed to have greater tolerance for plant phenols, but these animals might have problems from inhibition of essential digestive microflora by tannins, or from microbial breakdown of phenols to more absorbable forms. That microbial breakdown must be considered is indicated by the fact that *d*-catechin administered orally to the rat is metabolized to different end products depending on whether or not the microflora of the gut has been inhibited by antibiotics (Griffiths, 1964). Winery pomace and grape seeds are not particularly good feeds, but they appear to be acceptable, particularly by sheep. Up to about 10–12 kg pomace/day can be used by horses and 4–6 by cows. The high tannin content does limit the palatability of grape-seed meal, preventing its use as the sole feed, especially for cattle (Fabre, 1909; Cottier, 1924; Folger, 1940; Fabris, 1966; Ferrando and Catsaounis, 1966).

Since tannins are protein precipitants it is not surprising that toxic symptoms have been reported when they are injected into the body. Tannins, including tannic red wines, have been recommended as agents to combat diarrhea (Frey, 1908) and have some effect in fairly concentrated form in inhibiting pepsin and other digestive enzymes (Alquier, 1929; Emiliani, 1940). These effects evidently do not hinder normal digestion and could be beneficial in cases of hyperactivity.

A few other reports have appeared of specific effects of wine phenols. Masquelier (1961b) reports hypocholesteremic action by the cinnamic acid derivatives, catechins, leucocyanidins, and isoquercitrin of wine. Masquelier and Laparra (1967) find that an ethyl acetate extract of red wine containing cinnamic acid derivatives, purified and administered to the duodenum of a rat, gave a great increase in bile secretion. They believe the results have significance for man. Godlewski and Richet (1924) report that great diuresis resulted in a few minutes when anesthetized dogs received intravenous doses of about 1cc of fresh juice from white grapes. Juice from black grapes and juice that had been heated, stored, or dialyzed did not have this effect. Seeger (1967) reports an inhibitory effect on cancer cells by red wine and other red-pigment-bearing plant materials. Kabiev and Vermenichev (1966) report inhibition of growth of lymphosarcoma in mice by catechins and leucoanthocyanidins from plants.

A very serious but extremely rare condition, called the Marchiafava-Bignami syndrome after those who first described it, is characterized by degeneration of the corpus callosum, the nerves connecting the cerebral hemispheres. The condition is so rare (only about 50 cases have been described) that if it were not so apparently pathologically definite it would probably not be worth considering (Lolli, 1941; Thompson, 1956). It appears to have been recognized only in men 35 to over 80 years old, of Italian background, who had been alcoholics or heavy drinkers, some since childhood. The interest here is in the observation that the primary beverage was red wine. The patients also were generally of low income level, with inadequate food intake. The etiology is obscured by the fact that a similar condition has been induced in dogs by prolonged administration of high levels of alcohol alone. Thompson (1956) states it may occur without alcohol ingestion at all. This is the only condition which has been described in humans of which we are aware that appears to bear any suggestively specific relation to wine phenols, and its incidence appears to be less than 1 per 1,000,000 population. It is related to prolonged and excessive consumption and other inadequacies, and is still of uncertain cause. Pasteur was evidently correct in his enthusiastic endorsement of the healthfulness of wine.

Pasteur had in mind primarily, no doubt, the microbiological healthfulness and sanitary nature of wine. In recent years there has been renewed interest in factors other than alcohol which prevent microorganisms pathogenic for humans from growing or even living in wine. Such ancient practices as mixing water with wine for children probably grew from the fact that contaminated water was purified by the wine. Experimentation by adding typhoid germs and others to wine and noting their death rate led to the conclusion that wine contains factors other than alcohol that are bactericidal. Draczynski (1950-1951) found that white wine killed added *Escherichia coli* in 20 min to 2 hr. and found red wine about equally effective with equal alcohol and acid. Jenkins *et al.* (1957) note that practically every phenolic compound possesses some antibacterial properties. Recent studies on wine's antibacterial properties have centered on the phenolic substances as the agents active beyond the effects of alcohol, pH, etc.

The two most active groups studying this effect have been Masquelier and co-workers, in Bordeaux (Masquelier and Jensen, 1953b,c; Masquelier, 1957, 1958, 1961a; Fairbairn, 1959; Masquelier and Delaunay, 1965), and Powers and co-workers, in Georgia (Pratt *et al.*, 1960; Powers, 1964; Wheeler *et al.*, 1967). Masquelier and Jensen (1953b,c) showed that the bactericidal principle of red wine for *E. coli* was adsorbed by activated charcoal and could be eluted by acetone. The eluent contained anthocyanin and other polyphenols. A fresh extract of red grape skin was not antibacterial, but 5 minutes of boiling with 10% HCl made it so. Ether extractables from wine were not appreciably active. The activity was attributed to hydrolysis and partial demethoxylation of malvidin and other wine anthocyanins. Isolated malvidin had a phenol coefficient of about 33.

It was found that the bactericidal effect of red wine was higher in wines of intermediate age. Red wine had a relative bactericidal effect of 11 units at one year, a maximum of 19 units at 9 and 10 years, 12 units at 46 years, and 6 units at 82 years. Again, this was attributed to anthocyanidin production with aging. Teinturier wines were often lower in effect than other reds, though higher-colored. Heat treatment of wines soon lowered the bactericidal effect. The bactericidal effect of wine was very weak compared with the effects of commercial antibiotics. Dilution of red wine by more than about ¾ with water prevented killing of the small inocula used, 100 bacteria/ml. Synergistic effects were observed, however, with penicillin, hydroxytetracycline, and chloramphenicol, with the effect detectable at a dilution of 1 part wine with 14 parts water if contact time was increased.

Phenols other than anthocyanidins are antibacterial in wine. Masquelier and Delaunay (1965) chromatographed an ethyl acetate extract of wine on paper and tested the chromatogram for antibacterial effect. An antibacterial zone corresponded in mobility to hydroxycinnamic acid derivatives, and known samples of *p*-coumaric, ferulic, caffeic, gallic, protocatechuic, and chlorogenic acids were all antibacterial at 400–1400 mg/liter. It was noted that the freeing of *p*-coumaric acid from acyl combination with anthocyanins increased the antibacterial activity of wine. These effects of wine are not considered therapeutically interesting, but in the present as well as historical sense may be important from the viewpoints of food- and water-borne diseases, public health, sanitation, and epidemic control.

Gram-positive organisms seem more affected than others, a general finding with phenols. The gram-positive organisms have a "mucocomplex" external layer made up of amino sugars and amino acids. Gram-negative microorganisms, on the other hand, generally appear to have lipoprotein outsides (Stanier *et al.*, 1963). The greater interaction of mucoprotein than of lipoprotein with plant phenols would seem reasonable given the hydrogen-bonding and relative polarity properties of the grape phenols.

Pratt *et al.* (1960) found that anthocyanins isolated by paper chromatography from grapes and strawberries stimulated early bacterial growth but decreased the maximum growth obtained at concentrations of 5–500 mg/liter. Different synthetic anthocyanidins and grape and wine anthocyanins appear to have different effects on different organisms, but, at least in some cases, sizable inhibition zones are produced on agar plates inoculated with *E. coli*, *Staphylococcus aureus*, and *Lactobacillus casei* (Powers *et al.*, 1960; Hamady *et al.*, 1961). Degradative changes in heated pigments were great, and various fractions were found with different, though not always lower, antibacterial effects. In a number of other studies the same group investigated various aspects of the inhibitory effects of anthocyanins on microorganisms and specific enzymes, often including samples isolated from grapes (e.g., Powers, 1964; Somaatmadja *et al.*, 1964; Wheeler *et al.*, 1967; Carpenter *et al.*, 1967). Oxidation-reduction phenomena, metal chelation, and protein binding appear to play roles in the inhibitory effects, at least under certain conditions. Somaatmadja *et al.* (1965) found that 2 different grape leucocyanidins and a leucodelphinidin produced inhibition zones on inoculated agar plates and inhibited different microorganisms to different degrees.

The major significance of these results appears to be that a number of the phenolic substances of wine, perhaps nearly all, can produce bactericidal effects and that much of the effect appears related to protein-bonding ability. The increasing effect of wine after some, but not too much, aging appears to be reinterpretable in the light of more recent findings of anthocyanin loss in wine with aging. The unchanged anthocyanins are generally missing from wine within 5 years, and incorporation into "polymers" appears to be the main reaction. It is postulated that the maximum antibacterial effect, at about 8–10 years, noted by Masquelier's group correlates more reasonably with the gradual formation of more small oligomeric "tannin" molecules by reaction of anthocyanins and, probably, other phenols.

The lesser activity in young wine would be due to the low content of the optimum size or type of oligomer, and a lower activity later would be due to too large size, precipitation, etc. Basically the same idea was suggested by Swain with regard to Masquelier's studies (Fairbairn, 1959). The apparently more specific effects of pure anthocyanins (not in wine's complex solution) are difficult to relate to wine, but, considering the high instability of pure natural flavylium derivatives, such reactions might be involved even there. Successive repetitions of paper chromatography of the same pigment lead rapidly to most of the pigment staying at the origin, apparently oxidized and polymerized. In enzyme inhibition by tannins an important factor is molecular size of the tannin, particularly when diffusion is also a factor, and oxidation plays a role since the quinoid forms appear to bind irreversibly to enzymes whereas the phenol form is competitive (Wheeler *et al.*, 1967). It would be very interesting to compare wines made from white grapes by fermentation on the skins and aging as for red wine. An antibacterial effect would be predicted at least nearly as high as for red wines, and in proportion to the flavonoid tannin content. Some antibacterial activity for usual white wine beyond its alcohol content has been already noted and attributed to the low level of anthocyanogens present.

## B. Processing and Phenolic Interactions

### 1. Phenols Related to Fermentation and Associated Microorganisms

The possible relationships of natural phenols to the inhibition of grape pathogens and pests have already been discussed, as has the inhibition of animal-pathogenic bacteria. Remaining to be considered are the effects on yeasts and bacteria of the phenolic substances in or added to grape juice, musts, and wines.

Practically every phenol derivative possesses some ability to inhibit microorganisms (Jenkins *et al.*, 1957). An old process is the addition of preservatives, including phenols, to inhibit fermentation or microbial change in fruit juices and wines. The former use of salicylic acid in wine making has already been mentioned. Among 20 food preservatives discussed by Oka (1964), 13 were phenol derivatives: salicylic acid and its esters, *p*-hydroxybenzoates, phenol- and phenyl-substituted phenol, *o*-cresol, thymol, eugenol, guaiacol, naphthols, and three quinones. The concentrations that produced 99% killing in 2 hours of contact with a baker's strain of *Saccharomyces cerevisiae* ranged from 0.8 m*M*, for *n*-butyl-*p*-hydroxybenzoate, to 50 m*M*, for guaiacol and phenol; the quinones were even more active. The phenols seemed to inhibit by a generally similar surface-adsorption mechanism, since the amount adsorbed by the yeast from inhibitory solutions was fairly constant, 15–22 mmoles per kg of yeast cells. However, Bosund (1962) indicates that phenols can act by specifically uncoupling oxidative phosphorylation in the terminal oxidation of pyruvate via acetate.

Quinoids, especially those of the vitamin K series, can be used at doses of about 20 mg/liter to inhibit the growth of yeasts in grape juice and, usually at somewhat lower doses, to prevent secondary fermentation in sweet table wines, as shown by Verona (1952), Ribéreau-Gayon and Peynaud (1952), Pratt *et al.* (1948), Athanassiadis *et al.* (1956), Caspar and Yang (1957), and Yang *et al.* (1962). Unlike the simpler phenols, cell penetration is a factor in inhibition with quinones, according to Oka (1964). Starkov (1958) found that 2-methyl-1,4-naphthoquinone ($K_3$) suppressed yeast growth in wine in the presence of oxygen but had little effect on completely anaerobic fermentation. During fermentation, the yeast inactivated the quinone by reducing it to the hydroquinone. Such action promoted early onset of fermentation, presumably by furnishing a reducible substrate when natural hydrogen-acceptors were still low. Rasulpuri *et al.* (1965) found that 2-methyl-4-amino-1-napthol ($K_5$) inhibits oxidation of ethanol by *Saccharomyces cerevisiae* and decreases the free sulfhydryl groups in the yeast cell, probably those involved in dehydrogenase activity.

The phenols naturally present in grapes or wine have been reported to inhibit yeast growth. Carles and Nivière (1897) reported that "Jacquez" and other highly colored grape varieties were difficult to ferment completely to dryness, owing to inhibition of the yeasts by the anthocyanin and tannin present. Rosenstiehl (1902) found that apple tannins inhibited yeast's ability to ferment cider, but the cells were not killed and were able to reproduce. He found that grape juice prepared free of solids yet with a high anthocyanin content produced a similar effect, and concluded that anthocyanins and tannins behaved similarly in inhibiting fermentation by yeasts. The inhibitory effect was rather small, however, even though the yeast was immediately contacting a high initial pigment level, contrary to the usual situation in vinification of red wines, where pigment and tannin are leached from the solid grape tissues. The yeast became dyed with the pigment and yet would take up more if placed in fresh juice or wine. If portions of the red juice were fermented in sequence with the same yeast, the fermentation became more and more sluggish as the yeast became more and more colored. By the fifth transfer, some weeks (compared with the usual few days) were required for the sugar to disappear, and by that time a considerable increase in the total volume of yeast cells had occurred. The results appear to be explained by inhibition of the yeast by adsorption of pigment and tannin on the cell surface, inhibition in proportion to the extent or thickness of the surface covering, and gradual recovery as new surface was created by reproduction.

Comrie (1930) attributed sluggish attenuation (removal of the last sugar) in beer fermentations to incomplete formation and precipitation of phlobatannin during wort boiling, with the phenols remaining in solution and inhibiting the yeast. Flanzy (1935a) noted that fermentation to completion in wines from red grapes tended to be poor compared with that in white grape wine, and that the "Carignane" variety, noted for high color in his region, was especially troublesome. Experimental removal of most of the anthocyanin (including about 15% of the total tannin) from such wine with activated carbon allowed normally rapid fermentation of added sugar. Eliminated as causes of slow fermentation were sulfur dioxide content, minerals, sugar level, and temperature. It was concluded that excessive pigment caused the inhibition of fermentation to complete dryness, and that it was important to accustom yeast to high levels of phenols before fermenting such musts. If an unnecessarily high content of tannins, anthocyanins, and similar phenolic compounds was avoided, no nutritional supplementation was required to achieve complete fermentation. Lüthi (1957) reported that apple and pear juices with excess tannin often ferment poorly, but that fermentation is usually rapid if this tannin is removed by gelatin fining or if an extra nitrogen source is added.

Turbovsky *et al.* (1934) reported that wine yeasts, wild yeasts, and vinegar bacteria were not appreciably inhibited by the levels of tannin usual in wine production. Even added tannin of 20,000 mg/liter, only retarded alcoholic fermentation. They found that an unidentified malolactic fermenter was inhibited at about 4000 mg tannin/liter, but such wine was too astringent for human consumption. Peynaud and Dupuy (1964) have reported the separation and partial identification of specific phenolic fractions in wine which inhibit lactic-acid-producing microorganisms. The most active fraction was inhibitory only in acid solution, and was removed from wine by gelatin precipitation. Since red wines generally undergo malolactic fermentation more readily and more frequently than do white wines, it appears that relatively high total phenolic content might favor (Anon., 1962), or at least not inhibit, the bacteria that produce lactic acid in wines. An unidentified inhibitory principle for lactic acid bacteria in green olives was present at levels paralleling total phenol content. It was destroyed by raising pH to 10, but was enhanced by sodium chloride (Fleming and Etchells, 1967).

Thus, the phenols present in grapes and wine appear to be significantly inhibitory to the organisms normally important in wine only if associated with other suppressive conditions, such as low nutrient level or high alcohol level. This idea is reinforced by experience with sparkling wine fermentations. Sparkling wines are produced by a second fermentation (nutrients likely to be low) of a wine (10% or so alcohol) to carbon dioxide concentrations (pressures) approaching those that inhibit fermentation completely. It is commonly noted that the secondary fermentation is slower and more difficult to bring to completion in red sparkling wines than in white. Goldman (1963) found that red base wine required 38 days, and did not reach as high $CO_2$ pressure as did white wine, which fermented under the same conditions in 16 days. Although phenolic content may not be the only factor, e.g., red wines tend to be slightly higher in alcohol, it certainly seems to contribute.

Schanderl and Staudenmayer (1961) reported that certain white wines, particularly wines imported into Germany, when converted to sparkling wine did not referment easily and were subject to browning and other problems. The unsatisfactory wines had in common a relatively high phenol content. A simple test was devised that could be used to eliminate such wines in advance even if their high phenol content was partially masked by high $SO_2$ content. This test consisted of bringing the wine to pH 7 with alkali, which in a few minutes produced oxidation to gray to black-violet colors in the unsuited high-phenol-base wines, and to yellow or light color in acceptable wines. The color was proportional to the phenol content, but was also influenced by the iron content of the wine. In further development (Schanderl, 1962) of this "pH 7 test," the effect of added ferric ion on color aided the interpretation. Although wines from white "hybrid" grapes tended to give dark colors in this test, the overlap with pure *Vitis vinifera* wines, particularly those produced with pomace contact, was sufficient that they could not be distinguished reliably by this test. The ability of high-phenol wines to fix $SO_2$ and to form quinones by oxidation were believed to be factors in unsatisfactory champagnization.

Schanderl (1962) found that fermentation by wine yeast was halted at a lower total $CO_2$ yield when 2000 mg tannin/liter was present than in a control without added tannin. At 4000 mg tannin/liter the fermentation was delayed, but ultimate $CO_2$ yield was not suppressed further. With apiculate yeasts, however, fermentation was prevented almost completely with either level of tannin. Gallic acid, caffeic acid, and quinic acid at 100 mg/liter slightly stimulated fermentation by wine yeast in synthetic culture solution free of initial alcohol, but suppressed and delayed fermentation compared with that of the control when 6% ethanol was included in the medium. He concluded that, depending on the yeast strain, the inhibitory influence of phenols of white wine is usually not significant until alcohol content exceeds about 8%. Thus it is likely to be of practical importance only in sparkling wine, toward the end of fermentation of high-alcohol wines, or in flor-sherry production. Marcilla *et al.* (1939) showed that adding tannic acid to sherry originally containing total phenols of 70 mg/liter inhibited development of the aerobic yeast film stage of the flor-sherry process. Additions to a total as low as 160 mg/liter definitely delayed yeast film development. Film-forming sherry yeasts were noted by Florenzano (1952) as tolerating only small amounts of tannin. This may be one reason that red wines are seldom treated by a yeast-film aerobic process. Such treatment of red wine has been accomplished, however, by, for example, Cantarelli (1955, 1958), and again the inhibitory effects of wine phenols seem to be overcome relatively easily. Production of acetaldehyde during aerobic yeast growth decreased tannin and pigment by causing precipitation so that the inhibition effect should diminish with time in this process.

Sikovec (1966a,b) investigated the effect of chlorogenic, isochlorogenic, caffeic, gallic, and ellagic acids at 200 mg/liter, and mixtures of them at 200 and 800 mg/liter, on the wine yeast *Saccharomyces cerevisiae* strain Epernay and the "sulfite yeast" *S. ludwigii.* Studies were made of both alcoholic fermentation in model solutions at added alcohol levels of 0, 6, and 11%, and of aerobic respiration. With the wine yeast fermentation initially at zero alcohol, the control reached the greatest $CO_2$ yield by the twelfth day and continued slightly above all those with phenols. The 800-mg/liter level of mixed phenols reduced $CO_2$ yield to about half that of the control. Several of the phenols, notably isochlorogenic acid and the 200-mg/liter mixture, gave earlier onset of $CO_2$ production than the control, but were rapidly overtaken. At 6% and at 11% initial alcohol, isochlorogenic acid stimulated $CO_2$ production over that of the control. Most of the other phenols were similar to the control or slightly inhibitory, with ellagic acid tending to be the most inhibitory. With *S. ludwigii* in the absence of added ethanol, the control fermented most completely; with 11% added alcohol, all samples were considerably suppressed. At 6% alcohol the phenol mixture and part of the individual phenols stimulated fermentation over that in the alcohol-only control sample, approaching that of the zero-alcohol zero-phenol sample.

Caffeic acid was apparently considerably more inhibitory for both yeasts than were the chlorogenic acids. This is believed to explain the greater difficulty in refermenting older wines, in which the original ester forms would have hydrolyzed to free caffeic acid. Fermentation decreased the phenol content, and it was suggested that the yeast partly assimilated phenols as carbon sources. Growth in the presence of phenolics tended to produce distorted cell forms, which were less vigorous and were killed more easily than normal cells. When cell multiplication was the measure of inhibition rather than $CO_2$ yield during alcoholic fermentation, the effects were somewhat similar except that the phenols were more regularly inhibitory for the wine yeast than for the controls. Aerobic growth ($O_2$ consumption) by the wine yeast was stimulated by these phenols in the absence of alcohol and inhibited in the presence of it. The effects on *S. ludwigii* were both stimulatory and inhibitory, depending on the phenol and the alcohol level.

In summary, it appears that the natural phenol derivatives of grapes or wine, and to some extent those likely to be added, are definitely inhibitory but that inhibition is generally of such a low order as to be of practical significance only in special cases involving combinations of detrimental factors. There does appear to be a degree of selective inhibitor effect, depending upon the phenol involved, the yeast type, and prior adaptation or associated cultural conditions.

The adsorption of phenolic substances by yeast cells as observed by Rosenstiehl (1902) has already been mentioned. He described the removal of anthocyanin as a dye adsorbed to the yeast cell wall. Coudon and Pacottet (1901b) added tannin (presumably tannic acid) at 2000 mg/liter to red grape juice and to crushed whole red grapes. After fermentation, racking, and subtraction of the amount of tannin present in a similar portion of must without added tannin, only about 200 mg/liter of the added tannin was left in solution; the remaining 1800 mg/liter had been removed in the lees along with the yeasts and other solids. Preparation of wine from red juice by fortification gave a total phenol content of 1240 mg/liter whereas fermentation of the same juice gave a similar color level but 1020 mg/liter total phenol (Berg and Akiyoshi, 1957).

Both more color and considerably more tannin were obtained by mixing the corresponding white and red wines than by mixing white juice and red grapes before fermentation (Ough and Amerine, 1962). The difference is interpreted as arising from the double total yeast population in contact with the tannin in the sample diluted before fermentation. Other quantitative studies of precipitation of leucoanthocyanins or tannins from wines along with the yeast after fermentation have been reported (e.g., Wucherpfennig *et al.*, 1964; Cantarelli and Peri, 1964a).

Milisavljevic (1967) reported that about 25% of the leucoanthocyanins were removed from wine left in contact with yeast, and that, depending partly upon the strain of yeast, much of the enotannin, catechin, and protocatechuic acid may be removed from wine by yeasts. Chevalier *et al.* (1961) found that yeast behaved like a dispersed insoluble protein, adsorbing tannin from fermenting beer in proportion to the number of cells present and the tannin concentration. Treatment of the yeast with polyvinylpyrrolidinone solution displaced some of the tannin, and yeast cells so treated regained adsorptive capacity for beer tannin. Autolysis of the yeast cell seems to be associated with release of part of any adsorbed phenol back into the solution (Masschelein, 1963). Different amounts of tannin were adsorbed by the same weight of top, bottom, and baker's yeast.

Campbell (1966) reviewed the structure and composition of the yeast cell wall in relation to brewing. The cell wall is normally about 30% protein, and the structure is apparently conducive to complexing with larger polyphenols. Such complexing is a factor in flocculation and early settling of yeast from fermentation media (Ito, 1967a,b). Masschelein and Haboucha (1965) studied a number of factors relating to the adsorption of beer phenols by yeasts. The physiological condition of the yeast affects the mannan-protein composition of the cell wall, and therefore the adsorption capacity. Particularly, yeasts adsorbed more tannin at the end of fermentation or if heated. The anthocyanogens appeared to be adsorbed preferentially among the beer phenols, but if the anthocyanogens were already complexed with wort proteins, yeasts did not displace the proteins. Although it is not possible to interpret the available data very precisely, it appears that 10 g of live yeast may adsorb about 1 g of beer anthocyanogens. This would appear to put yeast's saturation capacity at a logical point above particulate nylon (about 1 g/10 g dry weight), and near that of soluble proteins such as gelatin (about 11 g tannin adsorbed at "saturation" per 10 g dry weight) (Singleton, 1967a,b). Since about 10 g of wet weight of yeast is produced in fermentation of a liter of wine, the potential removal of 1000 mg/liter of tannin by yeast adsorption may well explain a major part of the drop in red wine total phenol from the predicted total of extractable phenol of red grape must (Singleton and Draper, 1964). The fact that white and rosé wines have low and very similar total phenol content and very little true tannin compared with red wine is partly the result of the excess precipitation capacity of the yeast for tannin in relation to the amount present in the musts of these wines.

Yeasts may cause the precipitation of phenols in another manner—the production of aldehyde. Acetaldehyde is produced at several hundred mg/liter in wine by the flor-sherry process, which involves aerobic yeast growth. This process is not usually applied commercially to red wines. However, Cantarelli (1955, 1958) showed experimentally that the high phenol and anthocyanin content of red wine is converted rapidly to discolored and insoluble products by reaction of the aldehyde produced by the aerobic yeast with the phenols. Similar reactions may occur in white wines, but the initial relatively low phenol content and the processing conditions result in a desirable light-colored, clear product.

Many yeasts can be adapted to utilize phenols as carbon sources. Zimmermann (1958) found that *Saccharomyces cerevisiae* reduced the phenol content of a medium from 500 mg/liter to zero in about 130 hours, which was more rapid than with *Rhodotorula glutinis* or *Candida tropicalis*. Leibnitz *et al.* (1960) found that a wine strain and another strain of *S. cerevisiae* required more adaptation to phenol, used it more slowly, and were more sensitive to levels above about 100 mg/liter than were *Torulopsis utilis*, *T. dattila*, *R. glutinis*, and an unidentified yeast. Cresols, xylols, and orcinol were not used by these yeasts (*Saccharomyces* not tested), but catechol, resorcinol, and hydroquinone were. Rieche *et al.* (1962) were able to adapt *T. utilis* to grow in 5000 mg/liter of a phenol mixture including cresols. Wase and Hough (1966) found that adaptation of *Debaryomyces subglobosus* to phenol was rapid in continuous culture but not in batch culture. Adaptation was lost rapidly on other media, and regained as slowly as before.

The study by Harris and Ricketts (1962) emphasized the variation in ability to use phenol by genus and strain of yeast. They found that the ability to grow on catechol or resorcinol generally paralleled the ability to utilize phenol, although a strain of *S. pastori* used catechol but not phenol or resorcinol. Their study and that by Jayasankar and Bhat (1966) indicated that phenol was metabolized via catechols. Phloroglucinol was generally utilized poorly if at all by the yeasts tested. The conversion of natural plant phenols was indicated by a high incidence of phenol-adapted yeasts in retting solutions from coconut husk conversion to coir fiber. Martins and Drews (1966) found that *R. glutinis* metabolized phenols in the manner to be expected from bacterial studies (Ribbons,

1965), namely phenol conversion to catechol, *cis-cis*-muconic acid, β-ketoadipic acid, and succinic acid. The specific radioactivity was notably high in tyrosine and phenylalanine in cells grown on phenol labeled with radioactive carbon although all amino acids were labeled.

Westlake and Spencer (1966) found that some strains of *Pullularia* and *Cryptococcus* yeasts utilized rutin, quercitrin, taxifolin, and naringin. Apigenin and catechin were apparently not used readily. Media with complex nitrogen sources aided in degradation of the flavonoids. Products of yeast-cell-free enzymatic oxidative degradation of rutin were glucose, rhamnose, protocatechuic acid, phloroglucinol carboxylic acid, and carbon monoxide. All of these actions of yeasts on phenols were, of course, under obligatory aerobic conditions, since oxygen is consumed in the reaction. Therefore, any such conversions of phenolic substances of grapes or wines would be limited to conditions such as on the fruit skin in the vineyard, at the very beginning of the fermentation, or in aerobic growth (as during the flor-sherry process).

Rossi and Singleton (1966a) found that the many phenolic substances present in an extract of ripe grape seeds were still detectable after fermentation. Many similar observations have been made in that chromatograms of beer and wine still show much the same phenolic substances as were present in the raw materials. Additional substances often appear, but there is no evidence for a major conversion of phenols to nonphenols or severely degraded products of the type that appear possible from aerobic metabolism. Aerobic yeast growth on high-phenol wines (red wines) has not been studied enough from this viewpoint for the detection of such changes. Cantarelli (1955, 1958) showed a decrease of about 20% in the phenol estimated by a permanganate index of a red wine after 80 days under an aerobic yeast film. Chromatograms of the pigment show changes; they are interpreted, however, as resulting from phenol-aldehyde and hydrolytic reactions. No appreciable degradation of the phenolic pigments to colorless products is indicated.

The possibility exists that degradation of phenols of grapes may be caused by aerobic growth of pathogenic and saprophytic molds on the clusters. To the extent that affected berries or clusters are included in musts, the wine might be affected. Coudon and Pacottet (1901c) showed that weak and off-colored red wines resulted from infection of the grapes by *Botrytis cinerea*. Skins taken from healthy grapes and grapes on the same clusters infected either naturally or artificially gave extracts in 10% alcohol that were like medium red wine for the healthy grapes but were yellow to colorless for the botrytized. The tannin content of berries attacked by *Botrytis* was reduced 15% for slightly attacked berries and 17.5–18.6% for severely attacked. *Penicillum glaucum* attack reduced the tannin of the berry to half its original value. Moldy vintage in the winery appeared colored at crushing, but the pigment rapidly discolored and precipitated, yielding a brick-red or tile-colored wine. *P. expansum*, on the other hand, reportedly increased tannin and bitterness in wine from infected grapes (Milisavljevic, 1957). Poux (1966) reports that phenoloxidase content is about four times as high in moldy grapes as in healthy ones from the same cluster. The effect of molds, *B. cinerea* particularly, upon the destruction of anthocyanins, browning, and lowered total phenol content has been attributed to the action of oxidative enzymes produced by the action of the mold. The anthocyanins are affected more than the wine's tannin. In one instance (Sutidize, 1962), severe infection reduced pigment 78% and tannin only 1.1%. Bolcato and Lamparelli (1962), however, reported that such oxidative precipitation reactions could be caused by growth of *B. cinerea* in the absence of detectable peroxidase or phenolase. Bolcato *et al.* (1964a,b) reported that chromatograms of grapes with and without *Botrytis* infection showed qualitative differences in the anthocyanins and other flavonoids. Sakamoto (1959, 1960) reported degradation by *Penicillium* sp. of the flavonoid structure to products that included gallic acid or protocatechuic acid and phloroglucinol carboxylic acid. Westlake and co-workers (Westlake, 1963; Westlake *et al.*, 1961; Westlake and Spencer, 1966; Simpson *et al.*, 1963; Child *et al.*, 1963) made a detailed study of the degradation of flavonoids by inducible enzymes of molds, particularly *Aspergillus flavus.* They showed that, if the flavonoid has an accessible 3-hydroxyl, carbon monoxide is produced from that 3-carbon. The depside of the B-ring acid esterified to phloroglucinol carboxylic acid results, and is then hydrolyzed. Generally, methoxylation of any of the phenolic groups except at C-7 reduces the utilization of these compounds by fungi, and a C2-3 double bond appeared to be required, by *A. flavus* at least. The esterase that hydrolyzes the depside is readily able to hydrolyze tannic acid but not simple esters such as ethyl butyrate. Nelson and Ough (1966) inoculated "Thompson Seedless" grape clusters with *B. cinerea, A. niger, Rhizopus arrhizus, A. flavus,* and *Acetobacter roseus.* After the grapes were thoroughly infected they were blended with 60–90% healthy grapes and converted to wine. The total phenol content by Folin-Denis assay was not changed appreciably except by *A. flavus,* which raised the apparent tannic acid content from that of the control (100–200 mg/liter) to 800–1100 mg/liter. The abnormal phenol produced was not identified, but was associated with an orange color in the wine. Schuller *et al.* (1967) reported having detected traces of aflatoxin, a phenol derivative, in 2 of 33 wines tested. The wines were white wines from European areas where late harvesting may be practiced, and the aflatoxin they reported presumably arose from infection of the grapes with *A. flavus*.

In contrast to the aerobic systems just described, alcoholic yeast fermentations are noted for their ability to perform many reductive reactions. Several examples seem pertinent to our discussion. Mayer (1914) found that fermenting yeast would reduce salicylaldehyde to saligenin. Róna (1914) reduced cinnamaldehyde to cinnamyl alcohol with yeast, and cinnamic acid can be reduced by yeast to cinnamaldehyde (Chen and Peppler, 1956). Pauly and Feuerstein (1929) produced a conversion of coniferaldehyde to coniferyl alcohol in high yield with yeast, and the *p*-glucoside was similarly reduced intact. Lüers and Mengele (1926) and Neuberg and Simon (1926) showed that a series of quinones were all reduced to the hydroquinone form during yeast fermentation. Yeasts combine aromatic aldehydes with pyruvic acid or acetaldehyde, giving, for example, l-*p*-methoxyphenylacetyl carbinol from *p*-anisaldehyde (Neuberg and Liebermann, 1921). Such carbonyl compounds may be reduced to the corresponding secondary alcohol (Neuberg and Nord, 1919) or converted back to the appropriate benzoic acid derivative (Shimazu, 1950). Dakin (1914) used fermenting brewer's yeast to convert phenylglyoxal to *l*-mandelic acid and benzoylcarbinol. Fischer and Wiedemann (1934, 1935) showed that, in a series of cinnamyl ketone analogs, the ketone function could be reduced by yeast, leaving the double bond, or that both could be reduced.

The reducing effects of yeast fermentation are important during wine production in several ways that do not involve phenolic substances directly. Phenols are involved directly in the effects of fermentation on color, but the potential effects of yeasts on phenols of wine would appear to be greater than have yet been defined. It has been known for a long time that active fermentation of an oxidized must or wine will remove oxygen, decrease color, and remove oxidized flavors. As noted by Mathieu (1919), however, the bleaching of anthocyanins is unfavorable in red wines. "Fining" with yeasts freshly removed from an active fermentation is used commercially, particularly in Germany, to lighten color and improve white wines for sparkling-wine production. Schanderl (1951, 1959, 1964) emphasized that the yeast must have been actively fermenting if lowering of the oxidation-reduction potential of the treated wine is to be rapid. The effect is presumably one of reducing quinoid brown substances to a colorless phenolic form, although another factor may be adsorption of polymeric brown substances.

Reichel (1937) reported repeated reduction of anthocyanin or anthocyanidin to a colorless form in the presence of enzyme systems such as liver or yeast alcohol dehydrogenase, and reversion to the colored form in air. Reichel and Buckart (1938) reported, however, that intact yeast reduced anthocyanidin to catechin, which was converted to an amorphous insoluble condensation product. This would appear to be a further argument against the occurrence of anthocyanidins in young wines. Vetsch and Lüthi (1964) showed that anthocyanin color in wine was decreased as much as 38% during malolactic fermentation, and afterward partially regenerated. The color decrease was attributed to the action of *Leuconostoc* species on citric acid, producing reduced nicotinamide adenine dinucleotide, the active reducing agent. The anthocyanin decolorization can be visualized as occurring by reductive addition of hydrogen to the carbonium ion at C-2 to form the same type of colorless intermediate as is postulated as the C3-4 unsaturated dehydration product of a 3,4-flavandiol leucoanthocyanidin. Singleton *et al.* (1964a) found that arresting the fermentation of wines of the port type with aldehydic spirit gave about 20% higher anthocyanin color at 2–3 months. The difference diminished with time. They interpreted their results as effects of a Baeyer aldehyde-phenol condensation, but another factor may be that acetaldehyde would serve as a hydrogen acceptor alternative to anthocyanin as the yeast is halted with a high level of reduced coenzymes.

The presence of "medicinal" or creosotelike flavors is noted rarely in wines. It had always been assumed that these represented contamination. Chen and Peppler (1956) found that a mutant strain of *Saccharomyces cerevisiae* produced more volatile styrene from cinnamaldehyde. Products have recently been found in beer which appear to arise from similar yeast conversions of constituents common to wine also. Brumsted *et al.* (1965) found that a wild strain of yeast would produce phenolic off-flavor traceable to conversion of ferulic acid to a product identified as 4-vinylguaiacol, detectable by taste at 50 mg/liter or more. Vanillin was also converted to 4-methylguaiacol, but this did not seem to contribute to off-taste at the levels present in beer. Halcrow *et al.* (1966) reported that several strains of brewer's yeast and two of *Aerobacter* decarboxylated ferulic acid to 4-ethylguaiacol, the amount of which correlated directly with off-flavors in the beer. Steinke and Paulson (1964) found that whisky grain fermentations produced *p*-ethylphenol, 4-ethylguaiacol, their vinyl analogs, and *p*-methylphenol from *p*-coumaric acid, ferulic acid, and vanillin.

An interesting antioxidant phenolic compound with physiological activity has been isolated from alcoholic extracts of brewer's yeast by Forbes *et al.* (1958), and its structure proved by Meisinger *et al.* (1959) as 2-(6-hydroxy-2-methoxy-3,4-methylenedioxyphenol)-benzofuran:

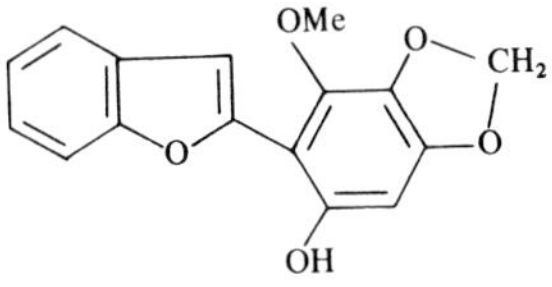

Although only 0.5 to 2 mg was found per pound of dry yeast, at least it illustrated the type of complex phenol metabolite which can be produced by yeast and could presumably occur in wine. The production of tyrosol in wine by the action of yeast has already been discussed. It only need be reiterated that tyrosol can be produced in wine from nonphenolic substrates and thus augment the total rather than simply modify the form of the phenols contained in the wine.

Molds, especially *A. niger* and some *Penicillia*, are, of course, noted for their ability to produce tannase, hydrolyze tannic acid, and use the glucose and gallic acid as carbon sources when suitably adapted (Knudson, 1913a,b). Biddle and Kelley (1912) reported that *S. cerevisiae* would hydrolyze gallotannin and remove optical activity from the solution. Nierenstein (1925) reported that the yeast liberated glucose and some gallic acid from tannic acid during anaerobic fermentation, but the major product was a tannin which was extractable into ethyl acetate and precipitable by gelatin and yet apparently was not *m*-digallic acid. In any case, yeast can apparently produce esterase or tannase capable of hydrolyzing gallates (Kim, 1939). Yeasts will hydrolyze glycosides such as amygdalin, sinigrin, and esculin (Bokorny, 1916; Barnett and Swain, 1956), and may contribute to the liberation of aglycones in wine. However, most of the phenolic glycosides appear to survive alcoholic fermentation without extensive hydrolysis by yeast action.

Carles and Pech (1960) and Carles *et al.* (1961) have theorized that bacterial hydrolysis and degradation of malvidin glucoside to gallic acid, and of other flavonoids to protocatechuic acid, account for the gradual appearance of these acids in wine. However, no satisfactory experimental verification has yet been obtained of bacterial demethylation of malvidin in wine.

Several of the possible reactions of lactic acid bacteria on phenols of wine have already been mentioned, and many of the reactions that yeasts can accomplish can also occur with these bacteria. Wallen *et al.* (1959) note that many microorganisms can convert quinic acid to protocatechuic acid. Phillips *et al.* (1956) and Whiting and Carr (1957) found that lactic acid bacteria in cider and perry (a fermented liquor from pears) hydrolyzed chlorogenic acid and converted quinic or shikimic acids to dihydroshikimic acid. Caffeic acid could be converted to dihydrocaffeic acid, and this decarboxylated to ethyl catechol. Metabolism of *p*-coumaric acid to unidentified products also occurred. Similar reactions in wine would be expected. The decarboxylation of cinnamic acids in grain mashes by bacteria and yeast has been mentioned (Steinke and Paulson, 1964).

## 2. Aging and Related Reactions

At the end of fermentation, wine is usually racked away from the yeast lees and subjected to one or more of a number of possible processes to produce clear wine and stabilize that clarity sufficiently for satisfactory marketing. Depending on the wine type, style, and quality level sought, the wine may be aged for a period in wooden cooperage and, after final preparation for bottling, aged in the bottle at the winery or within distribution channels for another period of greater or lesser length. With some wines, and ordinarily with grape juice, the least change possible would be desired once the product has passed the required minimum of clarification, etc., and been placed in the container in which it is to be sold. The reactions of phenols with which one would be concerned in these instances would be primarily color changes, such as gain of undesirable brownness or loss of desirable red color, with perhaps associated precipitation. Production and marketing practices would be geared to protecting the products from heat, light, oxygen, and other potentially harmful influences and to shortening the time between packaging and consumption.

With other wines, however, desirable effects on quality may be produced by as much as 3 or 4 years of aging in 50-gal oak barrels and, perhaps, as much as 10 years of additional storage in corked glass bottles. During this time the phenols are an important, perhaps the most important, class of constituents. They are important because the proper level of their own properties of color and flavor must be present at the end of the processing and aging, but they are also important as reactants and controllers in reactions which produce or spoil the desired effects of aging. The effects desired may be several, but in general may be characterized as increased complexity in flavor and color, coupled with maintained or improved balance or harmony. The color, for example, will shift from the stark purplish red of the young wine to the more fiery, slightly amber of the usual matured wine, and eventually to the tawny of the long-aged wine. Complexity in flavor is a positive value, as it is in all esthetics (Singleton and Ough, 1962). Aging can produce more harmonious yet more complex flavor, by a slow loss of obtrusive flavors (including astringency) and a gain of additional flavor nuances while retaining much of the originally desirable flavor.

High total phenol, for example, would tend to protect anthocyanins over the long period that may be necessary to allow the formation of a certain new odorous constituent. Direct participation of phenols and phenol oxidation products in the formation of new flavors in wine has been reported, and certainly they are involved in the oxidative reactions which are minimized in juices and some wines, maximized in some sherries and other wines, and held to some desired intermediate level in still other wines. The oxidation and browning reactions are discussed in the next section, but some oxygen contact is obviously inescapable in aging, particularly barrel aging, with its not infrequent transfers and rackings and the ullage resulting from evaporation and diffusion through the wood.

Many reports can be found on observations, relative color analyses, and total phenol analyses before and after processing wine or in a series of similar samples aged for different periods, etc. Unfortunately, what we would really like to know—the specific amount of each individual phenolic substance and how it participates and changes in wine in each type of storage reaction—requires much more detailed knowledge than is now available. The more nearly complete understanding of the qualitative and quantitative phenolic make-up of grapes and wines is too new and too complex to have been integrated into the body of knowledge we would like to have. Some information is available, however, and more rapid development is anticipated. Since many of the components are similar but the storage so much longer, wine studies should help clarify reactions in other foods from plants (and vice versa).

With aging, part of the "tannin" of red wines and some white wines precipitates, along with other substances, particularly potassium bitartrate. New wine is supersaturated with this salt, and its removal to levels that will remain in stable solution is one of the usual operations in winemaking. This tartrate stabilization is often accomplished by nearly freezing the wine and, after the tartrate has crystallized, filtering the precipitate from the cold solution. More or less phenolic substance is removed at the same time, depending partly on the wine's total phenol content and partly on previous treatment (Mestre, 1942; Cassignard, 1963). Three white ports, prepared with a range of total phenol content overlapping the range of red wine, dropped in 10 days at about 20°F from 1047 to 713, 711 to 533, and 186 to 175 mg/liter (Okolelov and Kotlyarenko, 1951). Note that the decrease was 32% for the high-tannin wine and only 6% for the low-phenol wine, indicating that the smaller juice phenols were less susceptible to precipitation than the larger flavonoid condensed tannins from grape solids.

The potassium bitartrate which wine will retain and the rate at which it drops from supersaturation to equilibrium is affected by other constituents. Polyphenols tend to delay precipitation and increase tartrate-holding capacity (Pilone and Berg, 1965; Klenk and Maurer, 1967; Wucherpfennig and Ratzka, 1967). Total phenol content, however, correlates poorly with this effect, and the addition of various seed tannin fractions has little effect (Balakian and Berg, 1968). The complexing of tartrate anions with anthocyanins appears to be the major factor in increasing the tartrate-holding capacity of red wine. Only a specific, though not clearly identified, part of the pigment appears to be effective, and tannins, although ineffective alone, appear to interact with the anthocyanins to increase their effectiveness.

Phenolic and pigmented hazes may form in chilled red wines even when tartrate or other precipitates are not formed. Such a precipitate was found to be, when dried, 8% protein, 24% water-soluble tannin plus undetermined amounts of insoluble tannin, and 11% ash, but only a trace of potassium bitartrate (Rentschler and Schaeppi, 1957). Techniques are available for distinguishing these pigment-tannin precipitates from other precipitates in wine (e.g., Tanner and Vetsch, 1956). The solubility of phenols is greatly affected by their tendency to associate and to hydrate (Johnson *et al.*, 1967). Tannins and phenols have a high degree of mutual solubilization, often appearing more soluble in concentrated than in dilute solution (White *et al.*, 1952). The phenolic hazes that are formed in clear red wine when it is chilled usually redissolve when it is warmed. If such cold-sensitive hazes are removed, they are likely to form again with more aging, and the phenolic precipitates formed during the aging of red wine appear relatively similar in composition and properties to those which can be brought out by chilling (Cassignard, 1963; de Almeida, 1962; Ribéreau-Gayon and Peynaud, 1935).

Red wines may deposit phenolic precipitates when warmed, but usually do not unless warming is prolonged and oxidation is involved (Weger, 1965). White wines, however, when heated (and when aged), frequently develop a turbidity that is predominantly protein but includes a portion (4% in one instance) of phenolic substance (Koch and Bretthauer, 1957). Occasionally the tannin level in white juice may be high enough to cause precipitation (Prillinger, 1952). In such an instance and in the red wines which form chill-hazes of phenolics, adding the proper amount of gelatin will precipitate the nearly insoluble phenolics and leave the wine clarified and stable. The formation of reversible chill-haze and permanent oxidation haze has been extensively studied in beer, and both proteins and phenolics play a role (Clark, 1960; Curtis, 1966, 1968).

The balance in amounts and relative molecular sizes of protein and proteinlike substances versus tannins and oligomeric phenols appears to govern the formation and composition of precipitates of this type in both wine and beer. The protein-tannin reaction and its significance in natural and induced precipitations in wine and beer have been reviewed (Ribéreau-Gayon and Peynaud, 1934–1935; Singleton, 1967a,b). Typically, untreated white wines contain some relatively large and incipiently insoluble proteins that may stay in solution unless denatured by heat, etc. The phenols of young white wine are limited to small, quite soluble molecules, as already discussed. As oxidation or other reactions convert the phenols to larger molecules, however, they can soon bind to and precipitate the protein. White wine, in this sense, is often deficient in tannin capable of precipitating protein. White wines in large containers are observed to clear first near a source of air contact producing phenolic polymers by oxidation, or near the walls of wooden containers furnishing tannin (Ribéreau-Gayon, 1939).

Red wines, on the other hand, are essentially devoid of protein precipitable by tannin and have an excess of relatively large tannins that, as they complex and polymerize, will precipitate. Beer tends to be intermediate in that most of the large protein molecules appear to be coagulated and removed during wort boiling, as do most of the large tannins, either at that time or with the yeast after fermentation. It would appear that a larger phenolic unit would need to form before the surviving beer proteins are caused to precipitate. Obviously the protein-tannin ratio in precipitates from these beverages could range from 0.0 to 1.0, and is likely to be protein-rich in white wine, phenol-rich in red wine, and intermediate in beer. When freshly fermented, racked, and filtered, these beverages are ordinarily only temporarily clear, probably not brilliantly so, and are colloidally metastable. The amount and composition of the precipitates that form with time are, however, so dependent on conditions of aging and the individual lot of wine that the processor is more interested in fining as a means of producing relatively stable and brilliant clarity.

Fining, by the addition of proteins, such as egg white, gelatin, and casein, to cause the precipitation and removal of colloidal solids, is a very old practice for obtaining stable clarity in wine. The fact that proteins may not precipitate from white wine unless tannic acid is added has also been known for a long time, as has the fact that excess astringency can be removed from red wines by gelatin fining or similar treatment (Ribéreau-Gayon and Peynaud, 1934-1935). Recently polyamides, insoluble and soluble poly(vinylpyrrolidinone), and similar synthetic substances have been used instead of proteins. When such materials are used to treat wine they complex with phenolic substances, largely by hydrogen bonding between the phenolic hydroxyls and the carbonyl oxygens in the protein or protein-mimicking polymer (Singleton, 1967a,b). Since astringency in either the leather or flavor sense appears to be in direct relation to protein-binding capacity, it is to be expected that astringent phenolics would be precipitated. Goldstein and Swain (1965) have determined astringency by precipitating tannin by enzyme protein and measuring the effect by residual enzyme activity.

Soluble fining agents, when added in amounts adjusted to produce satisfactory precipitation, usually remove total phenol equal to the weight of the agent added, but shifted either way by a factor of two or so, depending on which component is most limiting (e.g., Singleton, 1967a,b; Ough, 1960; Maveroff, 1949; Ribéreau-Gayon and Peynaud, 1934-1935; Moreau and Vinet, 1937). Color, from either white or red wine, is always partly removed also, but ordinarily cannot be removed completely by this means. It is possible with soluble fining agents, particularly poly(vinylpyrrolidinone), to exceed the dose that causes precipitation, instead producing solubilization of incipient tannin precipitates.

Insoluble agents acting as phenol adsorbents generally have much lower capacity for phenol removal than similar soluble agents, owing to less surface available for interaction with the phenolic molecules (Singleton, 1967a,b; Iwano, 1967; Rossi and Singleton, 1966b; Cantarelli, 1963b; Peri and Cantarelli, 1963; Wucherpfennig and Bretthauer, 1962a,b). Their relative affinity for different phenolics may be quite different also, owing to different manner of application (such as granular material in columns) and the relationships between surface availability, diffusion, and phenol molecular weight (Cantarelli, 1963b). Data by Rossi and Singleton (1966b) agree with previous findings that catechins are not precipitated from wine by soluble proteins, but show that they are readily adsorbed by polyamide or insoluble poly(vinylpyrrolidinone) chromatographic columns. These columns had a higher adsorption capacity for an ethyl-acetate-soluble preparation of oligomeric anthocyanogenic tannins from grape seeds and a considerably smaller affinity for large polymeric tannins from grape seeds. Gelatin in solution, conversely, precipitated slightly more of the large polymers than the small polymer tannins from wine.

Gramshaw (1968), with beer, found that the substances adsorbed on insoluble polyamide or poly(vinylpyrrolidinone) included leucoanthocyanidins that were monomeric, dimeric, trimeric, and mixed catechin-flavandiol oligomers. Poly(vinylpyrrolidinone), compared with the polyamide (nylon-66), appears to have a relatively high capacity to remove catechins from beer or wine, and less to remove large tannins (Gramshaw, 1968; Rossi and Singleton, 1966b). Since the catechins appear strongly related to browning, and the larger tannins to astringency (Rossi and Singleton, 1966a,b), a degree of separate manipulation of these properties appears possible.

The relationship of wine color, particularly the relative brown (420 mμ) and red (520 mμ), to the precipitation and color change during aging and fining is very interesting. Cold storage followed by filtration to remove phenolic precipitates lowers the color of red wine perhaps 10–25%, and this precipitate (or wine bottle-crust) still seems to contain flavylium ion units, as indicated by electrophoresis (de Almeida, 1962; Mestre, 1942; Ribéreau-Gayon and Peynaud, 1935; Ribéreau-Gayon, 1936). If red wine that is a few years old is protected from oxygen and dialyzed, the portion passing through the membrane will be only a little less colored but will remain clear when chilled, whereas the original wine or the polymeric nondialyzable part will cloud (Ribéreau-Gayon and Peynaud, 1935; Ribéreau-Gayon, 1936). The color that will not dialyze and will precipitate, the polymeric anthocyanin-bearing portion, is browner and has a higher 420/520-mμ absorbance ratio than the pigment remaining in solution (Mareca and Escudero, 1966; Mestre, 1942). The visible color shifts from a more tawny color to a more lively ruby. Gelatin fining has similar effects as a rule.

Gelatin lowers color intensity less in young red wines than in older red wines. Ough (1960) removed slightly more tannin and considerably more anthocyanin color with soluble poly(vinylpyrrolidinone) than with gelatin. The two agents, about equally, shifted the dominant wave length slightly toward longer wave lengths (less brown). Pallavicini *et al.* (1963), working with young red wines with a tendency to brown enzymatically, found that decolorization always accompanied fining with poly(vinylpyrrolidinone), but a dose that left 84–90% of the total phenol behind removed only about 1/200 of the red color and decreased the browning tendency. Nylon has similar effects (Klenk and Maurer, 1963). White wines also generally undergo a decrease in brown and browning tendency from treatments of this type.

Bentonite, a montmorillonite clay with some cation-exchange capacity, removes some anthocyanin color from wine. In young wines bentonite removed 10–20% of the color and gelatin 5–10%, but in 6-year-old wines bentonite removed 5–10% and gelatin generally more, up to 20% (Bergeret, 1963). Gelatin tended to shift the color away from brown, whereas bentonite tended to shift toward brown. It appears that bentonite preferentially takes up the free red flavylium salt, and gelatin the yellow (polymerized) form (Bergeret, 1963; Ferenczi, 1966).

Bentonite takes up anthocyanin monoglucosides more than diglucosides (Kain, 1967). It is, however, an active adsorbent that when properly swelled, has a high effective surface and removes at least some of the nonionic color from white wines and improves the stability of wines to phenolic precipitation (Weger, 1965). Bentonite is an excellent adsorbent for protein removal from wine and, as such, may influence phenolic composition by adsorbing tannin secondarily on its protein-coated surface and by removing polyphenol oxidase and other enzymes that would attack phenols in must (Saywell, 1934; Rankine, 1962; Tagunkov, 1966).

Ion-exchange resins may remove anthocyanins by cation-exchange reactions, but many resins, including anionic and nonionic ones, remove some of the phenols, including brown and red pigments, by adsorptive mechanisms. Grasso (1952) removed 48% of the color of red wine and 65% of white wine color with columns of the phenolic resin Duolite S-30. The same resin has considerable ability to adsorb the tannins of white wine (Colagrande and Corradini, 1967). Peterson and Caputi (1967) found that white wine polyphenols, as an example, were reduced from 220 to 130 mg/liter by Duolite C-25 cation exchange with sodium ion, to 80 mg/liter with hydrogen-ion exchange, and to 55 mg/liter with hydrogen exchange followed by Duolite A-7 hydroxyl anion-exchange acidified immediately afterward. Of course, phenols are very subject to oxidation in alkaline solution. With a sulfonated polystyrene resin in the sodium ion form, red wines treated at 100 liter/kg of resin in a column lost 3 to 20% of their total phenol (Begunova *et al.*, 1965).

To complete the discussion of adsorbents and fining agents, brief mention must be made of activated carbon. Owing to its partial graphite-leaflet structure, active carbon has a high affinity for benzene rings, and phenols constitute the predominant benzene derivatives in grape and most other food products from plants. Activated carbon is, however, a very active adsorbent, and therefore will adsorb desirable odors and other relatively nonpolar substances from wine when enough is added to make major reductions in the phenolic composition (Singleton, 1964). Active charcoal can adsorb about 10–30% of its own weight, and about 80 lb/1000 gal is required to decolorize a red wine fully, perhaps 20–40 lb/1000 gal to remove tannin from a red wine for pharmaceutical use (Ross, 1942), and under 9 lb/1000 gal to remove slight blemishes in haze or color, a more logical practice from a modern wine viewpoint. Carbons are very different in capacity, and somewhat different in specificity for red versus brown color, odors, etc. (Singleton and Draper, 1962). In general, brown color is adsorbed more readily than red color, but the relative activity of a series of carbons in decolorizing red wines of widely differing types is essentially constant. An example of a situation calling for use of charcoal would be a faint anthocyanin pink in a "Pinot noir" must to be vinified "en blanc" to sparkling wine (Geoffroy and Perin, 1966).

The effect of storage on anthocyanins and their color has been studied in grape juice and a number of other red, tart fruit juices, as well as in wine. Nebesky *et al.* (1949) found grape juice anthocyanins more stable to light than isolated currant or strawberry pigment. Low oxygen and low storage temperature are important for anthocyanin retention. Sastry and Tischer (1952a,b) found that the tannin of grape juice conferred important protective action on anthocyanin against the destructive effect of ultraviolet irradiation. Although fairly heat-stable on a short-time high-temperature basis, over an hour at 250°F produced considerably lower yield of anthocyanin in juice. Ponting *et al.* (1960) studied heat degradation of grape juice anthocyanin and found a linear relation between the logarithm of the time and temperature necessary for a 10% destruction of anthocyanin color. At 80°C about 80 min was required to produce the 10% decrease—a difference large enough to be visible in a 1-cm layer of the juice. Heating gave less browning in proportion to red color loss than did usual storage.

Studies of degradation of anthocyanins in other fruit juices (e.g., Dalal and Salunkhe, 1964; Daravingas and Cain, 1965, 1968) generally agree with and extend the data available from grape juice. Hydroxymethylfurfural and furfural hasten the deterioration of the pigments. The increased production of these compounds from sugars is one of the reasons that sugar and elevated storage temperature lower pigment retention. The $Q_{10}$ for pigment destruction appears to be about 2–3. Juice pigment is more stable than isolated pigment in buffer at juice pH, but the difference disappears at lower pH. The maximum stability for cyanidin-3-diglucoside is about pH 1.8 (Daravingas and Cain, 1968). The pigment is presumably more stable as pH is lowered because the carbinol base is less stable than the flavylium salt, but slightly less stable again below pH 1.8 owing to hydrolysis to the unstable aglycone. Cyanidin is very unstable, with only 20% retained after 1 day under nitrogen at 50°C and pH 3.25, and 10% under oxygen. The glycoside under the same conditions retained about 20% of its color for 14 days.

Hydrolysis of anthocyanins during wine aging has been claimed to be the source of the precipitate of phenolic pigment in old wine, the anthocyanidin being less soluble than the glycoside (Ferré and Michel, 1937). However, this conclusion was based upon noting an increase (maximum

400–600 mg/liter) in a reducing substance interpreted as sugar during aging. Since acidic hydrolysis did not seem to be able to account for the hydrolysis (e.g., more acid wines were generally slower to precipitate pigment), an enzyme was thought responsible. It is not clear that carbohydrate contribution from wooden-barrel storage was eliminated as a possible source of reducing sugar. The addition to red musts of acids, including sulfuric, gave more anthocyanin color without affecting the total phenol content (Ough *et al.*, 1967). The 525-m$\mu$ absorbance increase, measured for both at pH 3.3, was 23% for 1.5 g added $H_2SO_4$/liter, compared with that of untreated wine. After 8 months of aging, the color retention remained greater in the acid-treated samples than in the controls.

The hydrolysis of glycosides in general, and of flavonoids in particular, is greatly influenced by stereochemical factors (Feather and Harris, 1965; Harborne, 1965). Therefore, that flavonols are found as free aglycones in wine does not necessarily require similarly important hydrolysis of anthocyanins during aging (Ribéreau-Gayon, 1963d). Kaempferol-3-glycoside has a half-life of 1.8 min at 100°C in 2 *N* HCl, while cyanidin-3-glucoside has 15 min (Harborne, 1965). Time for complete hydrolysis of cyanidin-3-glucoside was estimated to be 30–45 min. Assuming a temperature coefficient of 2–3, it would require about 3 to 90 days to hydrolyze half of the cyanidin-3-glucoside in 2 *N* HCl at a fairly warm wine storage temperature, 68°F. No doubt the much lower acidity of wine would also be a large factor, but if the effect remained proportional to that with flavonols it seems that anthocyanin hydrolysis can probably occur during the long times involved in storage and aging. More study is needed.

Accumulation of anthocyanidins in grape juice or wine is another matter. All of the studies already mentioned indicate that, once free, the natural anthocyanidins are very unlikely to exist unchanged for any length of time, particularly if oxidation is possible. In fact, a fungal enzyme with the ability to hydrolyze anthocyanins to aglycones, which then spontaneously become colorless without appreciable browning, has been recommended for partial decolorization of berry wines that are too dark (Huang, 1955, 1956; Yang and Steele, 1958).

Another type of degradation that has been postulated during the aging of red wines is removal of the methyl groups from methylated anthocyanins such as the malvidin derivatives. As wine aged for 2–3 years in the cask, the relative methoxyl content of the pigment fraction declined from 8–10% to 3–4% (Genevois, 1951; Ribéreau-Gayon, 1933). It appears, however, that preferential reaction and precipitation of the less soluble, highly methylated pigments could account for the observation without invoking splitting of the ether linkage (Ribéreau-Gayon, 1963d). Vogt (1958) reports that the total methoxyl content of the wine plus deposit in very old wine seems unchanged. Durmishidze and Khachidze (1952) found that pigment methoxyl by analysis correlated with colorimetric pigment content in a young teinturier wine, 1280 versus 1250 mg/liter, but a comparable wine 20 years old retained 545 mg/liter by methoxyl estimation, and only 75 mg/liter by color. Robinson *et al.* (1966b) found malvidin and peonidin derivatives the most stable to color loss when added to white wine, and delphinidin derivatives least, among 10 anthocyanins tested.

The anthocyanin color intensity decreases with sufficiently long aging in barrels or bottles, and the hue tends to become brick red regardless of original color shade (Berg and Marsh, 1956), but the picture is complicated by the fact that the intensity often passes through a minimum and then rises for a time in the early stages of aging. The time sequence evidently depends upon the wine and the storage conditions. Valuiko *et al.* (1964) found the minimum apparent anthocyanin content at 3 months, followed by recovery to about the original level, whereas total phenol trended steadily downward. Under anaerobic storage, however, anthocyanin content apparently exceeded the original content after one year of aging (Valuiko, 1965b). This increase was attributed to anthocyanidin production from leucoanthocyanidin, but this is doubted for various reasons already discussed and because the leucoanthocyanidin content of white wine ordinarily stays relatively constant during storage without conversion to pink wine (e.g., Nachkov, 1967).

One source of decreased visible anthocyanin which can cause reappearance with time is the interaction of $SO_2$ with the pigment. With this agent added, anthocyanin release from the skins is aided but maximum color occurs before the end of fermentation, and the color exhibited by red juice or wine bears an inverse relation to the amount of $SO_2$ added (Berg and Akiyoshi, 1958, 1962; Harvalia, 1965). The nature of the reaction involved has been clarified by Jurd (1964a). It does not involve hydrolysis of the glycoside, reduction, or a chalcone intermediate, as previously postulated. Instead, the negative bisulfite ion evidently adds directly to the positive flavylium ion at position 2 (or perhaps 4), producing a colorless flavene sulfonic acid analogous to the carbinol base. The reaction is reversible, and the position of equilibrium depends on the concentrations of the flavylium ions, the bisulfite ions, and the hydrogen ions. Lowering the pH shifts the equilibrium to the colored free flavylium form because the bisulfite ion concentration is lowered and has more effect than the opposed increase of the flavylium ion. Any lowering of the effective bisulfite ion concentration in wine at a given pH, such as by oxidation or by complexing with acetaldehyde, will increase visible anthocyanin color so long as no other reaction occurs.

Berg (1963) and Berg and Akiyoshi (1962) have considered the total red pigment in red wine as composed of anthocyanins in reversibly colorless forms (carbinol bases, flavene-sulfonic acid), anthocyanins in associated or polymeric forms (brick red but not pH-responsive), and pH-responsive free anthocyanins in the colored form. The leucoanthocyanidins and the colorless and brown degradation products of anthocyanins and other phenols appear to participate in forming the polymerized pigment and, of course, in regulating the amount of unchanged phenols and anthocyanins remaining at any one time. Only about 14% of the anthocyanin was in a colored form in freshly prepared wines, about 18% at 3 months of aging, and about 41% after 52–64 months. Fermentation of 9 wines produced an average decrease of 18% in the total anthocyanins, of 20% in the colored forms, of 35% in visible color intensity, and of 34% in total potential pigment including leucoanthocyanins. Three months of aging produced relatively little further change in visible color, but a 28% decrease in total anthocyanins was countered by a 32% increase in the fraction which was colored and the total potential pigment decreased 25%. The increase in the colored fraction during aging was the result of association or polymerization which evidently held the anthocyans in colored (if somewhat brick-red) but unreactive form. The results also are considered to cast doubt on the hydrolysis of anthocyanins in wine during aging and to suggest some degree of reversibility of the polymerization of the red and brown pigments.

That anthocyanidins became incorporated into phenolic polymers, even polymers produced from their own degradation products, can be inferred from plots of anthocyanidin color loss with time in juice and in model solution. Daravingas and Cain (1968) show, for example, that the anthocyanidin remaining in a solution drops rapidly and then appears to remain constant at least for the limited time tested. The constant level was about 20% of the original color under nitrogen and 10% under oxygen. These data suggest that some of the flavylium ion is incorporated in its colored form into a reaction product, presumably a polymer, which then protects the "anthocyanidin" from further reaction. That oxygen lowers the amount of such protected product could be attributed to the greater destruction of the anthocyanidin in competition with this incorporation into polymers. Similar arguments can be made for the anthocyanins because they usually appear to approach some finite red color level asymptotically with time, rather than zero anthocyanin color. The apparent stable color level is affected by reaction conditions.

Somers (1966b, 1967a,b) separated by gel filtration the polymeric tannin fraction from a 1964 wine. This fraction was red and accounted for roughly half of the total absorbance at 535–540 mμ. The color was responsive to pH changes, although much less so than free anthocyanins. Acid hydrolysis released several anthocyanidins from this polymer and converted some leucocyanidin present. Sugar equivalent to about 8% of the polymer was released by hydrolysis, indicating that the anthocyanins were incorporated as glucosides. As noted previously, the molecular weight was predominantly 2000–5000 but was estimated to range up to 50,000.

The red polymers may even occur in grapes, or at least may form very readily during isolation. This polymeric anthocyanin-bearing tannin or phlobaphene increases in amount as wine ages, and the free anthocyanin content drops essentially to zero by the end of barrel aging, or early in bottle aging (Ribéreau-Gayon and Stonestreet, 1965; Ribéreau-Gayon and Nedeltchev, 1965). The nature of the polymer and the linkage involved can only be conjectured at present. Since oxidation appears to be involved and brown polymers appear similar except for color, possible structures are considered briefly in the next section. These results, however, explain a problem that has been troublesome: that an older wine which still appears quite red usually cannot be shown to contain anthocyanins. The pigment remains at the origin in the common solvents during paper chromatography. The wine's spectrum may still show a maximum near 520 mμ and a minimum near 420 mμ, and hydrolysis in boiling hydrochloric acid may still give glucose and most of the expected aglycones (Somers, 1966a, 1967a,b; Ribéreau-Gayon, 1957). It appears that these polymers which seem to form so easily and are such a nuisance in anthocyanin research may be very important in red color survival in wine, and probably in other products.

The polymers that form with time in wine are evidently primarily phenol-derived. The formation of an anthocyanin-bearing polymer which is appreciably red rather than brown and is of such a size that it remains in solution in high amounts would appear to depend, under good conditions of wine storage, upon the presence of relatively small reactive tannins and the ratio of anthocyanins to tannins. Certain grape varieties are noted for relatively poor, and other for relatively good, "stability" of red wine color (Olmo, 1952; Amerine and Winkler, 1947; Bernon and Nègre, 1939). Generally the teinturier varieties, although high in original pigment, are subject to great losses in red wine color (Amerine and Winkler, 1947). The varieties that have relatively stable color, such as "Petite Sirah" and "Ruby Cabernet" (Amerine and Winkler, 1947; Olmo, 1952), tend to have relatively high tannin and astringency, also. The fact that teinturiers generally have low total phenol other than anthocyanins has been obscured by the fact that their very high pigment content contributes to the total phenol analysis. Two young dark red wines, each with a total phenol content of 2000 mg/liter, might have 1300 mg/liter of pigment in one case and 500 mg/liter in the other. Blending the high-pigment wine with the lower would probably still give lowered stability, owing to insufficient tannin oligomers to complex with the pigment. "Pinot noir," at least in California, is noted for color instability and lacks acylated pigments (Webb, 1964). It is often relatively low in tannin also. The unstable blue precipitates from "Concord" juice involve complexes of acylated pigments and of metals and pigments (Ingalsbe *et al.*, 1963). Further study of these relationships promises important advances in processing of grape juices and wines and their storage stability.

Aging, particularly in wooden barrels, may have other effects on phenolics in wine. The qualitative and a few quantitative aspects have already been considered, but a few more comments seem required. Wine and brandy will evaporate through the wood of the barrel or vat in which they are aged. The rate of this evaporation depends on surface-to-volume ratio of the container, temperature, relative humidity, etc. All poorly diffusible and nonvolatile materials in the wine would be concentrated in proportion to the evaporative loss, which in small cooperage is about 3–4% per year (e.g., Kayser, 1897; Mathieu, 1907; Wüstenfeld and Luckow, 1933; Valaer, 1939). Total phenol content would then be expected to rise 10–15% in a 3–4 year period of barrel aging, provided no precipitation occurred.

The wood of the container contributes more or less extractives to its contents, depending on its previous treatment. Obviously, a large container has less surface per unit of capacity, and therefore contributes less extract per gallon of contents than a small container. Equally obviously, a short sojourn in the container, or previous extraction by other batches, will reduce the amount contributed. Singleton and Draper (1961) found that a sample of commercial, mixed white oak heartwood chips contained 71.5 g of solids exhaustively extractable by water and alcohol per kg dry wood. In the wine range of alcohol concentrations, about 35–40 g/kg were extracted in 16 days, and this extracted solid assayed 43% phenol calculated as gallic acid or tannic acid. Their data showed that a flavor change from chip extract in port wine was detectable before a significant change in total phenol content. However, the surface of a new, untreated 50-gallon barrel can be estimated from these figures to contribute roughly 100 mg of gallic-acid-equivalent phenols per liter of wine per millimeter of wood depth extracted.

Cotea and Tirdea (1959) reported that a white wine with 327 mg phenols per liter rose to 1392 mg/liter in an untreated barrel, and only to 356 mg/liter in an old (used) barrel. In slightly larger 225-liter Bordeaux barrels, 4 years of storage gave about 200 mg of tannin to each liter of wine, most of it in the first year (Ribéreau-Gayon and Peynaud, 1934-1935). Coudon and Pacottet (1901b) calculate that a new barrel would contribute about 400 mg/liter of phenol to wine in 3 years of aging, and found that a well-used barrel added 42 mg/liter in 1 month. Individual barrels produced differences of about 2-fold. Nelson and Wheeler (1939) reported periodic analyses for 6 wines held 2 years in vats, but did not give complete details of storage conditions. A few wines appeared to increase in tannin, but all but one were lower at the end of the period than just after fermentation. Three red wines dropped only 10% or less; 2 white or rosé wines dropped about 25% and 1 increased about the same amount. It appears that precipitation reactions tend to outweigh evaporation and extraction effects in wine aged for appreciable time unless, perhaps, it is a white wine in a new barrel.

Comparable figures are available for the tannin (total phenol) content of distilled beverages which obtain all of their phenolic substances from the wooden container. Tolbert *et al.* (1943) found 30–492 mg/liter tannin in a series of commercial brandies, and presented experimental data indicating that approximately 2.8 mg of total phenol as tannic acid was extracted per square centimeter of barrel surface, or roughly 280 mg/liter of brandy in 2 years in a new 50-gallon barrel. Similar levels are indicated from other reports on brandy, whisky, and other spirits, although reuse of barrels and the fact that charred barrels are used for bourbon whisky affect the results (Valaer, 1941; Liebmann and Scherl, 1949; Barikyan, 1960; Otsuka *et al.*, 1965).

The components that may be extracted from barrels during aging include sugars at about 1 g/liter calculated as glucose but including xylose, arabinose, fructose, and rhamnose (Otsuka *et al.*, 1963; Skurikhin, 1957). The important phenolic components appear to be hydrolyzable tannins and lignin. Brandies aged 2–20 years had total phenols by gravimetric methods of about 160–890 mg/liter, including tannin by methoxyl values of about 50–170 mg/liter (Skurikhin, 1959). Wine levels of alcohol will extract less lignin and more tannin from barrel oak in proportion to each other, and about 60% as many total solids as brandy (Skurikhin, 1966; Singleton and Draper, 1961). Ethanolysis and oxidation appear to be involved in solubilization of lignin from wood and conversion of lignin to aromatic aldehydes such as vanillin (Baldwin *et al.*, 1967; Egorov and Egofarova, 1965; Underwood and Filipic, 1964; Skurikhin, 1962). The amount of vanillin and vanillinlike substances in brandy aged more than 4 years is about 70–90 mg/liter, and sherry aged in wood has enough to affect the flavor favorably (Skurikhin, 1962, 1967).

### 3. Oxidation, Browning, and Related Aspects

Oxidation and its control have been primary considerations in winemaking, processing, and aging since even before Pasteur's classic demonstrations that, with a little oxygen, red wines matured normally but, with very little more, they became cloudy and brown. Wines may contain readily oxidized substances such as ascorbic acid, perhaps other endiols, added sulfur dioxide, traces of ferrous ion, etc. When the amounts of these substances found in wine are considered, however, it appears that the bulk of the readily oxidized substance of red wine is clearly included in the phenolic substances. The phenols of white wine are also important in oxidation, and the fact that these wines are naturally changed so much more readily and drastically by limited oxidation is evidently mainly because of their lower content of phenolic substances. Phenols of fruits are commonly determined, of course, by methods that depend upon their greater ease of oxidation than other major juice constituents, such as reducing sugars, hydroxyacids, etc. The ability of most wines to take up appreciable oxygen can be completely removed by removing the phenols by selective adsorption or precipitation (Ribéreau-Gayon, 1933).

Grapes contain polyphenol oxidase, and oxidative enzymatic browning occurs in the damaged berry or very rapidly in grape juice and must. Nonenzymatic browning occurs in wines and raisins and may be deliberately sought or not, depending on the specific product. The subjects of oxidations and brownings in foods are broad and complicated. The general properties of these reactions and some specific reaction systems are reasonably well known, but the interaction of substrates and the complexity of the reaction products have prevented complete understanding in systems as heterogeneous as wine, and more attention has been paid to nonphenolic nonenzymatic browning (Stadtman, 1948; Joslyn and Ponting, 1951; Hodge, 1953; Mapson and Swain, 1961; Corse, 1965; Mason, 1965).

The rapid uptake of oxygen and associated browning of freshly crushed grapes is primarily enzymatic and can be greatly decreased by heating to destroy the enzymes (Hussein and Cruess, 1940). The amount of oxygen consumed rapidly by 1 kg of grapes very shortly after crushing in a closed flask is about 20–200 mg (Nilov, 1964). The enzyme mainly responsible is a polyphenol oxidase, and a number of its properties have been reviewed (Casas, 1967; Deibner and Rifai, 1963a,b). The grape enzyme seems very similar to other plant phenolases (Pridham, 1963). It is a copper protein (Bayer *et al.*, 1957) readily inhibited by sulfur dioxide and adsorbed by protein precipitants such as bentonite. The optimum pH is above the pH of grape juice, about 5.5. The browning associated with its action on catechol is prevented by coupled oxidation of ascorbic acid, which is not oxidized alone. This evidently results from reduction by ascorbic acid of the quinones, which cause the browning, as fast as the enzyme forms them. This helps explain why the naturally low ascorbic acid content of grapes seldom survives into wine but may if the must is heavily sulfited (Ournac, 1966).

The enzyme has high oxidizing activity for *o*-dihydroxybenzene derivatives, but only about 1/10 as much activity with *p*-diphenols (Tagunkov, 1966). The "cresolase" activity has been stated to be low, but monophenols are found to be oxidized.

Although the enzyme is active in 10–20% alcoholic solution (Hussein and Cruess, 1940), it and other enzymes are not usually detectable in wines after fermentation (e.g., Pallavicini, 1966). A major reason for this loss in enzymes is that they are associated with the particulate matter in grape must, and phenolase is reduced to zero or a few percent of the original content as juice is centrifuged or filtered (Diemair *et al.*, 1960). The phenolase is not inactive in the precipitate, however. If grape solids are left in wine, phenolase activity has been reported for at least 18 months in those solids but not in the wine (Durmishidze, 1950).

Phenolase and the other enzymes are evidently associated with grape solids, however, only because they are precipitated by the tannin released at the same time the grapes are crushed (Bell *et al.*, 1965; Pridham, 1963; Porter *et al.*, 1961). If, during disruption of the cells, a substance, such as a polyamide, is added which will complex with the tannins, polyphenolase and other enzymes may be found in the solution (Pridham, 1963). To the extent that such agents remove phenols, of course, browning and oxygen uptake are also inhibited by lowered substrate concentration. Soluble polyphenol oxidase from grape leaves and berries has been produced by homogenizing the tissue with caffeine, by Drawert and Gebbing (1967a,b), or with nylon, by Cassignard (Anon., 1966a). The total activity was more than half again as large after such treatment, showing that the adsorbed enzyme is reduced in activity. For this reason some of the data upon relative concentration in various tissues or varieties may be questioned; the apparent activity differences might reflect differences in tannin rather than in enzyme content.

There nevertheless appear to be varietal differences. Cassignard (Anon., 1966a) found that a white variant of a red variety had more than

4 times as much phenolase activity as the normal red counterpart, whereas other white grapes were above and below red varieties. Poux (1966, Anon., 1966a) found relative phenolase activities of berry color variants of "Carignane" to be 54 for black, 24 for pink, and 10 for white. Similarly, "Grenache" berry phenolase activities were 21 for black, 13 for pink, and 30 for white. Grape phenolase apparently participates in the browning of sun-dried raisins, but more activity survives in shade-dried raisins (Hussein *et al.*, 1942). Radler (1964) found that dipping solutions which speed drying give lighter-colored raisins, presumably because of nonspecific phenolase inhibition by high sugar content. He also found a mutant grape which contained no phenolase, but normal phenol content, and produced light-colored raisins. Of course, molds, particularly *Botrytis*, produce phenol oxidases, as previously mentioned, that appear to carry over into wine.

The grape enzyme is usually localized in the skin and the solids of the pulp, although varieties differ (Pallavicini, 1967). Relative activity per unit of fresh weight was about 15 in skins, 1.5 in pulp plus juice, and 0.5 in juice (Ivanov, 1967). Cassignard (Anon., 1966a) found that seeds of "Semillon" had high phenolase activity, but "Merlot blanc" did not.

The question is not yet clearly answered whether phenolase activity is ever important in further oxidative browning of wine beyond the period of crushing and fermenting. That it is not required for this reaction is indicated by the oxidation and browning that may occur in bentonite-fined or pasteurized wines. Of course, the brown color already enzymatically produced in the earlier stages may be a problem in the wine. The products of oxidation of phenols include polymers that inactivate and precipitate phenolase or other proteins, so that the method of storage which best preserves phenolase includes protection from air as well as a low temperature. Under these conditions, perhaps half of the activity may be retained for 5 months (Casas, 1967) if the product is not fined. Even here there is some question of nonenzymatic oxidation and browning being confused with enzymatic activity, because the products and reactants are difficult if not impossible to distinguish in wine.

Drawert and Gebbing (1967a) found that the reactivity of grape phenolase was similar to that of potato with several substrates, but was highly active with *p*-coumaric acid whereas the potato enzyme was inactive. The grape phenolase appeared slightly less active with *d*-catechin, dopa, pyrogallol, and gallic acid, and slightly more active with protocatechuic acid. They report that the grape enzyme plus chlorogenic acid or *p*-coumaric acid gave a violet color when proline was added. This is very interesting because proline is the most common "amino" acid in wine. "Pinking" of white wines has been observed occasionally. The color has been assumed to be from leucoanthocyanin conversion, but this new observation is worthy of study in this connection.

Rossi and Singleton (1966a) found that spraying particulate grape phenolase on paper chromatograms of grape-seed phenols gave very rapid browning with *d*-catechin, *l*-epicatechin, and *l*-epicatechin gallate, but much slower browning with oligomeric and polymeric flavolans. In solution the catechin group was also more heavily oxidized, as indicated by the production of about four times the proportion of dark-brown precipitate from the catechin fraction as from the other fractions. The catechin brown products contributed absorbance at 440 m$\mu$ that was 6–20 times as much in relation to the amount of phenol left unprecipitated as that of other fractions tested. Since caffeic acid derivatives are generally known for producing low browning when oxidized and this appears true for wine (Berg, 1953a,b), the catechins appear to be the most significant, though not the only, potential source of enzymatic browning in wines.

Nonenzymatic oxidation and browning of wine is sufficiently rapid that wines allowed free access to air rapidly become oxidized, maderized (somewhat like baked or aged sherry or madeira wine in flavor and color), and spoiled from the standpoint of a table wine or port. These types of wines must be protected from this excessive oxidation, and, if completely sealed away from oxygen, wine or grape juice is preserved with minimal change even when other storage conditions are somewhat adverse (Tressler and Pederson, 1936; Pederson *et al.*, 1941). For example, total phenol in samples of a port wine stored 6 months was 1330–1640 mg/liter in completely full hot-filled bottles whether stored cool, hot, or in sunlight. Comparable treatment of bottles ¼ empty and subject to oxidation gave 1270–1180 mg/liter, off color, and other changes.

Some oxidation seems necessary, however, for many of the best reputed, most complex, aged wines. With red wines particularly, a certain (though variable and not yet well defined) level of oxidation appears beneficial. Although excellent rosé and white table wines are made with the lowest oxygen contact commercially possible (very low indeed), in some styles of these wines a low level of oxidation may be desirable and help produce diversity and complexity of products. Of course, in these light wines, oxidation is particularly likely to produce undesirable appearance and flavor changes. Maderization is deliberate in still other wines of the dessert and appetizer class.

Ribéreau-Gayon (1933) reviewed wine oxidation and made one of the first broad and thorough studies of it. From his data a wine would be expected to absorb about 30 mg $O_2$/liter per year in barrel storage under average conditions, plus an additional small amount picked up during transfers from container to container, etc. This level represents 3–4 times the amount of oxygen in wine saturated with air. Considering that red wines may benefit from 3–4 years in barrels, it appears that acceptable total oxygen consumption could be over 120 mg/liter. Even in the corked bottle there is some evidence for oxygen penetration—about 1 mg/liter annually—and the headspace and bottling may allow considerably more unless special techniques are used. Joslyn (1935) reports 40–170 mg $O_2$/liter absorbed in the total period of cask aging of red wines, and finds that faster aging may be produced by warmer storage and the introduction of 30–60 mg $O_2$/liter. Sherry production by baking (several weeks at 130°F) requires a certain minimum of oxygen, estimated to be about 55 mg $O_2$/liter (Joslyn, 1935; Singleton *et al.*, 1964b). Preobrazhenskii and Kharin (1965) recommended 200–300 mg $O_2$/liter for such wine over a 2-month period.

Dubaquie (1925) emphasized that the optimum degree of aging and oxidation varies for different wines, and that under some conditions prolonged reaction with 10 mg $O_2$/liter was detrimental. The wines less subject to damage by oxidation were those high in tannin, and a certain unidentified fraction of the total pigment plus tannin was most affected by oxygen. Laborde (1918) found that different wines would absorb very different amounts of oxygen, 3–30 mg $O_2$/liter in some experiments under constant conditions, but that relatively constant amounts were absorbed in relation to the amount of precipitate formed—about 20 mg/g precipitate. The precipitate was believed to be a reaction product between acetaldehyde and the tannins. In other tests, Laborde (1897, 1908-1919) reported that the total absorption of oxygen by wines subjected to "casse"—mold-enzyme-oxidized precipitation—was 70–160 mg/liter, with $CO_2$ production equivalent to about half the volume of the oxygen used. Healthy wine gave only slightly less total uptake, but less $CO_2$ was produced. Clarens (1921) showed that the uptake of oxygen by a wine and by the tannins precipitable from it by zinc were about the same under alkaline conditions.

Rossi and Singleton (1966a) found that the oxygen uptake in different wines was approximately proportional to the total phenol content, but was catalyzed by light, involved $CO_2$ release, and was too slow and prolonged for convenient Warburg manometry. The oxygen taken up by the wines in a given period was higher with higher pH, and oxidation appeared to be related to proportion of phenolate ion present. Red wines that were very alkaline rapidly took up roughly 400 mg $O_2$/liter, and white wines took up 70 mg $O_2$/liter. The oxygen consumed per unit phenol appeared greater for white wines than for red. All phenolic fractions from grape seeds took up considerable oxygen under alkaline conditions, and the more polymeric forms took up more in proportion to their color value in Folin-Ciocalteu assay. Color production by nonenzymatic oxidative browning of these seed fractions in carbon-decolorized (acidic) wine was great for the catechin fraction, but was scarcely detectable after 39 days for the oligomeric leucoanthocyanin fraction or the polymeric condensed tannin fraction.

Durmishidze (1955) showed that malvidin was not oxidized appreciably in comparison with *d*-catechin in the presence of phenolase, as would be predicted from its methylated structure, but in the presence of hydrogen peroxide it did take up oxygen much more rapidly than did catechin. He corroborated the observations, though not the conclusions, of previous Russian work claiming to show a thermostable peroxidase in wine and an oxygen uptake of wine primarily dependent on a readily extracted aglycone fraction (presumably including catechins and the smaller tannins).

There have been many studies of oxidation-reduction potentials in grapes and wine (e.g., Casas, 1965; Mareca, 1962). In general it appears that wines are very weakly poised, and no high concentration is normally present of any substance that is in readily reversed oxidation-reduction equilibrium. Catalytic amounts of ferrous-ferric, cuprous-cupric, dihydroxymaleic-diketosuccinic, ascorbic-dehydroascorbic, and possibly other systems exist or have been postulated in wine, but their content is evidently very low as a rule. Their main function appears to be in serving as transfer and regulatory systems for the major process of oxygen consumption which is phenol-quinone based. It appears to us that the more highly reductive endiols, etc., would be consumed at wine pH before appreciable quinone could form (Rodopulo, 1953; Tarr and Cooke, 1950). As quinones form they are largely converted to products of further oxidation or polymerization, and in wine do not appear to develop a significant poising level of reversible hydroquinone-quinone equilibrium. Oxidation-reduction titrations (Monties, 1967; Ribéreau-Gayon and Gardrat, 1957) verify that the apparent potentials have a polyphenol relationship, that the curves produced by catechin and chlorogenic acid are similar to those of wine, and that young wines have more oxidizable phenols than old wines. There does not seem to be reproducible "steps" in the oxidation-reduction titration curves by which one can distinguish individual oxidation or reduction substrates in wine.

The browning of white wine, as a process to be avoided, has been studied by many workers. Berg (1953a,b) and Berg and Akiyoshi (1956a) showed it to be apparently nonenzymatic, developed tests for browning tendency in white wines, showed important varietal differences, indicated correlation of the varietal tendency to brown with ultraviolet absorption spectra of the wine, and showed that total phenol content correlated inconsistently with tendency to brown. The indication from their work has been discussed: that browning in white wine is opposed by a higher proportion of caffeic acid derivatives with relatively low other phenols. This correlates with the adaption from beer findings of polyamide as a removal agent for not only the brown pigment but enough of the phenolic precursors to delay further browning (de Villiers, 1961). The use of proteins, particularly casein, for similar purpose had been known for some time. He (de Villiers, 1961) showed that leucoanthocyanins were removed quantitatively from the treated wines, and that the treatment reduced absorbance at 275 mμ slightly more than that at 325 mμ, although paper chromatograms of the phenols did not seem changed and caffeic acid was elutable from the nylon.

A number of other workers have studied the effect of phenol-complexing adsorbents on brown formation in white wine. Removal of part of the phenol by gelatin, polyamide, carbon, or other agent decreases browning without eliminating it. Specifically, critical changes in the qualitative phenol content are difficult to identify, although the most adsorbable fraction of white wine includes catechins as well as leucoanthocyanins (Wucherpfennig and Bretthauer, 1962a,b; Wucherpfennig and Kleinknecht, 1965; Fuller and Berg, 1965; Pallotta *et al.*, 1966a,b; Iwano, 1967).

Caputi and Peterson (1965) found that wines from some varieties browned only with oxygen whereas others browned at room temperature when oxygen was excluded. Phenol adsorbents sometimes increased as well as decreased the tendency of different white wines to brown, and accelerated storage tests did not always correlate with long-term normal storage browning. Two types of browning seem to be involved in white wines: oxidative polymerization of phenols; and nonoxidative reactions possibly involving nitrogen compounds and carbonyl compounds (Peterson and Caputi, 1967).

Anthocyanins, catechins, and leucoanthocyanins are all extremely reactive in oxidation and browning of the type occurring in wine (Cantarelli, 1963a). The synthetic polyamide and poly(vinylpyrrolidinone) polymers and proteins are about equally effective in decolorizing white wines before and after maderization, but carbons are relatively more effective before than after (Cantarelli, 1963a), suggesting that the brown polymers are too large for many of the pores of carbon (Singleton, 1964). Other workers (Cantarelli and Peri, 1964b; Cantarelli, 1966) have studied the protection of wine from browning, and the course of browning. White wine that was browned in an air stream at elevated temperature for 50 hours or more developed an absorption maximum at about 445 m$\mu$. This maximum correlated directly with the amount of flavanols in the wine.

The deliberate oxidation of wine to sherry or other maderized products by heat or long aging, coupled with some oxidation, presumably involves the same kind of reactions, though to a different degree. Rodopulo (1950, 1953) has reported the production of about 10 mg/liter of quinones during active oxidation of wine, but disappearance upon cessation of active oxidation. Relatively high polyphenol levels are preferred for white wines in order to produce the proper color and the desired maderized flavor. The addition of ethylacetate grape phenolic extracts or longer time on the skins is reportedly beneficial to maderized wine quality because of more total phenol and, particularly, more flavonoids in the wine (Politova-Sovzenko, 1950; Muraki *et al.*, 1960). Heating in barrels may be advantageous partly because extra tannin is extracted from the wood under these conditions (Kul'nevich, 1957a,b). About 500 mg/liter of tannic substances in wine to be maderized and 200–300 mg $O_2$/liter have been recommended. A high nitrogen content from the grapes or from yeast autolysate has also been recommended (Berg and Preobrazhenskii, 1960; Preobrazhenskii, 1963; Preobrazhenskii and Kharin, 1965). Those Russian workers used heating at 140°–153°F (rather high, from American experience) for about 2 months, and included 3–12% sugar in the wine so that the wine also would have sugar-derived products, caramel, hydroxymethylfurfural, etc. Even dry wines have some unfermentable sugar and, with prolonged heating, produce furfural.

The phenols of the wine undergo oxidative polymerization, with analysis declining for both the phloroglucinol and total phenolic group (Kul'nevich, 1957a,b). Analyses by Masuda and Muraki (1963) indicate that, allowing for the considerable evaporative concentration produced during storage of wine in barrels for 50 days at about 120°F, the major volatile substances decreased, polymeric phenols appeared to increase slightly, and smaller phenols and amino nitrogen decreased noticeably. Heating a high-tannin Kakhetia wine for 3 days at about 150°F in the absence of oxygen produced a small decrease in total phenol (from 2170 to 2070 mg/liter) (Nanitashvili, 1957). After long aging of Kakhetia wine none of the catechins were present, although wine of intermediate age (about 14 years) was reported (Nutsubidze, 1958a) to contain *d*-catechin, 1-epigallocatechin, and their racemates. When wine was maderized by heating and oxidation, gallocatechin was oxidized first, and *d*-catechin was epimerized and partially converted to gallocatechin, according to Nutsubidze (1958b).

There have been a number of reports on the reactions of amino acids in wine to form volatile carbonyl compounds via Strecker degradation, which could explain increased ammonia and carbon dioxide release noted in oxidation. It is clear that volatile aldehydes important to flavor and brown pigments can arise from reactions between sugar and amino acids (Reynolds, 1965). The formation of such products by interaction of amino acids with the quinones produced by polyphenolase has been suggested as the source of wine bouquet, particularly in aged sparkling wine (Politova-Sovzenko, 1957).

The nonenzymic formation of odors from amino acids held in model solutions under madeira baking conditions (up to 280 days at about 130°F) was modified by the presence of tannin but not affected much by oxygen (Nilov and Ogorodnik, 1967; Ogorodnik and Nilov, 1966; Kazumov, 1961). Most of the aroma and melanoidin pigments produced in baked wine, however, appear to come from sugar amino acid interaction, and the tannin level did not seem to change significantly in nitrogen-related oxidative flavor change (Kishkovskii and Potii, 1964; Kishkovskii *et al.*, 1966; Zinchenko, 1964).

It is believed that the production of aldehydes in wine in normal aging under relatively cool conditions is not because of reactions between quinone and amino acids. Convincing evidence has been supplied by Agabal'yants and Glonina (1966), who showed that oxidation of 5 Kakhetia and 5 European-style wines did not change amino acid content appreciably even though aldehyde was produced and added radioactive valine was recovered 98.9% unchanged. They concluded that such deamination of amino acids as does occur does not account for flavor changes in the wine, and that it goes on at about the same rate with or without oxidation, suggesting that it is not quinone-related. Amino acids may influence champagne bouquet and flavor in other ways, such as production of more tyrosol and other flavorous higher alcohols during the refermentation process (Rodopulo, 1964).

The increase in volatile aldehydes, predominantly acetaldehyde, during the production of sherries of the fino type is about 500 mg/liter. This increase results from aerobic metabolism of ethanol by yeast grown either as a film on the surface of the wine or in aerated submerged culture. In the film process a tannin content in white wine greater than 300 mg/liter impaired the accumulation of aldehydes, and the flavor was considered poorer (Motalev *et al.*, 1962). If large yeast inocula and aeration are used to produce high levels of aldehyde in a short time (600 mg/liter in 4 days), a tannin-rich white wine is preferred (Martakov *et al.*, 1967). The product can be blended with about equal amounts of wine that has been gradually oxidized (800 mg $O_2$/liter in 45 days at 130°F) to simulate aged fino sherry. If red wines are processed by aerobic yeast fermentation, much of the pigment and tannin is changed and precipitates from acetaldehyde condensation, and the wines become aged in appearance (Cantarelli, 1955, 1958; Iñigo and Bravo, 1960). When the process is halted at the correct stage, the wine can simulate aged red table wine and remain stable in the bottle.

The reaction of aldehydes with phenols has been known, and the effect of acetaldehyde on phenols in wine studied, for a very long time. In a series of papers, Trillat (1909) and others (Laborde, 1918; Fallot, 1909b) showed that acetaldehyde formation during wine aging was presumably from air oxidation of ethanol, that it combined with and precipitated tannins and pigments of red wine, and that aldehyde-tannin products were sometimes associated with bitterness. The bacterial production of acrolein, and the combination of acrolein with tannin to produce very bitter products, have been mentioned (Rentschler and Tanner, 1951; Herrmann, 1963a).

The reaction of the aldehyde with the wine phenols was soon understood as an example of Baeyer-type condensation, which leads to the phenol-formaldehyde resins, etc. (Fallot, 1909b). The reactive aldehyde is confined to condensation with the phloroglucinol ring of flavonoids under acidic low-temperature conditions, with the reactivity of *d*-catechin and formaldehyde at a minimum at about pH 4.5 (Hillis and Urbach, 1959). Addition of hydrogen peroxide to red wine causes a rapid increase in acetaldehyde content with little change in tannin (Joslyn and Comar, 1941). In time, however, the aldehyde decreases and the tannins and pigments partially precipitate. If acetaldehyde is added to red wine it decreases rapidly to a relatively constant low level. Acetaldehyde or formaldehyde at 25–100 mg/liter precipitates 50–65% of the total phenol (2200 mg/liter) from a red wine in a few days (Valuiko, 1965a). Further precipitation did not occur with longer storage. Reaction of the anthocyanins was somewhat slower, but the reaction was more complete and the original anthocyanin level (317 mg/liter) could be reduced to zero with sufficient aldehyde.

These data suggest that the residual nonprecipitable phenol is nonflavonoid (no phloroglucinol). If the 317 mg/liter of anthocyanin and the 65% aldehyde-precipitable tannin is subtracted from the total phenol, the difference, 450 mg/liter, is about what would be postulated as the sum of the caffeic acid and other derivatives not flavonoids in rich red wine. The lack of accumulation of appreciable acetaldehyde in slow aging of red wine, compared with its sizable accumulation in white wine (Mareca, 1962; Mareca and de Campos, 1957a,b), would be explained by the low level of phloroglucinol derivatives in white wine. Free aldehyde also arises in wine from gradual oxidation of the bisulfite, which releases the acetaldehyde bound during fermentation, which then may react with phenols, etc. (Pisarnitskii, 1965).

It has been shown that ports fortified with alcohol containing acetaldehyde retain more anthocyanin color than if purer alcohol is used (Singleton and Guymon, 1963; Singleton *et al.*, 1964a). It was postulated that this color retention was the result of acetaldehyde condensation to the A-ring and stabilization without destroying the flavylium structure or color and without producing a sufficiently large polymer to cause insolubility. A higher proportion of the pigment was adsorbable and less reactive after aldehyde addition. The difference between the samples diminished with time, perhaps owing to greater precipitation of the condensation product and the formation of similar products in the other wine. Mentioned earlier as another possible factor was the preferential biochemical reduction of acetaldehyde over anthocyanins.

Similar reactions evidently occur with other aldehydes. The vanillin-flavanol color reaction is an example, and its specificity lends credence to the idea that the Stiasny reaction for condensed tannin can be extended to distinguishing flavonoids from other phenols in solutions like wine. Furfural reacts with tannins, including grapevine tannin, and when furfural was added to tea catechins no catechins remained in the solution and black pigments resulted (Bokuchava and Skobeleva, 1957). Nutsubidze (1958b) believed that oxidation of catechins to quinones produced acetaldehyde from ethanol and that this was a key reaction in wine aging and explained precipitation, color changes, polymerization, etc.

The significance and interpretation of these observations on wine aging, oxidation, and browning are not entirely obvious, and no complete single picture of these reactions is likely in such a complex group of systems. Some tentative conclusions, drawn from these wine studies and from selected other studies on phenolic compounds under similar conditions, nevertheless seem worthwhile.

That a major share of the phenol and pigment in red wines becomes polymeric with aging seems to be a fact, and it becomes necessary to explain the nature(s) of such polymers. Aged wine appears to be a mixture of products of rather vigorous oxidation, ethanol yielding acetaldehyde, for example, in the presence of apparently much more easily oxidized substances such as delphinidin or petunidin glucosides. This, too, needs explanation.

It is generally known today that autoxidation of ascorbic acid may actually produce extra oxidation of foods instead of the protection from oxidation expected from a readily oxidized "antioxidant." This is because hydrogen peroxide is produced along with dehydroascorbic acid, and if insufficient ascorbic acid is present to consume the peroxide it is available for mischief (e.g., Chapon and Urion, 1960; Prillinger, 1963).

It is not so well known among food scientists that oxidation of hydroquinones produces the same effect, but, in fact, this is one way of manufacturing hydrogen peroxide. Hathway and Seakins (1955, 1957a) and Hathway (1958) showed that *d*-catechin and other analogs capable of forming quinones autoxidized at 35°C in phosphate buffer at pH 4–8, with formation of hydrogen peroxide. The requirement for any metal catalyst was not indicated, and since the reaction also took place in a nonpolar solvent a free-radical mechanism was indicated. Production of the hydrogen peroxide was summarized as:

$$H_2Q + O_2 = Q + H_2O_2$$
$$H_2Q + H_2O_2 = Q + 2\,H_2O.$$

Obviously, the amount of hydrogen peroxide available for other oxidations could be up to the equivalent of the amount of quinone produced.

Ethanol is readily oxidized to acetaldehyde by hydrogen peroxide (e.g., Chauvin, 1909), and this would appear from data of Joslyn and Comar (1941) although oxidation of bisulfite to free bound acetaldehyde could also be claimed in wine. It would appear that oxidation of about 600 mg/liter of gallic acid equivalent *o*-diphenol in wine could produce about 100 mg/liter of acetaldehyde, and, considering the regeneration of hydroquinones by condensation and hydroxyl insertion reactions, much more should be possible. These levels seem consistent with the wine analyses reported.

Hydrogen peroxide may also account for a number of other products. Visintini-Romanin (1967) found a fluorescent degradation product of malvidin in withered grapes and mature red wine. It appeared to be malvone, derived from malvin by hydrogen peroxide oxidation as reported by Karrer and de Meuron (1932). Jurd has shown that anthocyanins and related compounds undergo reactions with hydrogen peroxide in mildly acidic solvents to yield disruption of the heterocyclic ring, giving benzofurans and ring-opened derivatives (e.g., Jurd, 1966b, 1964c). Involvement of the heterocyclic ring of anthocyanins in such reactions with hydrogen peroxide may explain not only its decolorization in fruit juices and wine, but also that the oxidation rate depends upon the carbinol base percentage (e.g., Sondheimer and Kertesz, 1952; Lukton *et al.*, 1956; Jurd, 1964d).

In the presence of traces of iron, which are naturally present in wine, hydrogen peroxide is a very potent oxidant, Fenton's reagent. It has been believed that the active unit was a hydroxyl radical, but it now appears that an organic peroxide such as peracetic acid would probably be involved in wine (Brodskii and Vysotskaya, 1963). Organic peroxides related to tannin oxidation have been reported in wine, but not well studied (Kul'nevich, 1957a). Semiquinone radicals produced in the presence of oxygen can react to give peroxides and hydroperoxides.

These very active oxidants derived from or related to hydrogen peroxide are capable of "inserting" hydroxyl groups *ortho* or *para* to a preexisting phenolic hydroxyl (e.g., Bate-Smith, 1957; Turney, 1965). This would appear to explain Nutsubidze's (1958b) initially surprising finding of the production of gallocatechin from catechin during the maderization of wine, and indicate a mechanism for replenishing the supply of *o*-diphenols during wine oxidation. When the same operation is performed enzymatically, it is found that the group or atom displaced by the entering hydroxyl is not usually freed to the solution but migrates to an adjacent carbon (Guroff *et al.*, 1967). This is very attractive as a mechanism for nonenzymatic reactions as well, because it would explain a number of findings with food phenols. The production of gentisic acid during "saponification" of grapes could come from hydroxylation and oxidation of *p*-coumaric acid or kaempferol, for example (Bate-Smith, 1957). The hydroxylation of caffeic acid and the production of esculetin on paper chromatograms or in wine might reasonably, from steric or further oxidation considerations, be limited to esculetin and not the alternative isomer, which has not been found in practice. Heimann and Andler (1965) found that *p*-methoxyphenol was converted to methanol and *p*-benzoquinone enzymatically. This type of reaction could be involved in the reported demethoxylation of malvin in wine.

A number of possibilities appear to exist to explain the increases in molecular weight and polymerizations that the phenols of wine undergo with aging. The disappearance of acetaldehyde under wine conditions would seem to be explained by and limited to the aldehyde's electrophilic substitution at the nucleophilic carbons 6 or 8 of flavonoids. Since precipitation of phenols from wine does occur upon addition of acetaldehyde, it would appear that cross-linking with a second flavonoid may occur. Unless other activation mechanisms intervene, this type of cross-linking would appear to be limited to "head-to-head," A-ring through A-ring, polymerization.

Of course, polymerization by carbon to carbon, carbon to oxygen, and oxygen to oxygen linkages is an important reaction of the free radicals involved in phenol oxidation (Hassall and Scott, 1961; Scott, 1965). Although the reactions are probably of more than one type in wine, certain types of products appear most probable. Linkage in various natural and synthesized flavonoid "polymers" and related products has been under very active investigation, and several different types evidently occur (Geissman, 1962; Harborne, 1964a). Recent papers in important series are by Weinges *et al.*, (1968), Jurd and Lundin (1968), and Heyns *et al.* (1964). The natural dimeric anthocyanogens of fruits studied so far appear to fall into two groups (Weinges *et al.*, 1968). The best known group appears to be equivalent to the product of the carbonium ion produced by OH loss from a flavan-3,4-diol (leucocyanidin) in the 4 position linking to the nucleophilic 8 (or possibly 6) of a catechin molecule. The second group is evidently the same sort of dimer, with an additional link between the 7-O of the catechin moiety and the 2-C of the potential anthocyanidin moiety, or possibly the two bonds reversed, 2-C to 8-C, 4-C to 7-O.

A number of studies have shown the joining under mildly acidic conditions in aqueous solution of flavonoids with themselves and with other phenols, particularly of the phloroglucinol series. Jurd has shown (1967) that a flavylium salt (or, presumably, a natural anthocyanin) will link C-4 to C-6 (or C-8) of catechin in dilute acetic acid solution, forming a dimeric flavene that is then oxidized to the equivalent of a (red) catechin-anthocyanin dimer. A second mole of the flavylium salt serves as the oxidant and is reduced to a flav-2-ene. In dilute aqueous solution this product converted to a dihydrochalcone, but in an acetic acid solution it disproportionated into flavylium salt and flavan.

The possibility of similar linkage, so as to retain or reform the flavylium ion of natural anthocyanins and thus retain the visible red color, is further suggested by the structure of dracorubin, which is a natural orange-red (in acid) dimeric flavylium-flavan pigment from a palm (Jurd, 1967). The linkage to the 6 or 8 of an anthocyanin unit by electrophilic agents other than acetaldehyde, as previously suggested, may also lead to "anthocyanin-bearing tannins" as have been indicated in wine or grapes.

The dimerization reaction of 4–8-linked dimers can be looked upon as the reverse of the leucoanthocyanidin to anthocyanidin conversion. It need not involve air oxidation (Jurd, 1967), and if quinones are present they would affect the direction and ratio of products obtained (Jurd, 1966a). Geissman and Dittmar (1965) isolated an anthocyanogen from avocado seed. Its structure was shown to be a dimer with a readily broken covalent link between carbon 4 of a potential flavan-3,4-diol and carbon 6 or 8 of a catechin moiety. Furthermore, it was reported that this was a reversible reaction, and free flavan-3,4-diol would link with itself or catechin spontaneously to form polymers or dimers, respectively. Creasy and Swain (1965) showed the occurrence of a similar compound in strawberry leaves. They also obtained spontaneous dimerization between synthetic or natural flavan-3,4-diol monomers and catechin. They discussed evidence for possible incorporation of other flavonoids, lignin units, amino acids, etc., into such polymeric structures, and suggested that branching three-dimensional structures were possible.

It appears that reactions of these types could occur in wine and could account for the anthocyanin-bearing "tannins" in wine, changes in size and composition of the original grape tannins, and the production of monomeric and polymeric flavonoid derivatives not present in grapes. It seems significant that Masquelier *et al.* (1959) found the natural leucoanthocyanin polymers to decrease in size or amount as wine ages when we know that the fraction of anthocyanins or total phenols that is dialyzable shows increasing polymerization with time.

Other types of flavonoid linkage have been indicated to be important, such as chalcone formation and linkage from the former 2-C of the chalcone to a 6 or 8 position of another flavonoid (Freudenberg and Weinges, 1962, 1963). More complicated ring-opened products have also been described (Heyns *et al.*, 1964). Most of these products, proved or postulated, involve a phloroglucinol 6 or 8 nucleophilic site as one of the reactants. This would appear to involve greater stability to decoloration but greater tendency to brown of 3,5-diglucosides than of anthocyanidin-3-glucosides (Robinson *et al.*, 1966b). The 6 and 8 positions could be blocked by the 5-glucose, so polymerization would be prevented by that mode of linking. Protection of some of the anthocyanin against rapid browning and loss is thought to be possible by 6 or 8 attachment of another group. Garoglio (1968) found that grape anthocyanidin-3,5-diglucosides could be distinguished from 3-monoglucosides by failure of the diglucosides to precipitate readily with acetaldehyde.

Polymerization of catechin occurs alone at conditions that should obtain in wine, but only if oxygen is available (Hathway and Seakins, 1955, 1957a; Hathway, 1958). The product was produced more rapidly at pH 8 but was formed also at pH 4. It was quite different from the phlobaphene produced with strong heating at pH less than 2. It appeared to be identical to the polymer produced from catechin by phenolase (Hathway and Seakins, 1957b). The autoxidation polymer of catechin is evidently complex, but has been postulated to be a "head-to-tail" polymer with the 6 or 8-C of a catechin molecule linked to one of the B-ring positions of another moiety activated by quinoidal oxidation. Degradation and browning accompanies and complicates the reaction (Evelyn *et al.*, 1960).

The catechin tannin is described as red, but the absorption maximum was at 400 m$\mu$ at pH 6, and at 420–430 m$\mu$ at pH 8. Monomeric catechin quinone absorbed maximally at 370 m$\mu$. The development of 445-m$\mu$

absorbance by wine during vigorous oxidation has been described (Cantarelli, 1966). The brick red or orange of old wine undoubtedly results from such products in addition to the postulated anthocyanin-bearing polymers.

So far nothing has been said about the production of carbon dioxide during oxidation and phenolic browning of wine. One source might be oxidation of phenols, particularly those with pyrogallol units, to the point of ring fission and decarboxylation of the acids produced. This type of reaction occurs nonenzymatically, at least in alkaline solution and in the presence of iron and hydrogen peroxide (Grinstead, 1964). The possible source of $CO_2$ from wine phenol oxidation that appears most attractive to us is produced in the formation of purpurogallinlike derivatives. Dimeric flavonoid derivatives form enzymatically in tea oxidation (Roberts and Myers, 1960). They evidently constitute a family of related tropolone derivatives formed by the fusing of a pyrogallol ring to a catechol ring with the loss of one carbon as $CO_2$. The tea products have absorbance maxima near 380 $m\mu$, but also about 460 $m\mu$. The existence of such compounds appears possible in partly oxidized wine, and if found in appreciable amounts they would have "orange-red" color and perhaps taste properties. Although these substances are primarily enzymatic products in black tea, many phenolase actions occur without the enzyme, given the proper conditions.

Roberts (1962) deduced first a dimerization, tail-to-tail, of two gallocatechins, and then the B-ring of one condensing with gallic acid to give a benzotropolone or purpurogallin derivative. Further oxidation produced ring fission and $CO_2$ liberation from the benzene portion, leaving an acidic derivative, presumably the carboxy-substituted tropolone. Takino and Imagawa (1963a,b) produced a red glycoside, erycitrin, from pyrocatechol and myricitrin, by oxidation with either tea oxidase or potassium iodate. It had the structure shown, R = H, and similar derivatives were prepared from tea catechin and gallocatechin in which R = the remainder of *d*-catechin (Takino *et al.*, 1964).

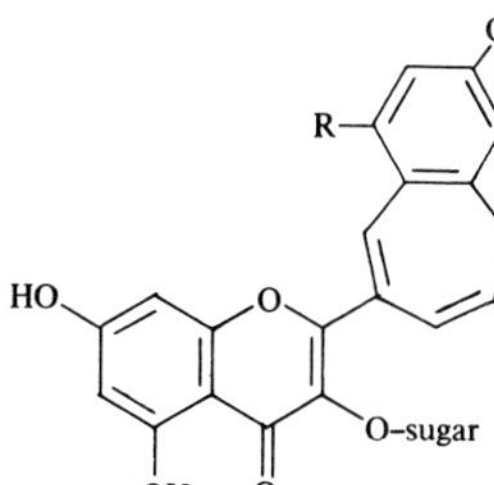

Purpurogallin derivatives have evidently not yet been reported in wine, but have been proposed to explain the color of aged brandy (Shpritsman, 1961).

The participation of amino acids in browning and oxidative reactions in wine has been indicated. Generally, the amino acids appear to act in connection with oxidation of the phenol to quinoid forms. Primary and secondary amines add to *o*-quinones in a position *para* to one of the keto groups and regenerate the phenol. Thus, temporarily, the effects of oxidation would be reversed and the browning quinone removed. The amino-substituted phenol would have lowered oxidation-reduction potential, however, and thus promote coupled oxidations of any *o*-diphenol of higher potential including its parent substance (Monties, 1966). Mercaptans such as cysteine behave similarly. Roberts (1959) found that cysteine in high concentration suppressed, and in low concentration promoted, the formation of oxidized purplish pigments with tea oxidase and catechin.

Phenolase oxidation does produce extra browning in the presence of many amino acids, and continuous consumption of the amino acid, with deamination and the production of aldehydes and carbon dioxide, is possible in juice, must, or wine if phenolase is present or if quinones are otherwise produced (Trautner and Roberts, 1950; Haider *et al.*, 1965; Clark and Geissman, 1948; Burton *et al.*, 1963).

Rinderknecht and Jurd (1958) noted that phloroglucinol browned poorly alone but produced a 483-m$\mu$ maximum, whereas with glycine added the reaction was rapid at pH 7–8 and gave a 443-m$\mu$ maximum absorbance. Oxidized white wine developed 445 m$\mu$ absorbance (Cantarelli, 1966). Resorcinol and catechol did not brown with glycine under the same conditions. The phloroglucinol-glycine browning was suggested to occur via a Schiff's-base link of the amino group to the triketo form of phloroglucinol, enolization, and cyclization to an indole. Since epicatechin did produce yellowing with glycine, a possible contribution of chalcone or other free *m*-trihydroxy (keto) derivatives of flavonoids in this reaction appears possible. In the suggested reaction sequence there appears to be no reason that oxidation would be required nor any reason that other amino acids could not participate. The contribution of amino acids in browning with phenols could be envisioned as oxygen-dependent if similar reactions with *o*-quinones were occurring, and oxygen-independent if with *m*-triketo derivatives. Burton *et al.* (1963) found that resorcinol did brown with glycine in their tests, and, of course, catechol browns when oxidized to the quinone, though more rapidly in the presence of amino acids. The difference in the results with resorcinol would presumably be due to oxidative versus relatively anaerobic conditions, as suggested by the lack of a catechol reaction in the earlier tests. The phloroglucinol reaction was evident at pH 3, though much suppressed. It was essentially prevented by bisulfite additions, which had greater effect on the glycine reactions than browning without glycine.

The possible complexity of browning reactions in wine is suggested by the many possible reaction products making up the polymers and other brown products we have been discussing. Data of Wright *et al.* (1960) on a tobacco polymer showed the presence of 18 amino acids (hydrolyzable), caffeic acid, quinic acid, quercetin, glucose, rhamnose, rutinose, and a molecular weight of 20,000–30,000. The brown pigments in wine may be similarly complex, and quinone-amino or mercaptan covalent linking is one reason that oxidation haze involving protein is more "permanent" than the hydrogen-bonded chill-haze formed with phenols and proteins.

## 4. Miscellaneous Reactions

The reversible partial reduction of anthocyanins is suggested by the bleaching effect of anaerobic yeast fermentation (Berg, 1963) or malolactic fermentation (Vetsch and Lüthi, 1964) and the later regeneration of much of the color by mild oxidation. The reaction appears to be rather analogous to the decolorizing of anthocyanins by sulfite, as clarified by Jurd (1964a). Flavylium to flavene reduction would appear to be involved (Jurd, 1966a, 1967). Reductive decolorization by ascorbic acid has been reported (Nebesky *et al.*, 1949). When wine is strongly $\gamma$-irradiated in bottles, both brown and red colors are bleached without appreciable change in total phenol content (Singleton, 1963; Tsetskhladze *et al.*, 1957). This appears also to be the result of reduction, since hydrogen is generated by the irradiation. Undoubtedly other free radical reactions occur; volatile aldehyde increases. Reduction of cyanidin to catechin and other analogous reactions are, of course, possible, and simple oxidative regeneration would not then be possible.

Bisulfite ion in wine serves to inhibit phenolases, bind acetaldehyde, and bleach anthocyanins, as we have seen. It may also have an important though largely undetermined role in liberating lignin into wine from the cask or cork, and thus adding phenols. Similar effects may influence the formation or breakdown of polymeric phenols (Quesnel, 1964).

Tannins seem to act in binding part of diethylpyrocarbonate, which is added to wine as a sterilant. The reaction is evidently carbethoxylation of phenolic OH groups (Paulus and Lorke, 1967; Hennig, 1962).

A startling problem, when it occurred in wines (now rare because of less iron contamination in the modern winery), was the production of inky blue wines and dark precipitates from the phenol-ferric color formed (Carles, 1919; Rentschler, 1963).

# VI

# Research Needs

When qualitative composition is compared with quantitative and total phenol content of grapes and wine, there does not appear to be a major portion unidentified. Some qualitative identifications need to be buttressed by more detailed studies, however, and some uncertainties remain. Particularly, the oligomeric and polymeric condensed tannins in grapes and their reaction products in wine need to be studied and defined.

The individual phenols need to be isolated or synthesized in sufficient quantities that their contributions to sensory properties and reactions in juice and wine can be more easily determined. Much remains to be done in determining the effects of various agronomic and processing variables on both the specific classes and individual phenols. We need to be able to follow the changes in each and all phenols and determine a "phenol balance" after controlled development or reaction periods. Silation and gas chromatography may be the answer, but a number of techniques appear to be ready for development and application to this problem.

The roles of the phenols in the plant itself appear to be much more important than has generally been realized. There are already practical examples of the use of qualitative phenol composition as an indicator of the genetic background of the grape or young wine. Little practical use has been made so far of the large quantitative and perhaps qualitative differences in phenolic composition related to weather, ripeness, and other growing conditions. With more basic data it is believed that such information will be very useful to the grower-processor.

The biochemistry of the synthesis of phenols is being developed, but practically nothing is known of the mechanisms of breakdown in plants of complex compounds like the condensed tannins or even the anthocyanins. The rapid responses to environment and tannin disappearance in tissues of seeds or fruit during germination or development indicate that important mechanisms exist for mobilization of these compounds. Why do the catechins occur as gallates and not as glycosides, when the reverse is true of other flavonoids? Is epicatechin gallate the long-sought pyrogallol analog of caffeic acid in the plant? Why are the anthocyanins of grapes highly methylated and the other flavonoids not? Many such problems cry for solution, and the grape seems to be a worthy plant in which to study them. It has a generally high content of nearly all the most common forms with relatively simple glycosidation, etc.

The direct relationships of each phenol to flavor, color, and processing reactions need to be clarified. Better means of retaining desired color and avoiding off-color would seem to be available if present hypotheses could be verified and used toward that end. Complex as the potential phenol reactions in grapes and wines are, the basic reaction types and products appear different enough that the roles of the alternatives should be determinable. Probable dividends of such study would be better color, longer shelf life, quicker aging, and optimum quality.

**ACKNOWLEDGMENTS**

Support that helped make this study possible was supplied under the United States National Institutes of Health Grant ES-00300 and California Wine Advisory Board Contract V-13, and is gratefully acknowledged.

## REFERENCES

Accum, F. C. 1820. "A treatise on the adulteration of foods and culinary poisons; exhibiting the fraudulent sophistications of bread, beer, wine, spiritous liquors, tea, coffee, cheese, pepper, mustard, etc., etc., and methods of detecting them." Abraham Small, Philadelphia.

Adkinson, J. 1913. Some features of the anatomy of the Vitaceae. *Ann. Botany London* **27**, 133–139.

Agabal'yants, G. G., and Glonina, N. N. 1966. Aminokisloty i okislennost vina. *Vinodelie i Vinogradarstvo SSSR* **26**, 9–16.

Agostini, A. 1964. Utilization of by-products of vines and wines. *Die Wynboer* (395), 13, 15–16.

Aizenberg, V. Ya., Davtyan, I. O., and Gulamiryan, A. K. 1966. Chemical composition of some grape varieties. *Izv. Sel'skokhoz. Nauk; Min. Sel'sk. Khoz. Arm. SSR*(6), 51–57; *Chem. Abstr.* **66**, 64500 (1967).

Akiyoshi, M., Webb, A. D., and Kepner, R. E. 1963. Major anthocyanin pigments of *Vitis vinifera* varieties Flame Tokay, Emperor, and Red Malaga. *J. Food Sci.* **28**, 177–181.

Albach, R. F., Kepner, R. E., and Webb, A. D. 1959. Comparison of anthocyan pigments of red vinifera grapes. II. *Am. J. Enol. Viticult.* **10**, 164–172.

Albach, R. F., Kepner, R. E., and Webb, A. D. 1963. Peonidin-3-monoglucoside in vinifera grapes. *J. Food Sci.* **28**, 55–58.

Albach, R. F., Kepner, R. E., and Webb, A. D. 1965a. Structures of acylated anthocyan pigments in *Vitis vinifera* variety Tinta Pinheira. Identification of the anthocyanidin, sugar, and acid moieties. *J. Food Sci.* **30**, 69–76.

Albach, R. F., Webb, A. D., and Kepner, R. E. 1965b. Structures of acylated anthocyan pigments in *Vitis vinifera* variety Tinta Pinheira. II. Position of acylation. *J. Food Sci.* **30**, 620–626.

Alexiu, A. 1965. Studiul comparativ al unor metode pentru caracterizarea cromatico a vinurilar rosii. *Lucrari Stiintifice. (Inst. Cerc. Hort.-Vit., Bucharest).* **9**, 499–511.

Alfieri, R., and Sole, P. 1966. Influence des anthocyanosides administrés par voie oro-perlinguale sur l'adaptoelectroretinogramme (AERG) en lumiere rouge chez l'homme. *Compt. Rend. Soc. Biol.* **160**, 1590–1593.

Alley, C. J., Goheen, A. C., Olmo, H. P., and Koyama, A. T. 1963. The effect of virus infections on vines, fruit, and wines of Ruby Cabernet. *Am. J. Enol. Viticult.* **14**, 164–170.

Alquier, J. 1929. Valeur biologique et hygienique du vin. *Rev. Viticult.* **70**, 361–366, 377–384, 393–399, 411–413; **71**, 5–11, 21–25, 261–265, 283–287, 293–298, 309–312, 327–332, 341–352.

Alston, R. E. 1964. The genetics of phenolic compounds. *In* "Biochemistry of Phenolic Compounds" (J. B. Harborne, ed.), pp. 171–204. Academic Press, New York.

Alston, R. E., and Turner, B. L. 1963. "Biochemical Systematics," p. 404. Prentice-Hall, Englewood Cliffs, New Jersey.

Alwood, W. B. 1914. Crystallization of cream of tartar in the fruit of grapes. *J. Agr. Res.* **1**, 513–514.

Alwood, W. B., Hartmann, B. G., Eoff, J. R., Ingle, M. J., and Sherwood, S. F. 1916. Development of sugar and acid in grapes during ripening. *U.S. Dept. Agr. Bur. Chem. Bull.* **335**, 1–28.

Amati, A., and Formaglini, A. 1965. Riconocimento di antifermentativi nei vini mediante cromatografia su strato sottile I. *Riv. Viticolt. Enol.* **18**, 387–395.

Ambrosi, H., van Niekerk, C. I., and Spies, H. T. 1966. Grape must separation in South Africa. *Die Wynboer* (416), 24–36.

Amerine, M. A. 1947. The composition of California wines at exhibitions. *Wines & Vines* **28**, (1), 21–23, 42–43, 45; (2), 24–26; (3), 23–25, 42–46.
Amerine, M. A. 1954. Composition on wines. I. Organic constituents. *Advan. Food Res.* **5**, 353–510.
Amerine, M. A. 1955. Further studies with controlled fermentations. *Am. J. Enol.* **6**, 1–16.
Amerine, M. A. 1956. The maturation of wine grapes. *Wines & Vines* **37**(10), 27–30, 32, 34–36, 38; (11), 53–55.
Amerine, M. A. 1965. "Laboratory Procedures for Enologists," p. 109. Univ. of California, Davis, California.
Amerine, M. A., and Bailey, C. B. 1959. Carbohydrate content of various parts of the grape cluster. *Am. J. Enol. Viticult.* **10**, 196–198.
Amerine, M. A., and De Mattei, W. 1940. Color in California wines. III. Methods of removing color from the skins. *Food Res.* **5**, 509–519.
Amerine, M. A., and Ough, C. S. 1957. Controlled fermentations. III. *Am. J. Enol.* **8**, 18–30.
Amerine, M. A., and Winkler, A. J. 1941. Color in California wine. IV. The production of pink wines. *Food Res.* **6**, 1–14.
Amerine, M. A., and Winkler, A. J. 1944. Composition and quality of musts and wines of California grapes. *Hilgardia* **15**, 493–673.
Amerine, M. A., and Winkler, A. J. 1947. Relative color stabilities of the wines of certain grape varieties. *Proc. Am. Soc. Hort. Sci.* **49**, 183–185.
Amerine, M. A., and Winkler, A. J. 1963. California wine grapes: composition and quality of their must and wines. *Calif. Univ. Agr. Expt. Sta. Bull.* **794**, 1–83.
Amerine, M. A., Ough, C. S., and Bailey, C. B. 1959. Color values of California wines. *Food Technol.* **13**, 170–175.
Amerine, M. A., Pangborn, R. M., and Roessler, E. B. 1965. "Principles of Sensory Evaluation of Food," p. 602. Academic Press, New York.
Amerine, M. A., Berg, H. W., and Cruess, W. V. 1967. "The Technology of Wine Making," 2nd ed., p. 799. Avi, Westport, Connecticut.
Amirdzhanov, A. G. 1962. Fall leaf coloring as an indicator of the carbohydrate balance in the grape vine. *Izv. Timiryazev. Sel'skokhoz. Akad.* (3), 224–227; *Chem. Abstr.* **58**, 12858 (1963).
Anderson, R. 1923. Concerning the anthocyans in Norton and Concord grapes. A contribution to the chemistry of grape pigments. *J. Biol. Chem.* **57**, 795–813.
Anderson, R. 1924. A contribution to the chemistry of grape pigments. III. Concerning the anthocyanins in Seibel grapes. *J. Biol. Chem.* **61**, 685–695.
Anderson, R. J., and Nabenhauer, F. P. 1924. A contribution to the chemistry of grape pigments. II. Concerning the anthocyans in Clinton grapes. *J. Biol. Chem.* **61**, 97–107.
Anderson, R. J., and Nabenhauer, F. P. 1926. A contribution to the chemistry of grape pigments. IV. The anthocyans in Isabella grapes. *J. Am. Chem. Soc.* **48**, 2997–3003.
Andreev, V. V. 1965. Rezhim izvlecheniya krasyashchikh veshchestv iz vinograda. *Vinodelie i Vinogradarstvo SSSR* **25**(3), 13–17.
Anonymous. 1959. Vitamin sources and their utilization, Collection 4. Vitamin P, its properties and uses. *Izd. Akad. Nauk SSSR. Moscow*; Trans. Artman, M., Israel Program Sci. Trans., Jerusalem (*U.S. Offic. Tech. Serv., Dept. Commerce*), pp. 1–216 (1963).
Anonymous, 1962. Die Erreichung der Säureharmonie. *Schweiz. Z. Obst. u. Weinbau* **71**, 120–123.
Anonymous, 1965. "Fruit, a review," pp. 4, 65–67. Commonwealth Economic Committee, London.
Anonymous. 1966a. Colloque oenologique de Montpellier, 1965. L'influence des phases préfermentatives de la vinification sur la qualité des vins. *Vignes et Vins* (Spec. No., May), 11–26, 39–149.

Anonymous. 1966b. Color additives. Grape skin extract (enocianina). *Federal Register* **31**, 4784.

Anonymous, 1967a. Annual statistical summary. *Wine Inst. Bull.* (1418), 1-21.

Anonymous. 1967b. Méthodes d'analyse et éléments constitutifs des vins. Compte-rendu de la 9e réunion de la Sous-Commission des méthodes d'analyses et d'appreciation des vins. *Bull. Offic. Intern. Vigne Vin* **40**(439), 934-972.

Anonymous. 1968a. "The Merck Index," 8th ed. (Stecher, P. G., ed.), p. 1713. Merck, Rahway, New Jersey.

Anonymous. 1968b. Fruit situation. *U.S. Dept. Agr. Econ. Res. Serv.* TFS-**166** (Jan.), 1-27.

Areshidze, I. V. 1963. Pigments and photosynthesis in anthocyanin-forming plants. *Fiziol.-Biokhim. Osnovy Pidvishchennya Produktivsnosti Roslin,* 440-444; *Chem. Abstr.* **60**, 9603 (1964).

Artyunayan, A. S., Dzhanpoladyan, L. M., Samvelyan, A. M., and Khachatryan, A. L. 1964. The effect of fertilizers on the increase of the content of anthocyanins and aromatic compounds in grapes and wines. *Tr. Armyansk. Nauchn.-Issled. Inst. Vinogradarstva, Vinodeliya i Plodovodstva* (6-7), 455-469; *Chem. Abstr.* **64**, 20575 (1966).

Asen, S., and Jurd, L. 1967. The constitution of a crystalline, blue cornflower pigment. *Phytochemistry* **6**, 577-584.

Ash, C. 1952. Reminiscences of pre-prohibition days. *Proc. Am. Soc. Enol.* **3**, 39-44.

Aso, K., Nakayama, T., and Maki, M. 1953. The bitter components in alcoholic drinks. I. The tyrosol content in saké. *J. Ferment. Technol.* (*Japan*) **31**, 43-47.

Aso, K., Nakayama, T., and Sato, A. 1956a. Studies on the utilization of grape. V. Color change of Concord grape juice by various treatments and pasteurization. *J. Ferment. Technol.* (*Japan*) **34**, 37-41.

Aso, K., Nakayama, T., and Sato, A. 1956b. Studies on the utilization of grape. VI. Color change of Concord grape juice during storage. *J. Ferment. Technol.* (*Japan*) **34**, 127-133.

Association of Official Agricultural Chemists. 1965. "Official Methods of Analysis," 10th ed., p. 957. A.O.A.C., Washington, D. C.

Astruc, H., and Castel, A. 1934. Etude du pouvoir dissolvant de l'anhydride sulfureux sur les matières colorantes et extractives du raisin. *Prog. Agr. Vit.* **101**, 519-523.

Athanassiadis, C. M., Caspar, J. N., and Yang, H. Y. 1956. Controlling secondary fermentation in wine with vitamin K5. *Wines & Vines* **37**(11), 47-48, 50.

Averna Sacca, R. 1906. "I tannini nell'uva e nel vino," p. 240. U. Hoepli, Milan.

Bailey, L. H. 1935. "The Standard Cyclopedia of Horticulture," 3 Vols., p. 3639. Macmillan, New York.

Baker, G. A., and Amerine, M. A. 1953. Organoleptic ratings of wines estimated from analytical data. *Food Res.* **18**, 381-389.

Baker, R. T., and Smith, H. G. 1906. Vitis opaca, F. v. M., and a chemical investigation of its enlarged rootstock (tuber). *J. Roy. Soc. New S. Wales* **40**, 52-60.

Baker, G. A., Ough, C. S., and Amerine, M. A. 1966. Sensory scores and analytical data for dry white and dry red wines as bases for predictions and groupings. *Am. J. Enol. Viticult.* **17**, 255-264.

Balakian, S., and Berg, H. W. 1968. The role of polyphenols in the behavior of potassium bitartrate in red wines. *Am. J. Enol. Viticult.* **19**, 91-100.

Baldwin, S., Black, R. A., Andreasen, A. A., and Adams, S. L. 1967. Aromatic congener formation in maturation of alcoholic distillates. *J. Agr. Food Chem.* **15**, 381-385.

Baragiola, W. I., and Godet, C. 1914. Chemisch-analytische Untersuchungen über das Reifen von Trauben und über die Entwicklung des daraus gewonnenen Weines. *Landwirtsch. Jahrb.* **47**, 249-302.

Barikyan, Kh. G. 1960. Ob uskorenii stareniya kon'yachnogo spirta. *Vinodelie i Vinogradarstvo SSSR* **20**(6), 12-14.

Barnes, E. O. 1940. Biochemical studies of some varieties of apples, plums, and grapes grown in Minnesota. *Minn. Univ. Agr. Expt. Sta. Tech. Bull.* **143**, 1–35.

Barnett, J. A., and Swain, T. 1956. Splitting of Aesculin by yeasts. *Nature* **177**, 133–134.

Bastianutti, I., and Romani, B. 1961. Detection of cinnamic acid and of *o*-, *m*-, and *p*-chlorobenzoic acids in wines. *Boll. Lab. Chim. Provinciali (Bologna)* **12**, 172–178; *Chem. Abstr.* **56**, 6474c (1962).

Bate-Smith, E. C. 1954a. Flavonoid compounds in foods. *Advan. Food Res.* **5**, 261–300.

Bate-Smith, E. C. 1954b. Leuco-anthocyanins. I. Detection and identification of anthocyanidins formed from leuco-anthocyanins in plant tissues. *Biochem. J.* **58**, 122–125.

Bate-Smith, E. C. 1956. Chromatography and systematic distribution of ellagic acid. *Chem. & Ind.* **1956**, R-32-R33.

Bate-Smith, E. C. 1957. Oxidation of plant phenolics. *Nature* **179**, 1283–1284.

Bate-Smith, E. C. 1958. The contribution of phenolic substances to quality in plant products. *Qualitas Plant. Mater. Vegetabiles* **3–4**, 440–455.

Bate-Smith, E. C. 1962a. Plants and polyphenols. *Recent Advan. Food Sci., Papers, Glasgow 1960* **1**, 205–211.

Bate-Smith, E. C. 1962b. The phenolics constituents of plants and their taxonomic significance. *J. Linnean Soc. London* **58**, 95–173.

Bate-Smith, E. C. 1963. Usefulness of chemistry in plant taxonomy as illustrated by the flavonoid constituents. *In* "Chemical Plant Taxonomy" (T. Swain, ed.), pp. 127–139. Academic Press, New York.

Bate-Smith, E. C., and Lerner, N. H. 1954. Leuco-anthocyanins. II. Systematic distribution of leuco-anthocyanins in leaves. *Biochem. J.* **58**, 126–132.

Bate-Smith, E. C., and Metcalfe, C. R. 1957. Leuco-anthocyanins. III. The nature and systematic distribution of tannins in decotyledonous plants. *J. Linnean Soc. London* **55**, 669–705.

Bate-Smith, E. C., and Ribéreau-Gayon, P. 1959. Leucoanthocyanins in seed. *Qualitas Plant. Mater. Vegetabiles* **5**, 189–198.

Bate-Smith, E. C., and Swain, T. 1953. Identification of leucoanthocyanins as tannin in foods. *Chem. & Ind.* **1953**, 377–378.

Bate-Smith, E. C., and Swain, T. 1965. Recent developments in the chemotaxonomy of flavonoid compounds. *Lloydia* **28**, 313–331.

Bayer, E. 1957. Aromastoffe des Weines. Die Carbonsäureester des Weines und der Traube, *Vitis* **1**, 34–41.

Bayer, E. 1958. Anwendung chromatographischer Methoden zur Qualitatsbeureilung von Weinen und Mosten. *Vitis* **1**, 298–312.

Bayer, E., Born, F., and Reuther, K. H. 1957. Über die Polyphenoloxydase der Trauben. *Z. Lebensm.-Untersuch. Forsch.* **105**, 77–81.

Bazilevskii, V. V. 1964. O novoi tekhnologii prigotovleniya krasnykh vin. *Vinodelie i Vinogradarstvo SSSR* **24**(8), 6–8.

Begunova, R. D. 1953. Dinamika nakopleniya krasyashchikh veshchestv v yagodakh vinograda pri ikh sozrevanii. *Tr. Vses. Nauchn.-Issled. Inst. Vinodeliya i Vinogradarstva, Magarach* **4**, 169–174.

Begunova, R. D., Zakharina, O. S., and Rokhlenko, I. S. 1965. Use of cation exchange for stabilization of red wines. *Tr. Vses. Nauchn.-Issled. Inst. Pivo-Bezalkogol'n. i Vinodel'chesk. Prom.* (12), 93–98; *Chem. Abstr.* **65**, 2965 (1966).

Bell, T. A., Etchells, J. L., and Smart, W. W. G., Jr. 1965. Pectinase and cellulase enzyme inhibitor from Sericea and certain other plants. *Botan. Gaz.* **126**, 40–45.

Benard, P., and Jouret, C. 1962. Comparaison entre trois techniques de vinification en rouge. *Vignes et Vins* (108), 25–31; (109), 7–11.

Benvegnin, L., and Capt, E. 1936. Composition dû moût de raisin à différents stades du pressurage. *Ann. Agr. Suisse*, 899–907.

Benvegnin, L., Capt, E., and Piguet, G. 1947. "Traité de Vinification," p. 575. Librairie Payot, Lausanne.

Berezenko, G. Z., and Avazkhodzhaev, M. Kh. 1965. Role of tannins in making salts harmless to the grapevine. *Vzb. Biol. Zh.* **9**(2), 82–83; *Chem. Abstr.* **67**, 8774 (1967).

Berg, H. W. 1940. Color extraction for port wine manufacture. *Wine Rev.* **8**(1), 12–14.

Berg, H. W. 1953a. Varietal susceptibility of white wines to browning. I. Ultraviolet absorption of wines. *Food Res.* **18**, 399–406.

Berg, H. W. 1953b. Varietal susceptibility of white wines to browning. II. Accelerated storage tests. *Food Res.* **18**, 407–410.

Berg, H. W. 1963. Stabilisation des anthocyannes. Comportement de la couleur dans les vins rouges. *Ann. Technol. Agr. Paris Numero Hors Ser.* I. **12**(1), 247–259.

Berg, H. W., and Akiyoshi, M. 1956a. Some factors involved in browning of white wines. *Am. J. Enol.* **7**, 1–7.

Berg, H. W., and Akiyoshi, M. 1956b. Effect of contact time of juice with pomace on the color and tannin content of red wines. *Am. J. Enol.* **7**, 84–90.

Berg, H. W., and Akiyoshi, M. 1957. The effect of various must treatments on the color and tannin content of red grape juices. *Food Res.* **4**, 373–383.

Berg, H. W., and Akiyoshi, M. 1958. Further studies of the factors affecting the extraction of color and tannin from red grapes. *Food Res.* **23**, 511–517.

Berg, H. W., and Akiyoshi, M. 1960. The effect of sulfur dioxide and fermentation on color extraction from red grapes. *Food Res.* **25**, 183–189.

Berg, H. W., and Akiyoshi, M. 1962. Color behavior during fermentation and aging of wines. *Am. J. Enol. Viticult.* **13**, 126–132.

Berg, H. W., and Marsh, G. L. 1956. Effects of heat-treatment of musts on the color characteristics of red wine. *Food Technol.* **10**, 4–9.

Berg, H. W., Filipello, F., Hinreiner, E., and Webb, A. D. 1955a. Evaluation of thresholds and minimum difference concentrations for various constituents of wines. I. Water solutions of pure substances. *Food Technol.* **9**, 23–26.

Berg, H. W., Filipello, F., Hinreiner, E., and Webb, A. D. 1955b. Evaluation of thresholds and minimum difference concentrations for various constituents of wine. II. Sweetness: the effect of ethyl alcohol, organic acid, and tannin. *Food Technol.* **9**, 138–140.

Berg, H. W., Ough, C. S., and Chichester, C. O. 1964. Prediction of perceptibility of luminous-transmittance and dominant wavelength differences among red wines by spectrophotometric measurements. *J. Food Sci.* **29**, 661–667.

Berg, V. A., and Preobrazhenskii, A. A. 1960. Faktory, opredelyayushchie maderizatsiyu vin v germeticheskoi tare. *Biokhim. Vinodeliya* **6**, 60–78.

Bergeret, J. 1963. Action de la gélatine et de la bentonite sur la couleur et l'astringence de quelques vins. *Ann. Technol. Agr. Paris* **12**, 15–25.

Beridze, G. I. 1950. Obshchaya kharakteristika vin kakhetinskogo tipa i sposoby ikh uluchsheniya. *Biokhim. Vinodeliya, Akad. Nauk SSSR.* **3**, 171–209.

Beridze, G. I. 1956. Novyĭ sposob prigotovleniya belogo kakhetinskoga vina. *Vinodelie i Vinogradarstvo SSSR.* **16**(1), 11–16.

Beridze, G. I. 1965. Sort i kachestvo vinograda-osnova vinodeliya. *In* "Povyshenie kachestva vinograda dlya tekhnicheskoĭ pererabotki" (T. G. Katar'yan), pp. 102–110. Pishchevaya Promyshlennost, Moscow.

Beridze, G. I., and Sirbiladze, M. G. 1964. Nitrogenous substances in vines and wines. *Probl. Evolyutsionnoi i Tekhn. Biokhim., Akad. Nauk SSSR. Inst. Biokhim.* 348–353; *Chem. Abstr.* **59**, 10735 (1965).

Berishvili, L. I. 1963. Use of ultrasound in processing grapes. *Novye Fiz. Metody Obrabotki Pishchevykh Produktov, Kiev* 248–251; *Chem. Abstr,* **62**, 7037 (1965).

Bernon, G., and Nègre, E. 1939. Nouvelles recherches sur les producteurs directs cultivés a l'École Nationale D'Agriculture de Montpellier. *Ann. École Natl. Agr. Montpellier* **25**, 295–352.

Beyer, G. F. 1945. Report on the spectrophotometric examination of wines. *J. Assoc. Offic. Agr. Chemists* **28**, 471–475.

Beythien, A. 1949. "Die Geschmackstoffe der Menschlischen Nahrung," p. 150. Steinkopff, Dresden.

Bhaskara, P., Murty, R., and Seshadri, T. R. 1939. Chemical composition of *Vitis quadrangularis* (Wall). *Proc. Indian Acad. Sci.* **9A**, 121–127.

Biddle, H. C., and Kelley, W. P. 1912. Tannic acid, ethyl gallate and the supposed ester of tannic acid. *J. Am. Chem. Soc.* **34**, 918–923.

Bigelow, W. D. 1895. Coloring matter in the California red wines. *J. Am. Chem. Soc.* **17**, 211–216.

Biol, H. 1967. Le traitement thermique de la vendage rouge: buts—techniques—résultats. *Vignes et Vins* (161), 29–31, 33–35; (162), 51, 53–55.

Biol, H., and Foulonneau, C. 1962. Sur certaines causes de la présence accidentelle dans les vins du diglucoside du malvidol, characteristique de certains hybrides. *Vignes et Vins* (110), 3–4.

Biol, H., and Foulonneau, C. 1963. Décoloration des vins rouges par certains sels de métaux lourds. *Ann. Technol. Agr. Paris* **12**, 27–38.

Biol, H., and Michel, A. 1961. Etude chromatographique des vins rouges issus de cépages réglementés. *Ann. Technol. Agr. Paris* **10**, 339–344.

Bioletti, F. T. 1905. Manufacture of dry wines in hot countries. *Calif. Univ. Agr. Expt. Sta. Bull.* **167**, 1–66.

Bioletti, F. T. 1906. A new method of making dry red wine. *Calif. Univ. Agr. Expt. Sta. Bull.* **177**, 1–36.

Bioletti, F. T. 1914. Amounts of wine and of by-products yielded by grapes in a California winery. *Ann. Rept. of Director, Calif. Sta. Rept.*, 192–193.

Bioletti, F. T. 1938. Outline of ampelography for the vinifera grapes in California. *Hilgardia* **11**(6), 227–293.

Birkofer, L., Kaiser, C., Koch, W., Donike, M., and Wolf, D. 1963. Acylierte Anthocyane. II. Cis-trans-Isomeric bei Acylanthocyanen. *Z. Naturforsch.* **18B**, 631–634.

Birkofer, L., Kaiser, C., Donike, M., and Koch. W. 1965. Acylierte Anthocyane. III. Konstitution von Acyl-anthocyanen. *Z. Naturforsch.* **20B**, 424–428.

Black, R. A., Rosen, A. A., and Adams, S. L. 1953. The chromatographic separation of hardwood extractive components giving color reactions with phloroglucinol. *J. Am. Chem. Soc.* **75**, 5344–5346.

Blyth, A. W., and Blyth, M. W. 1927. "Foods: their composition and analysis," 7th ed., p. 619. Griffin, London.

Bockian, A. H., Kepner, R. E., and Webb, A. D. 1955. Skin pigments of the Cabernet Sauvignon grape and related progeny. *J. Agr. Food. Chem.* **3**, 695–699.

Boettinger, C. 1901 Studien über Weinbildung. 4. Die im Wasser löslichen Bestandteile der Weintraubenblatter. *Chem. Ztg.* **52**, 6–8, 17–18, 24–25.

Böhm, K. 1959-60. Die Flavonoide. *Arzneimittel-Forsch.* **9**, 539–544, 647–653, 778–785; **10**, 54–58, 139–142, 188–192, 468–474, 547–554.

Boireau, R. 1888. Wines, their care and treatment in cellar and store, together with a short treatise on vinification. *Rept. Chief Executive Offic. Calif. State Board Viticult. Commissioners,* p. 148.

Bokorny, T. 1916. Emulsin und Myrosin in der Münchener Brauereipresshefe. *Biochem. Z.* **75**, 376–416.

Bokuchava, M. A., and Novozhilov, N. P. 1946. Taste qualities of various fractions of tea tannins and their significance in tea quality. *Biokhim. Chaĭnogo Proizvodstva* (5), 190–196; *Chem. Abstr.* **46**, 2712 (1952).

Bokuchava, M. A., and Skobeleva, N. I. 1957. The role and significance of aldehydes in the transformation of tannin substances at high temperatures. *Biochemistry* (USSR, Trans.) **22**, 952–956.

Bolcato, V., and Lamparelli, F., 1962. Le probabili basi biologiche della rottura ossidasica dei vini. *Agrochimica* **6**(4), 366–374.

Bolcato, V., Losito, F., Campanella, V., and Pallavicini, C. 1964a. Trasformazione delle sostanze coloranti dei vini in sequito a rottura ossidasica. *Riv. Viticolt. Enol.* **17**, 24–34.

Bolcato, V., Lamparelli, F., and Losito, F. 1964b. Azione della *B. Cinerea* e del lievito sulle sostanze coloranti dei mosti d'uva. *Riv. Viticolt. Enol.* **17**, 415–421.

Bonnet, A. 1902. Etude de la graine. *Ann. Ecole Natl. Agr. Montpellier* **2**, 73–134.

Bonnet, A. 1903. Recherches sur la structure du grain de raisin. *Ann. Ecole Natl. Agr. Montpellier* **3**, 58–102.

Bonnet, L. O. 1929. The color in grapes. *Calif. Grower* **1**(2), 12.

Booth, A. N., Masri, M. S., Robbins, D. J., Emerson, O. H., Jones, F. T., and DeEds, F. 1959. Metabolic fate of gallic acid and related compounds. *J. Biol. Chem.* **234**, 3014–3016.

Bosticco, A. 1966. I pigmenti antocianici dell'uva, quali fattori di accrescimento nel pollo (contributo sperimentale). *Ann. Fac. Sci. Agrar., Torino 1961-1962* **1**, 41–51.

Bosund, I. 1962. The action of benzoic and salicylic acids on the metabolism of microorganisms. *Advan. Food Res.* **11**, 331–353.

Boubals, D., Cordonnier, R., and Pistre, R. 1962. Etude du mode de transmission héréditaire du caractère "diglucosides anthocyaniques des baies" dans le genre *Vitis*. *Compt. Rend. Acad. Agr. France* **48**, 201–203.

Bourzeix, M. 1967. Le dosage des flavonols du vin et des extraits de raisin. *Ann. Technol. Agr. Paris* **16**, 349–355.

Bourzeix, M., and Baniol, P. 1966. L'isolement des flavonols du vin par chromatographie sur couche mince de cellulose. *Ann. Technol. Agr. Paris* **15**, 211–217.

Boutaric, A., Ferré, L., and Roy, M. 1937. Recherches spectrophotometriques sur la couleur des vins. *Ann. Fals. Fraudes* **30**, 196–209.

Brad, S. C. 1966. Action of fungal and pectolytic preparations on the coloring substances in grapes and red wine. *Rev. Ferment. Inds. Aliment.* **21**(2), 51–56.

Bradfield, A. E., and Bate-Smith, E. C. 1950. Chromatographic behavior and chemical structure. II. The tea catechins. *Biochim. Biophys. Acta* **4**, 441–444.

Breider, H. 1964a. Qualität und Resistenz. *Mitt. (Klosterneuburg) Ser. A.* **14**(1), 21–28.

Breider, H. 1964b. Untersuchungen über den Einfluss des Traubensaftes von Hybridenreben auf den Tierorganismus. *Weinberg Keller* **10**, 513–517.

Breider, H., Wolf, E., and Schmitt, A. 1965. Embryonalschäden nach Genuss von Hybridenweinen. *Weinberg Keller* **12**, 165–182.

Brodskii, A. I., and Vysotskaya, N. A. 1963. Study of the mechanism of hydroxylation of the aromatic ring with peroxides by means of $^{18}O$. *Khim. Perekisnykh Soedin., Akad. Nauk. SSSR. Inst. Obshch. i. Neorgan. Khim.* 249–251; *Chem. Abstr.* **61**, 5458 (1965).

Brown, S. A. 1963. The formation of natural coumarins. *Lloydia* **26**, 211–222.

Brown, S. A. 1965. Chemotaxonomic aspects of lignins. *Lloydia* **28**(4), 332–341.

Brown, W. L. 1940. The anthocyanin pigment of the Hunt muscadine grape. *J. Am. Chem. Soc.* **62**, 2808–2810.

Brumsted, D. D., Lautenbach, A. F., and West, D. B. 1965. Phenolic characteristics in brewing. III. The role of yeast. *Am. Soc. Brewing Chemists, Proc., Ann. Meet.* 142–145.

Brunet, R. 1912. L'analyse et la constitution du raisin. *Rev. Viticult.* **37**, 15–20.

Bujard, M. 1968. Die Maischeerwärmung, eine neue Technologie. *Schweiz. Z. Obst. Weinbau* **104**, 49–55, 88–91.

Bukharov, P. S., and Fertman, V. K. 1962a. Polarographic determination of gallic acid in tannins of wines and brandies. *Metody Issledovatel. v Vinodelii, Tsentr. Nauch.-Issledovatel. Lab. Vinodeliya i Sev. Vinogradarstva* (1), 34–40; *Chem. Abstr.*, **60**, 2298 (1960).

Bukharov, P. S., and Fertman, V. K. 1962b. Sravnitel'naya otsenka metodov opredeleniya dubul'nykh veshchestv v vinakh. *Metody Issledovatel. v. Vinodelii, Tsentr. Nauch.-Issledovatel. Lab. Vinodeliya i Sev. Vinogradarstva* (1), 41–48.

Bukharov, P. S., and Fertman, V. K. 1962c. Razrabotka metodiki opredeleniya ryadovykh-(1,2,3) i orto-(1,2) gidroksilov v dubil'nykh veshchestvakh vinogradnykh vin i kon'yakov. *Metody Issledovatel. v Vinodelii, Tsentr. Nauch.-Issledovatel. Lab. Vinodeliya i Sev. Vinogradarstva* (1), 49–55.

Burkhardt, R. 1963. Einfache und schnelle quantitative Bestimmung der kondensierbaren Gerbstoffe in weissen weinen und Tresterweinen. *Weinberg Keller* **10**, 274–285.

Burkhardt, R. 1965. Nachweis der *p*-Cumarylchinasäure in Weinen und das Verhalten der Depside bei der Kellerbehandlung. *Mitt.* (*Klosterneuburg*) **15A**, 80–86.

Burton, H. S., McWeeny, D. J., and Pandhi, P. N. 1963. Nonenzymic browning. Browning of phenols and its inhibition by sulfur dioxide. *Nature* **199**, 659–661.

Butler, W. L., and Siegelman, H. W. 1959. Conversion of caffeic acid to esculetin during paper chromatography. *Nature* **183**, 1813–1814.

Cahn, R. S., Ingold, C. K., and Prelog, V. 1956. The specification of asymmetric configuration in organic chemistry. *Experientia* **12**, 81–124.

Caldwell, J. S. 1925. Some effects of seasonal conditions upon the chemical composition of American grape juices. *J. Agr. Res.* **30**, 1133–1176.

Campbell, I. 1966. Yeast biochemistry. *Brewers Dig.* **41**(4), 78–83.

Cantarelli, C. 1955. Biochemical aspects of red wine accelerated aging by film-forming strains of *Saccharomyces oviformis*. *Biochim. Appl.* **2**, 167–190; *Chem. Abstr.* **51**, 7645 (1957).

Cantarelli, C. 1958. Invecchiamento naturale lento ed invecchiamento accelerato del vino in rapporto ai caratteri organolettici ed al valore biologico del prodotto. *Accad. Ital. Vite Vino, Siena, Atti* **10**, 192–227.

Cantarelli, C., 1963a. La stabilizzazione contro la maderizzazione mediante eliminazione dei leucoantociani. *Ital. Vinicola Agrar.* **53** (10), 335–337.

Cantarelli, C. 1963b. I trattamenti con resine poliammidiche in enologia. *Accad. Ital. Vite Vino, Siena, Atti* **14**, 219–249.

Cantarelli, C. 1966. Il colore dei vini bianchi. *Il corso nazionale di aggiornamento per enotechnici* (G. Battista Cerletti, Conegliano), pp. 261–283.

Cantarelli, C., and Peri, C. 1964a. The leucoanthocyanins in white grapes: their distribution, amount, and fate during fermentation. *Am. J. Enol. Viticult.* **15**, 146–153.

Cantarelli, C., and Peri, C. 1964b. Le cause e la prevenzione della maderizzazione dei vini bianchi. 2. Esame di alcuni trattamenti protettivi. *Vini d'Italia* **6**, 337–347.

Cappelleri, G. 1965. Risultati di un'indagine sulla ricerca della malvina in una serie di vini di *Vitis vinifera*. *Accad. Ital. Vite Vino, Siena, Atti* **17**, 153–159.

Cappelleri, G., Calò, A., and Liuni, C. S. 1964. Sull'eredità del carattere diglicoside del malvidolo nella sostanza colorante delle bacche del genere *Vitis* L. *Riv. Viticolt. Enol.* **17**, 284–302.

Cappelleri, G., Calò, A., and Liuni, C. S. 1966. Ulterione contributo allo studio dei pigmenti antocianici di alcune Ampelidee. *Accad. Ital. Vite Vino, Siena, Atti* **18**, 299–301.

Caputi, A., Jr., and Peterson, R. G. 1965. The browning problem in wines. *Am. J. Enol. Viticult.* **16**, 9–13.

Carles, J., and Alquier-Bouffard, A. 1962. Les acides organiques de la vigne et leurs gradients. *Compt. Rend.* **254**, 925–927.

Carles, J., and Lamazou-Betbeder, M. 1958. Principales différences des acides organiques et des acides aminés libres entre le jus de raisin et le vin. *Compt. Rend.* **247**, 2181.

Carles, J., and Pech, R. 1960. L'acide gallique et l'acide protocatéchique, produits de dégradation des anthocyanes du vin. *Compt. Rend.* **251**, 2764–2766.

Carles, J., Lamazou-Betbeder, M., and Pech, R. 1960. De quelques acides organiques de degradation trouvés dans le vin qui vieillit. *Bull. Soc. Franc. Physiol. Végétale* **6**, 80–81.

Carles, J., Lamazou-Betbeder, M., and Pech, R. 1961. Etat actuel de nos connaissances sur quelques actions bactériennes dans les vins. *Ann. Technol. Agr. Paris* **10**, 61–71.

Carles, P. 1919. Sur la casse bleue des vins. *Compt. Rend.* **169**, 1422–1423.

Carles, P., and Nivière, G. 1897. Influence des matières colorantes sur la fermentation des vins rouges très colorés. *Compt. Rend.* **125**, 452–453.

Carpenter, J. A., Wang, Y. P., and Powers, J. J. 1967. Effect of anthocyanin pigments on certain enzymes. *Proc. Soc. Exptl. Biol. Med.* **124**, 702–706.

Carpentieri, F. 1908. Color reactions of some hybrid grapes. *Staz. Sper. Agrar. Ital.* **41**, 637–656; *Chem. Abstr.* **4**, 1869 (1910).

Carpentieri, F. 1948. "Enologia Teorico-Pratica," 2 Vols., 14th ed., p. 1063. Fratelli Ottavi, Casale Monferatto.

Casas Lucas, J. F. 1965. Phénomènes d'oxydo-réduction au cours de la maturation et du vieillissement du vin et procédés pour les modifier. Rapport Espagnol. *Bull. Offic. Intern. Vigne Vin* **38**, 1406–1427.

Casas Lucas, J. F. 1967. Evolution des oxydases dans les moûts et les vins suivant les procédés de vinification et de traitement. *Bull. Offic. Intern. Vigne Vin.* **40**(442), 1369–1378.

Caspar, J. N., and Yang, H. Y. 1957. Vitamin $K_5$ as a preservative for wine. *Food Technol.* **11**(10), 536–540.

Cassignard, R. 1963. Emploi du froid dans le traitment des vins. *Ann. Technol. Agr. Paris Numero Hors Ser. I.* **12**(1), 331–341.

Castor, J. G. B. 1953a. The free amino acids of musts and wines. I. Microbiological estimation of 14 amino acids in California musts. *Food Res.* **18**, 139–145.

Castor, J. G. B. 1953b. The free amino acids of musts and wines. II. The fate of amino acids of must during alcoholic fermentation. *Food Res.* **18**, 146–151.

Chapon, L., and Urion, E. 1960. Ascorbic acid and beer. *Wallerstein Labs. Communs.* **23**, 38–44.

Chaudhary, S. S., Webb, A. D., and Kepner, R. E. 1968. GLC investigation of the volatile compounds in extracts from Sauvignon blanc wines from normal and botrytised grapes. *Am. J. Enol. Viticult.* **19**, 6–12.

Chauveaud, G. 1894. Sur les caractères internes de la graine des vignes, et leur emploi dans la détermination des espèces et la distinction des hybrides. *Compt. Rend.* **118**, 485–487.

Chauvin, A. C. 1909. Action of hydrogen peroxide on different alcohols and common spirits. *Mon. Sci.* **23**, 567; *Chem. Abstr.* **3**, 2726 (1909).

Chelle, J. L. 1925. L'acide salicylique et l'acide benzoique dans les vins normaux. *Ann. Fals. Fraudes* **18**, 134–148.

Chen, L. F., and Luh, B. S. 1967. Anthocyanins in Royalty grapes. *J. Food Sci.* **32**, 66–74.

Chen, S. L., and Peppler, H. J. 1956. Conversion of cinnamaldehyde to styrene by a yeast mutant. *J. Biol. Chem.* **221**, 101–106.

Chevalier, P., Chollot, B., Chapon, L., and Urion, E. 1961. Contribution des levures a la stabilité de la bière: leur pouvoir adsorbant sur les tannoides. *European Brewery Conv. Proc. Congr.* **8**, 246–259.

Chiancone, F. M. 1949. Esters of gallic acid. *Boll. Soc. Ital. Biol. Sper.* **25**, 627–629; *Chem. Abstr.* **44**, 8387 (1950).

Child, J. J., Simpson, F. J., and Westlake, D. W. S. 1963. Degradation of rutin by *Aspergillus flavus*. Production, purification, and characterization of an esterase. *Can. J. Microbiol.* **9**, 653–664.

Ciusa, W., and Barbiroli, G. 1967. La misura delle caratteristiche del colore nei vini. *Riv. Viticolt. Enol.* **20**, 433–458.

Clarens, J. 1921. Application des lois de la cinétique chimique à l'analyse quantitative: Du dosage fractionné des tannins en général et des tannins du vin, en particulier. *Bull. Soc. Chim. France* **29**, 837–852.

Clark, A. G. 1960. Nonbiological haze in bottled beers: a survey of the literature. *J. Inst. Brewing* **66**, 318–330.

Clark, W. G., and Geissman, T. A. 1948. The role of flavonoids and related substances in biological oxidations. *Conf. Biol. Antioxidant Trans* **3**, 92–112.

Clark-Lewis, J. W. 1958. Stereochemistry of catechins and related flavan derivatives. *In* "Current Trends in Heterocyclic Chemistry" (A. Albert, G. M. Badger, and C. W. Shoppee, eds.), pp. 40–50. Butterworth, London.

Clark-Lewis, J. W. 1962. Configurations of optically active flavonoid compounds. *Rev. Pure Appl. Chem.* **12**, 96–116.

Clore, W. J., Neubert, A. M., Carter, G. H., Ingalsbe, D. W., and Brummund, V. P. 1965. Composition of Washington-produced Concord grapes and juices. *Wash. State Coll. Agr. Expt. Sta. Tech. Bull.* **48**, 1–21.

Clore, W. J., Nagel, C. W., Brummund, V. P., and Carter, G. H. 1967. Grape variety responses and wine making trials in central Washington. *Wash. State Coll. Agr. Expt. Sta. Circ.* **477**, 1–11.

Coffelt, R. J., and Berg, H. W. 1965. Color extraction by heating whole grapes. *Am. J. Enol. Viticult.* **16**, 117–128.

Colagrande, O. 1961. Sul dosaggio del tannino nei vini. *Riv. Viticolt. Enol.* **14**, 323–331.

Colagrande, O., and Corradini, C. 1967. Effetto detannizzante sui vini bianchi delle resine adsorbenti. *Ind. Agr.* (*Florence*) **5**, 169–173.

Combes, R. 1922. La recherche des pseudo-bases d'anthocyanidines dans les tissues végétaux. *Compt. Rend.* **175**, 58–61.

Comrie, A. A. D. 1930. The brewing value of hop tannin. *J. Inst. Brewing* **36**, 307–311.

Cook, J. A., Lynn, C. D., and Kissler, J. J. 1960. Boron deficiency in California vineyards. *Am. J. Enol. Viticult.* **11**, 185–194.

Cook, M. T., and Traubenhaus, J. J. 1911. The relation of parasitic fungi to the contents of the cells of the host plants. I. The toxicity of tannin. *Delaware Univ. Agr. Expt. Sta. Bull.* **91**, 1–77.

Copeman, P. R. v.d. R. 1927. Some changes occurring during the ripening of grapes. *Union S. Africa Dept. Agr. Div. Chem., Sci. Bull.* **60**, 1–19.

Cornforth, J. W. 1939. The anthocyanin of *Vitis hypoglauca* F. v. M. *J. Roy. Soc. New S. Wales.* **72**, 325–328.

Corse, J. 1965. The enzymatic browning of fruits and vegetables. *Plant Phenolics Group N. Am. Proc. 4th Symp.* **1964**, 41–62.

Corse, J., Lundin, R. E., and Waiss, A. C., Jr. 1965. Identification of several components of isochlorogenic acid. *Phytochemistry* **4**, 527–529.

Corse, J., Ludin, R. E., Sondheimer, E., and Waiss, A. C., Jr. 1966. Conformation analyses of D-(—)-quinic acid and some of its derivatives by nuclear magnetic resonance. *Phytochemistry* **5**, 767–776.

Coste-Floret, P. 1901. "Les résidus de la vendage," p. 239. Masson, Paris.

Cotea, V., and Tirdea, C. 1959. Tratamente aplicate butoaielor noi de Stejar in vederea detaninizarii si influenta lor asupra calitatii vinului. *Gradina Via Si Livada* **7**, 31–35.

Cottier, H. 1924. Les pépins de raisins et l'alimentation des animaux. *Progr. Agr. Vit.* **82**, 179–182.

Coudon, H., and Pacottet, P. 1901a. Du tanin dans les vins. *Rev. Viticult.* **15**, 61–63.

Coudon, H., and Pacottet, P. 1901b. De l'influence du tanin sur la fermentation et la couleur des vins rouges. *Rev. Viticult.* **15**, 121–124.

Coudon, H., and Pacottet, P. 1901c. Le *Botrytis cinerea*, le tannin et la coloration des vins rouges. *Rev. Viticult.* **15**, 145–147.

Crawford, C. M., Bouthilet, R. J., and Caputi, A., Jr. 1958. A review and study of color standards for white wines. *Am. J. Enol.* **9**, 194–201.

Crawford, R. M. M. 1967. Phenoloxidase activity and flooding tolerance in higher plants. *Nature* **214**, 427–428.

Creasy, L. L., and Swain, T. 1965. Structure of condensed tannins. *Nature* **208**, 151–153.

Creasy, L. L., and Swain, T. 1966. Flavan production in strawberry leaves. *Phytochemistry* **5**, 501–509.

Cronenberger, L., Vallet, C., Netien, G., and Mentzer, C. 1955. The flavondiols of plant tissue cultures. *Compt. Rend.* **241**, 1161–1163.

Cross, F. B., and Webster, J. E. 1934. Physiological studies of uneven ripening of Concord grapes. *Proc. Am. Soc. Hort. Sci.* **32**, 361–364.

Cross, F. B., and Webster, J. E. 1935. Color development of Concord grapes associated with the number of leaves per cane. *Proc. Am. Soc. Hort. Sci.* **33**, 440–441.

Crowther, R. F. 1959–1960. Colour extraction during the fermentation of Concord grape. *Rept. Hort. Expt. Sta., Vineland, Ontario,* 88–93.

Crowther, R. F., Neudoerffer, N. C., and Bradt, O. A. 1964. Location effects on wines from the Concord grape. *Rept. Hort. Expt. Sta., Vineland, Ontario,* 112–113.

Cruess, W. V. 1920. Commercial production of grape syrup. *Calif. Univ. Agr. Expt. Sta. Bull.* **321**, 401–416.

Cruess, W. V. 1934. Fermentation of a red wine. *Fruit Prods. J.* **13**, 199–204, 207, 212, 214, 217.

Cruess, W. V. 1935. Some observations on wine tannin. *Wines and Vines* **16**(1), 5–7.

Curtel, G. 1904. Dé l'influence de la greffe sur la composition du raisin. *Compt. Rend.* **139**, 491–493.

Curtis, N. S. 1966. Head and haze—a review. *J. Inst. Brewing* **72**, 240–250.

Curtis, N. S. 1968. Tannins in brewing—today's picture. *Brewer's Dig.* **43**, 56–59, 64.

Dakin, H. D. 1914. The formation of benzoyl carbinol and other substances from phenyl glyoxal by the action of fermenting yeast. *J. Biol. Chem.* **18**, 91–93.

Dalal, K. B., and Salunkhe, D. K. 1964. Thermal degradation of pigments and relative biochemical changes in canned apricots and cherries. *Food Technol.* **18**, 1198–1200.

dal Piaz, A. 1885. "Die Verwertung der Weinruckstande. Praktische Anleitung zur rationellen. Verwerten von Weintrester, Weinhefe und Weinstein," p. 190. Hartleben, Vienna.

dal Piaz, A. 1892. The utilization of wine residues. *Calif. Bd. State Viticult. Commissioners, Appendix C., Ann. Rept.* **1891–1892**, 141–191.

Daravingas, G., and Cain, R. F. 1965. Changes in the anthocyanin pigments of raspberries during processing and storage. *J. Food Sci.* **30**, 400–405.

Daravingas, G., and Cain, R. F. 1968. Thermal degradation of black raspberry anthocyanin pigment in model systems. *J. Food Sci.* **33**, 138–142.

Das, D. N. 1963. Effect of tea and its tannins upon capillary resistance of guinea pigs. *Ann. Biochem. Exptl. Med.* (*Calcutta*) **23**(6), 219–222; *Chem. Abstr.* **59**, 13276 (1963).

da Silva, F. 1911. La vinification des vins de Madère et leur composition. *Ann. Fals. Fraudes* **4**, 57–65.

Davy, H. 1819. "Elements of Agricultural Chemistry," 2nd ed., p. 304. Hudson, Hartford, Connecticut.

de Almeida, H. 1962. Interprétation des phénomènes d'oxydo-réduction au cours du vieillissement du vin de "Porto" en bouteilles. *Bull. offic. Intern. Vigne Vin* **35**(371), 60–84.

Dean, F. M. 1952. Naturally occurring coumarins. *Progr. in Chem. Org. Natural Prods.* **9**, 225–291.

Dean, F. M. 1963. "Naturally Occurring Oxygen Ring Compounds," p. 661. Butterworths, London.

de Boixo, P. 1950. Observations sur le rôle de la radiation francaise dans la maturation du raisin faites en 1948 et 1949. *Bull. Offic. Intern. Vigne Vin* **23**(233), 37–41.

DeEds, F. 1949. Vitamin P properties in grapes and grape residues. *Proc. Wine Technol. Conf., Univ. Calif., Davis*, 48–51.

DeGori, R., and Grandi, F. 1955. Color curves of Brolio wine in Chianti of diverse ages. *Boll. Lab. Chim. Provinciali* (*Bologna*) **6**, 119–121; *Chem. Abstr.* **50**, 11605 (1956).

Deibner, L. 1952. Etat actuel du dosage direct des substances colorantes du raisin et du vins rouges d'apres les récents travaux. *Ann. Technol. Agr. Paris* **1**, 345–357.

Deibner, L., and Bourzeix, M. 1963. Sur l'extraction des composés polyphénoliques colorés des vins rouges a l'aide de l'acetate de plomb. *Ann. Technol. Agr. Paris* **12**, 287–312.

Deibner, L., and Bourzeix, M. 1965. Etat actuel de la detection des anthocyannes diglucosides dans les vins et les jus de raisin. *Ann. Fals. Expert. Chim.* **58**, 149–159.

Deibner, L., and Bourzeix, M. 1966. Über die Gesamtextraktion der Anthocyane aus den Schalen von roten Weintrauben. *Mitt.* (*Klosterneuburg*) **16**, 200–206.

Deibner, L., and Rifai, H. 1963a. Die Polyphenoloxydase im Traubensaft. I. Über die Bestimmung der Aktivität der Polyphenoloxydase mit Hilfe der elektrophotometrischen Metode mit Purpurogallin. *Mitt.* (*Klosterneuburg*) **13A**, 56–70.

Deibner, L., and Rifai, H. 1963b. Die Polyphenoloxydase. II. Veränderungen der Polyphenoloxydaseaktivität im Traubensaft während der Konservierung. *Mitt.* (*Klosterneuburg*) **13A**, 113–119.

Deibner, L., Bourzeix, M., and Cabibel Hugues, M. 1966. Sur la valeur analytique de l'-isolement des anthocyannes des vins et des jus de raisin au moyen des acetates de plomb et des résines échangeuses d'ions. *Ann. Fals. Expert. Chim.* **59**, 39–47.

Dekker, J. 1913. "Die Gerbstoffe. Botanische-Chemische Monographie der Tannide," p. 636. Gebrüder Borntraeger, Berlin.

Delaveau, P. G. 1964. Essai de localisation de la biogenèse des flavonoïdes dans le genre *Tropaelum*. I. Isolement de rutoside de *T. peregrinum* L. *Compt. Rend.* **258**, 318–320.

Delmas, J., Poitou, N., and Levadou, B. 1963. Les acides organiques dans la feuille de vigne. *Ann. Agron.* **14**, 951–965.

Denisova, T. V. 1965. The phenol complex in phylloxera-infested grape roots as a resistance factor. *Vestn. Sel'skokhoz. Nauki, Min. Sel'sk. Khoz. SSSR* **10**(5), 114–118; *Chem. Abstr.* **64**, 20214 (1966).

Dennison, R. A., and Stover, L. H. 1962. Blue Lake grape and its potential for processed juice. *Proc. Florida State Hort. Soc.* **75**, 281–283.

de Villiers, F. J. 1926. Physiological studies of the grape. I. The anatomy of the grape berry and the distribution of its primary chemical constituents. *Union S. Africa Dept. Agr., Sci. Bull.* **45**, 6–15.

de Villiers, J. P. 1961. The control of browning of white table wines. *Am. J. Enol. Viticult.* **12**, 25–30.

Deyoe, C. W., Deacon, L. E., and Couch, J. R. 1962. Citrus bioflavonoids in broiler diets. *Poultry Sci.* **41**, 1088–1090.

Dieci, E. 1967. Sull'enocianina tecnica *Riv. Viticolt. Enol.* **20**, 567–572.

Diemair, W., and Polster, A. 1967a. Über Gerbstoffe im Rotwein. I. Dünnschichtchromatographie und präparative Isolierung der Gerbstoffkomponente. *Z. Lebensm.-Unters. Forsch.* **134**, 80–86.

Diemair, W., and Polster, A. 1967b. Über Gerbstoffe im Rotwein. II. Eigenschaften und Isolierung. *Z. Lebensm.-Unters. Forsch.* **134**, 345–352.

Diemair, W., Janecke, H., and Krieger, G. 1951a. Über eine Methode der Gerbstoffbestimmung in der Rebe tnd im Wein. I. *Z. Anal. Chem.* **133**, 346–352.

Diemair, W., Janecke, H., and Krieger, G. 1951b. Über eine Methode der Gerbstoffbestimmung in der Rebe und im Wein. II. Das Schicksal des Gerbstoffes der Weinrebe während des Wachstums. *Z. Anal. Chem.* **133**, 353–359.

Diemair, W., Koch, J., and Hess, D. 1960. Über der Einfluss der schwefligen Saure und L-Ascorbinsäure bei der Weinbereitung. II. Inaktivierung der Polyphenoloxydase. *Z. Lebensm.-Untersuch. Forsch.* **113**, 381–387.

Dimotaki-Kourakou, V. 1967. Chauffage préalable de la vendage rouge. *Bull. Offic. Intern Vigne Vin* **40**(431), 50–81.

Draczynski, M. 1950–1951. Über die bakterizide Wirkung von Wein und Traubenmost, mit Bacterium coli als Test. *Wein Rebe, Jahrb. Weinbauwiss. Önologie, Deut. Wein-Ztg.*, 25–42.

Drawert, F. 1961. Über Anthocyane in Trauben, Mosten and Weinen, Methoden zur Anreicherung und Trennung der Farbkomponenten. *Vitis.* **2**, 288–304.

Drawert, F., and Gebbing, H. 1967a. Bound phenoloxidase and its release from plant cell structures. *Naturwissenschaften* **54**, 226–227.

Drawert, F., and Gebbing, H. 1967b. Sprühtest zur Prüfung der Substratspezifität von Phenoloxidasen. *Z. Naturforsch.* **22B**, 559–560.

Dreaper, W. P. 1910. Tannins. *In* "Allen's Commercial Organic Analysis" (W.A. Davis, and S. S. Sadtler, eds.), Vol. 5, pp. 1–104. McGraw-Hill (Blakiston), Philadelphia.

Drewes, S. E., and Roux, D. G. 1964. Stereochemistry of flavan-3,4-diols. *Chem. & Ind.* **1964** (43), 1799–1800.

Drews, B., Specht, H., and Schwartz, E. 1965. Die aromatischen Alkohole im Bier. *Monatschr. Brau.* **18**, 240–242.

Dubaquie, J. 1925. Oxidation et aération des vins en futailles. *Chim. & Ind. Paris* (Spec. No.), 606–607.

Dupuy, P., and Puisais, J. 1955. Contribution à l'étude des colorantes flavoniques chez certaines Ampelidées. *Compt. Rend.* **241**, 48–50.

Durmishidze, S. V. 1948a. Kolichestvennoe opredelenie enidina v vinograde i v vine. *Biokhimiya* **13**, 16–22.

Durmishidze, S. V. 1948b. Kolichestvennoe opredelenie tannina v krasnom vinograde i v vine. *Biokhim. Vinodeliya, Akad. Nauk SSSR.* **2**, 169–176.

Durmishidze, S. V. 1950. Polifenoloksidaza vinograda i ee rol' v tekhnologii vinodeliya. *Biokhimiya* **15**, 58–66.

Durmishidze, S. V. 1951. l'Gallokatekhin v sostave dubil'nykh veshchestv vinograda. *Dokl. Akad. Nauk. SSSR.* 77, 859–862.

Durmishidze, S. V. 1955. "Dubil'nye Veshchestva i Antotsiany Vinogradnoi Lozi i Vina," p. 323. *Izda. Akad. Nauk. SSSR,* Moscow.

Durmishidze, S. V. 1958. Vitamin P v vinograde. *Vinodelie i Vinogradarstvo SSSR.* **18**(2), 15.

Durmishidze, S. V. 1966. Biokhimicheskie protsessy pri pererabotke vinograda. *Tr. Akad. Nauk Gruz. SSR, Lab. Biokhim. Rastenii,* 252–273.

Durmishidze, S. V., and Khachidze, O. T. 1952. Fotometricheskii metod opredeleniya krasnykh krasyashchikh veshchestv v vinograde i vine. *Vinodelie i Vinogradarstvo SSSR.* 12(1), 18–20.

Durmishidze, S. V., and Nutsubidze, N. N. 1954. Khromatograficheskoe issledovanie dubil'nykh veshchestv vinogradnoi lozy. *Dokl. Akad. Nauk SSSR.* **96**, 1197–1199.

Durmishidze, S. V., and Nutsubidze, N. N. 1958. Antotsianovye pigmenty vinograda. *Soobshcheniya Akad. Nauk Gruzin SSR* **21**, 677–684.

Durmishidze, S. V., and Tsiskarishvili, T. P. 1950. Prevrashcheniya dubil'nykh veshchestv vinogradnoi grozdi v protsesse ee sozrevaniya. *Biokhim. Vinodeliya, Akad. Nauk SSSR.* **3**, 7–25.

Durmishidze, S. V., Urushadze, Z. D., and Ushakova, M. P. 1961. Semena vinograda kak istochnik bioflavonoidov. *Vinodelie i Vinogradarstvo SSSR* **21**(6), 8–10.

Durquety, P. M. 1958. La génétique des anthocyanes des raisins. *Qualitas Plant. Mater. Végétabiles* **3–4**, 500–507.

Dutoit, P., and Duboux, M. 1909. Quelques résultats de l'analyse physico-chimique des vins. *Bull. Soc. Vaudoise* **45**, 417–461.

Dzhanpoladyan, L. M., and Dzhanazyan, R. S. 1962. Prevrashcheniya uglevodov drevesiny duba v kon'yachnom spirte. *Vinodelie i Vinogradarstvo SSSR* **22**(4), 15–18.

Earle, F. R., and Jones, Q. 1962. Analyses of seed samples from 113 plant families. *Econ. Botany* **16**, 221–250.

Egorov, I. A., and Borisova, N. B. 1957. Aromatitcheskie al'degidy kon'yachnogo spirta. *Biokhim. Vinodeliya* **5**, 27–37.

Egorov, I. A., and Egofarova, R. Kh. 1965. Role of oak tree lignin in cognac-brandy production. *Prikl. Biokhim. i Mikrobiol.* 1(6), 680–683; *Chem. Abstr.* **64**, 11821 (1966).

Ehrlich, F. 1907. Über die Bedingungen der Fuselölbildung und über den Zussammenhang mit dem Eiweissabbau der Hefe. *Ber. Deut. Chem. Ges.* **40B**, 1027–1047.

Ekster, Ya. E. 1957. The connection between fat and tannin substances in grape seeds. *Tr. Odessk. Sel'kokhoz. Inst.* **8**, 89–92; *Chem. Abstr.* **53**, 507 (1959).

Emiliani, E. 1940. Influence on digestive processes of wine and some substances contained in it. *Ann. Chim. Appl.* **30**, 429–438; *Chem. Abstr.* **35**, 2976 (1941).

Érczhegyi, L., and Mercz, Á. 1966. A szölöfeldolgozás és sajtolás hatása a must minöségére. *Borgazdasag* 14(3), 84–88.

Erdmann, J. 1879. Ueber die Veränderlichkeit des Rothweinfarbstoffes. *Ber. Deut. Chem. Ges.* **11**, 1870–1876.

Esau, K. 1948. Phloem structure in the grapevine, and its seasonal changes. *Hilgardia* **18**, 217–296.

Esau, K. 1965. "Plant Anatomy," 2nd. ed., p. 767. Wiley, New York.

Esau, P. 1967. Pentoses in wine. I. Survey of possible sources. *Am. J. Enol. Viticult.* **18**, 210–216.

Esau, P., and Amerine, M. A. 1966. Quantitative estimation of residual sugars in wine. *Am. J. Enol. Viticult.* **17**, 265–267.

Escudié, P. 1939. Sur les tanins, en particulier sur celui de la vigne. *Rev. Viticult.* **91**, 175–177.

Evelyn, S. R., Maihs, E. A., and Roux, D. G. 1960. Condensed tannins. V. The oxidative condensation of (+)-catechin. *Biochem. J.* **76**, 23–27.

Fabre, J. 1909. Sur la valeur alimentaire du marc de raisin. *Progr. Agr. Viticult.* **52**, 362–368.

Fabre, R., Rougier, G., and Rougié, J. 1950. Sur la toxicité comparée du vin et de son eau-de-vie de premier bout. *Compt. Rend. Soc. Biol.* **144**, 1498–1499.

Fabris, A. 1966. Chemical and dietary characteristics of wine by-products. *Aliment. Anim.* **10** (11), 509–516; *Chem. Abstr.* **66**, 84732 (1967).

Fairbairn, J. W. 1959. "The Pharmacology of Plant Phenolics," p. 151. Academic Press, New York.

Fallot, B. 1909a. Sur la possibilité de distinguer chimiquement les vins de raisins blancs des vins de raisins rouges vinifiés en blanc. *Rev. Viticult.* **32**, 126.

Fallot, B. 1909b. L'aldehyde acétique dans le vin. *Rev. Viticult.* **31**, 586–590.

Feather, M. S., and Harris, J. F. 1965. The acid-catalyzed hydrolysis of glycopyranosides. *J. Org. Chem.* **30**, 153–157.

Feenstra, W. J. 1962. The genetics of flavonoid formation. *Planta Med.* **10**, 412–420.

Ferenczi, S. 1966. Studio comparitivo di alcune bentoniti nei rignardi della stabilizzazione proteica dei vini. *Ind. Agrar.* (*Florence*) **4**, 469–474.

Fernández, O., and Sanchez-Gavito, P. 1949. Filogenia quimica de la lignina. *Farm. Nueva Madrid* **14**, 553–561.

Ferrando, R., and Catsaounis, N. 1966. Étude de la digestibilité du tourteau de pépins de raisins chez le mouton. *Compt. Rend. Acad. Agr. France* **52**, 218–223.

Ferré, L. 1926. Autolyse de la matière colorante dans les raisins entiers soumis a l'action de la chaleur humide – application a la vinification des vins rouges. *Compt. Rend. Acad. Agr. France* **12**, 370–381.

Ferré, L. 1928. Vinification des vins de Bourgogne par chauffage prealable des raisins. *Rev. Viticult* **69**, 5–11, 21–29.

Ferré, L., and Michel, A. 1937. Hydrolyse de la matière colorante des vins. *Compt. Rend. Acad. Agr. France* **33**, 239–241.

Fessler, J. H. 1939. The composition of wine in California discloses a rich sectional variety in grape types and wine chemistry. *Wine Rev.* 7(1), 12–15.

Fessler, J. H. 1947. Some technical notes on tannin content in wines. *Wine Rev.* **15**(5), 12–13.

Fischer, F. G., and Wiedemann, O. 1934. Biochemische Hydrierungen. I. Über die Hydrierung ungesättigter α-Ketosäuren, Aldehyde und Alkohole durch gärende Hefe. *Ann. Chem., Liebigs* **513**, 260–280.

Fischer, F. G., and Wiedemann, O. 1935. Biochemische Hydrierungen. II. Über die Hydrierung ungesattigter Ketone durch gärende Hefe. *Ann. Chem., Liebigs* **520**, 52–70.

Fitelson, J. 1967. Paper chromatographic detection of adulteration in Concord grape juice. *J. Assoc. Offic. Agr. Chemists* **50**, 293–299.

Flanzy, M. 1935a. Sur une cause d'arret de la fermentation alcoolique dans la vinification et d'égrappage. *Rev. Viticult.* **83**, 217–222.

Flanzy, M. 1935b. Sur la vinification des vins rouges. *Rev. Viticult.* **83**, 315–319.

Flanzy, M. 1967. Quelques nouveaux procédés de vinification. *Bull. Offic. Intern. Vigne Vin* **40**(433), 283–300.

Flanzy, M., and Aubert, S. 1967. Observations sur la défécation des vins de *Vitis vinifera* et d'hybrides. Essai de differenciation des vins d'hybrides blancs. *Ann. Technol. Agr. Paris* **17**, 241–250.

Flanzy, M., and Causeret, J. 1952. Contributions a l'étude physiologique des boissons alcooliques. I. Etude comparée d'un vin et de l'alcool. *Ann. Technol. Agr. Paris* **1**, 227–240.

Flanzy, M., and Poux, C. 1958. Les possibilités de la microvinification, application a l'étude de la macération. *Ann. Technol. Agr. Paris* 7, 377–401.

Flanzy, M., Causeret, J., Hugot, D., and Guerillot, J. 1955. Contribution a l'étude des boissons alcooliques. II. Étude comparée de differents vins et de diverses techniques oenologiques. *Ann. Technol. Agr. Paris* 4, 359–380.

Flanzy, M., Puisais, J., Combescot, C., Demaret, J., Reynouard-Brault, F., and Igert, C. 1965. Contribution a l'étude physiologique des actions comparées du vin et de l'-éthanol, rôle du régime alimentaire, importance du mode de vinification. *Ann. Technol. Agr. Paris* **14**, 263–274.

Flanzy, C., André, P., Flanzy, M., and Chambroy, Y. 1967a. Variations quantitatives des acids organiques stables, non cétoniques, non volatils, dans les baies de raisin placées en anaérobiose carbonique. I. Influence de la temperature. *Ann. Technol. Agr. Paris* **16**, 27–34.

Flanzy, C., André, P., Flanzy, M., and Chambroy, Y. 1967b. Variations quantitatives des acides organiques stables, non cétonique, non volatils, dans les baies de raisin placées en anaérobiose carbonique. II. Influence de la durée d'anaérobiose. *Ann. Technol. Agr. Paris* **16**, 89–107.

Fleming, H. P., and Etchells, J. L. 1967. Occurrence of an inhibitor of lactic acid bacteria in green olives. *Appl. Microbiol.* **15**, 1178–1184.

Flerov, A. F., and Kovalenko, E. I. 1947. Effect of growth substances and alkaloids on the sprouting seeds of the grape. *Dokl. Akad. Nauk SSSR* **58**, 677–679; *Chem. Abstr.* **46**, 4723 (1952).

Florenzano, G. 1952. Film-forming wine yeasts with reference to sherry making and other applications. II. Zymotechnical properties. *Agr. Ital. (Pisa)* 7, 177–199; *Chem. Abstr.* **48**, 2980 (1954).

Folger, A. H. 1940. The digestibility of ground prunes, winery pomace, avocado meal, asparagus butts, and fenugreek meal. *Calif. Univ. Agr. Expt. Sta. Bull.* **635**, 1–11.

Folin, O., and Ciocalteu, V. 1927. On tyrosine and tryptophane determinations in proteins. *J. Biol. Chem.* **73**, 627–650.

Folin, O., and Denis, W. 1912. On phosphotungstic-phosphomolybdic compounds as color reagents. *J. Biol. Chem.* **12**, 239–243.

Fonzes-Diacon, and Laforce, 1926. L'acide salicylique conservateur des vins soumis a l'expertise. *Ann. Fals. Fraudes* **19**, 99–102.

Forbes, M., Zilliken, F., Roberts, G., and György, P. 1958. A new antioxidant from yeast. *J. Am. Chem. Soc.* **80**, 385–389.

Frater, G. 1924. The composition of ripe wine grapes from the government viticultural station, Paarl. *Union S. Africa, Dept. Agr., Div. Chem., Sci. Bull.* **32**, 1–30.

Fresenius, W., and Grünhut, L. 1899. Beiträge zur Kenntniss der Tresterwein. *Z. Anal. Chem.* **38**, 472–512.

Fresenius, W., and Grünhut, L. 1921. Beiträge zur chemischen Analyse des Weines. 21. Bestimmung des Gerbstoffs und Farbstoffs. *Z. Anal. Chem.* **60**, 406–417.

Freudenberg, K. 1924. Mitteilung über Gerbstoffe und ähnliche Verbindungen. 16. Raumisomere Catechine. IV. *Ann. Chem., Liebigs* **437**, 274–285.

Freudenberg, K. 1956. Catechins and related substances. *Sci. Proc. Roy. Dublin Soc.*, **27**, 153–160.

Freudenberg, K., and Weinges, K. 1960. Systematik und Nomenklatur der Flavonoide. *Tetrahedron* **8**, 336–349.

Freudenberg, K., and Weinges, K. 1962. Catechins and flavonoid tannins. *In* "The Chemistry of Flavonoid Compounds" (T. A. Geissman, ed.), pp. 197–216. Macmillan, New York.

Freudenberg, K., and Weinges, K. 1963. Zur Kenntis der Flavonoid-Gerbstoffe. *Ann. Chem., Liebigs* **668**, 92–96.

Freudenberg, K., Fikentscher, H., Harder, M., and Schmidt, O. 1925. Die Umwandlung des cyanidins in catechin. *Ann. Chem., Liebigs* **444**, 135–145.

Frey, E. 1908. Die Wirkung des Tannins auf Resorption und Sekretion des Dünndarms. *Arch. Ges. Physiol. Pflügers* 123, 491–514.

Frolov-Bagreev, A. M. 1948. Ispol'zovanie vinogradnykh semyan. *Vinodelie i Vinogradarstvo SSSR* **8**(8), 15–16.

Fuleki, T., and Francis, F. J. 1968. Quantitative methods for anthocyanins. 1. Extraction and determination of total anthocyanin in cranberries. *J. Food Sci.* **33**, 72–77.

Fuller, H. L., Cheng, S. I., and Potter, D. K. 1967. Detoxification of dietary tannic acid by chicks. *J. Nutrition* **91**, 477–481.

Fuller, W. L., and Berg, H. W. 1965. Treatment of white wine with nylon 66. *Am. J. Enol. Viticult.* **16**, 212–218.

Gadzhiev, D. M., and Abramov, Sh. A. 1962. O novykh metodakh polucheniya krasnykh stolovykh vin. *Vinodelie i Vinogradarstvo SSSR* **22**(1), 13–16.

Gaivoronskaya, Z. I. 1959. Tanidnye zapasy v kozhitse yagod kak dapolnitel'nyi faktor pri vybore pravil'nogo napravliniya ispol'zovania novogo sorta (vinograda). *Tr. Vses. Nauchn.-Issled. Inst. Vinodeliya i Vinogradarstva, Magarach* 7, 26–39.

Gaivoronskaya, Z. I. 1960. Izmenenie soderzhania dubil'nykh i azotistykh veshchestv v protsesse nastaivaniya i brozhenia susla na mzge vinograda belykh sortov. *Tr. Vses. Nauchn.-Issled. Inst. Vinodeliya i Vinogradarstva, Magarach* **9**, 145–152.

Gaivoronskaya, Z. I. 1963. Ekstrakt kak pokazatel' kachestva desertnykh vin i izmenenie ego v khode sozrevaniya vinograda. *Biokhim. Vinodeliya* 7, 43–57.

Galet, P. 1962. "Cépages et Vignobles de France," 3 Vol., p. 2891. Dehan, Montpellier.

Galston, A. W. 1967. Regulatory systems in higher plants. *Am. Scientist* **55**(2), 144–160.

Garino-Canina, E. 1924. Contributo all studio e alla determinazione delle sostanze tanniche e colorante dell'uva. *Staz. Sper. Agrar. Ital.* **57**, 245–274.

Garino-Canina, E. 1940–1941. The chemical composition of vine shoots. *Ann. Accad. Agr. Torino* **84**, 127–142; *Chem. Abstr.* **38**, 3688 (1944).

Garoglio, P. G. 1962. Nuovi metodi proposti per la ricèrca di sostanze coloranti derivate da ibride produttori ed altre attualità enochimiche. *Riv. Viticolt. Enol.* **15**, 187–189.

Garoglio, P. G. 1965. "La Nuova Enologia," 3rd ed., p. 1503. Inst. Industrie Agrarie, Florence.

Garoglio, P. G. 1968. Proposta di modifiche nella manualità del metodo rapido Dorier e Verelle per la determinazione del diglucoside malvosidico nei vini. *Riv. Viticolt. Enol.* **21**, 11–15.

Gärtel, W. 1954. Beerengrösse, Kernzahl und Mostgewicht beim Riesling. *Weinberg Keller* **1**, 51–58.

Gärtel, W. 1966. Untersuchungen über die bei der abschlussspritzung an Trauben haften bleibende Brühemenge. *Weinberg Keller* **13**, 463–471.

Gatet, L. 1939. Annales de physiologie et de physicochimie biologique. *Ann. Physiol. Physiocochim. Biol.* **15**, 984–1064.

Gautier, A. 1892a. Sur l'origine des matières colorantes de la vigne; sur les acides ampélochroïques et la coloration automnale des végétaux. *Compt. Rend.* **114**, 623–629.

Gautier, A. 1892b. Sur les acides Ampélochroïques. *Bull. Soc. Chim. France* 7, 823–829.

Gautier, A. 1911. Sur les mèchanismes de la variation des races et les transformations moléculaires qui accompagnent ces variations. *Compt. Rend.* **153**, 531–539.

Gavrish, G. A. 1966. Izmenenie sostava soka v protsesse uvyalivaniya vinograda. *Vinodelie i Vinogradarstvo SSSR* **26**(2), 26–29.

Geissman, T. A. 1962. "The Chemistry of Flavonoid Compounds," p. 666. Macmillan, New York.

Geissman, T. A. 1963. Leucoanthocyanins. *Plant Phenolics Group N. Am. Proc. 3rd Symp.*, 1–18.

Geissman, T. A., and Dittmar, H. F. K. 1965. A proanthocyanidin from avocado seed. *Phytochemistry* **4**, 359–388.

Gel'gar, L. L. 1967. Issledovanie protsessov otdeleniya vinogradnogo susla pressovaniem. *Tr. Vses. Nauchn.-Issled. Inst. Vinodeliya i Vinogradarstva, Magarach* **16**, 219–234.

Genevois, L. 1951. Matières colorantes et viellissement des vins. *Rev. Ferment. Inds. Aliment* **6**, 111–115.

Genevois, L. 1952. Les acides caféique et férulique chez les végétaux. *Bull. Soc. Chim. France* **19**, 241.

Genevois, L., and Peynaud, E. 1947. Chemical composition of sixteen varieties of peaches. *Rev. hort. Paris* **30**, 295–298; *Chem. Abstr.* **42**, 1675 (1948).

Gentilini, L. 1939. Contributo allo studio dell'analisi cromatografica applicata alle materie coloranti naturali del vino. *Ann. Chim. Appl.* **29**, 169–183.

Gentilini, L. 1941. Studio dell' analisi cromatografica applicata al riconascimento di alcuni dei pui comuni sofisticanti del colore dei vini rossi. *Ann. Staz. Sper. Viticolt. Enol. Conegliano* **10**, 59–72.

Gentilini, L. 1960. Indagini cromatografiche sull'acido salicilico naturale nei vini. *Riv. viticolt. e enol.* **13**, 376–378.

Gentilini, L. 1966. Applicazioni della luce di Wood in campo enologico. *Riv. viticolt. e enol.* **19**, 275–306.

Geoffroy, P., and Perin, J. 1966. La décoloration des moûts et des vins tachés (action du noir et des levures). *Le Vigneron Champenois* **87**, 416–426.

Gerasimov, M. A., and Smirnova, A. P. 1964. Khromatograficheskoe opredelenie katekhinov v vine, obrabotannom teplom. *Vinodelie i Vinogradarstvo SSSR* **24**(2), 14–17.

Getov, G. 1966. Izsledvane na s'stava na antotsianite v grazdeto na chepvenite perspektivin sortove lozi. *Grad. Lozar. Nauka (Sofia)* **3**, 537–545.

Getov, G., and Petkov, G. 1966. Nachweis der Anwesenheit von Malvin in Vitis-vinifera-Sorten. *Mitt. (Klosterneuburg)* **16**, 207–210.

Giashvili, D. S. 1966. Composition of a sweet pressed extract (infusion) and its uses. *Tr. Gruzinsk. Nauchn.-Issled. Inst. Pishch. Prom.* **2**, 3–9; *Chem. Abstr.* **67**, 10363 (1967).

Ginzburg, C., 1967. The relation of tannins to the differentiation of the root tissues in *Reaumuria palaestina*. *Botan. Gaz.* **128**, 1–10.

Gladwin, F. E. 1924. A correlation between color of grape leaves at time of foliation and fruit color. *New York Agr. Expt. Sta., Geneva, Tech. Bull.* **107**, 1–8.

Godlewski, H., and Richet, C., Jr. 1924. Pouvoir diurétique experimental du jus de raisin blanc en injection intraveineuse. *Compt. Rend. Soc. Biol.* **90**, 408–410.

Goheen, A. C., and Cook, J. A. 1959. Leafroll (red-leaf or rougeau) and its effects on vine growth, fruit quality and yields. *Am. J. Enol. Viticult.* **10**, 173–181.

Goldman, L. M. 1963. Factors influencing the rate of carbon dioxide formation in ferment-ed-in-the-bottle champagne. *Am. J. Enol. Viticult.* **14**, 36–42.

Goldstein, J. L., and Swain, T. 1963. Methods for determining the degree of polymerization of flavans. *Nature* **198**, 587–588.

Goldstein, J. L., and Swain, T. 1965. The inhibition of enzymes by tannins. *Phytochemistry* **4**, 185–192.

Golodriga, P. Ya., and Khe, P.-C. 1963. Rannespelost' vinograda i nekotorye biokhimicheskie pokazateli. *Tr. Vses. Nauchu.-Issled. Inst. Vinodeliya i Vinogradarstva Magarach* **12**, 74–83.

Golodriga, P. Ya., and Suyatinov, I. A. 1966. O nekotorykh faktorakh, vliyayushchikh na nakoplenie krasyashchikh veshchestv v yagodakh. *Vinodelie i Vinogradarstvo SSSR* **26** (4), 29–32.

Goodman, R. N., Király, Z., and Zaitlin, M. 1967. "The Biochemistry and Physiology of Infectious Plant Disease," p. 354. Van Nostrand, Princeton, New Jersey.

Gortner, W. A. 1963. Tissue hydroxy-cinnamic acids and soil moisture stress. *Nature* **197**, 1316–1317.

Gortner, W. A., and Kent, M. J. 1958. The coenzyme requirement and enzyme inhibitors of pineapple indoleacetic acid oxidase. *J. Biol. Chem.* **233**, 731–735.

Gramshaw, J. W. 1968. Phenolic constituents of beer and brewing materials. III. Simple anthocyanogens from beer. *J. Inst. Brewing* **74**, 20–38.

Grasso, S. 1952. Le nuove resine sintetiche decoloranti in enotecnia. *Riv. Viticolt. Enol.* **5**, 195–199.

Grebenyuk, S. M., Dotsenko, F. P., and Buz'ko, I. A. 1966. Ekstraktor nepreryvnogo deistviya dlya vinogradnoi vyzhimki. *Vinodelie i Vinogradarstvo SSSR* **26**(3), 44–47.

Griffith, J. Q., Jr., Krewson, C. F., and Naghski, J. 1955. "Rutin and Related Flavonoids: Chemistry, Pharmacology, Clinical Applications," p. 275. Mack Publ. Co., Easton, Pennsylvania.

Griffiths, L. A. 1964. Flavonoid metabolism. Identification of the metabolites of (+)-catechin in rat urine. *Biochem. J.* **92**, 173–179.

Grimmer, W., and Wauschkuhn, B. 1939. Beiträge zur Biochemie der Mikroorganismen. VIII. Der Abbau des Tyrosins durch verschiedene Bakterien. *Milchwirtsch. Forsch.* **20**, 110–130.

Grinstead, R. R. 1964. Metal-catalyzed oxidation of 3,5-di-*tert*-butylpyrocatechol and its significance in the mechanism of pyrocatechase action. *Biochemistry* **3**, 1308–1314.

Grohmann, H., and Mühlberger, F. H. 1954. Über die Unterscheidung zwishen natürlichen und zugesetzten Mengen an Vanillin in Dessertweinen und Aperitfs. *Deut. Lebensm. Rundschau* **50**, 183–186.

Grünhut, L. 1898. "Die Chemie des Weines," pp. 72–186. Enke, Stuttgart.

Guidry, N. P. 1964. A graphic summary of world agriculture. *U.S. Dept. Agr. Misc. Publ.* **705**, 1–64.

Guillemonat, A., Traynard, J. C., Triaca, Poite, and Roussel. 1963. Sur la constitution chimique du liège. IV. Structure de la phellochryséine. *Bull. Soc. Chim. France* **1963**(1), 142–144.

Guilliermond, A., Magenot, G., and Plantefol, L. 1933. "Traité de Cytologie Végétalé," p. 1195. Librairie E. Le Francois, Paris.

Guimberteau, G., and Portal, E. 1961. Contribution á la recherche de l'acide benzoique et des acides – phenols dans les vins. *Ann. Fals. Expert Chim.* **54**, 330–337.

Guinot, Y., and Menoret, Y. 1965. Le chauffage continu du raisin avant pressurage. *Compt. Rend. Acad. Agr. France* **51**, 866–872.

Guroff, G., Daly, J. W., Jerina, D. M., Renson, J., Witkop, B., and Udenfriend, S. 1967. Hydroxylation-induced migration: The NIH shift. *Science* **157**, 1524–1530.

Haider, K., Frederick, L. R., and Flaig, W. 1965. Reactions between amino acid compounds and phenols during oxidation. *Plant Soil* **22**, 49–64.

Halcrow, R. M., Glenister, P. R., Brumsted, D. D., and Lautenbach, A. F. 1966. Phenolic characteristics in brewing. IV. Off-flavors from biological sources. *Inst. Brewing (Aust. Sect.), Proc. Conv.* **9**, 273–279; *Chem. Abstr.* **67**, 81094 (1967).

Halperin, Z., and Goldman, M. L. 1950. Recovery of tartrates and other by-products from winery wastes. *Fruit Prods. J.* **29**, 148–153, 156–157.

Hamdy, M. K., Pratt, D. E., Powers, J. J., and Somaatmadja, D. 1961. Anthocyanins. III. Disc sensitivity assay of inhibition of bacterial growth by pelargonidin-3-monoglucoside and its degradation products. *J. Food Sci.* **26**, 457–461.

Handler, P., and Baker, R. D. 1944. The toxicity of orally administered tannic acid. *Science* **99**, 393.

Harborne, J. B. 1960. Genetic variation of anthocyanin pigments in plant tissues. *In* "Phenolics in Plants in Health and Disease" (J. B. Pridham, ed.), pp. 109–117. Pergamon Press, New York.

Harborne, J. B. 1962. Chemicogenetical studies of flavonoid pigments. *In* "The Chemistry of Flavonoid Compounds" (T. A. Geissman, ed.), pp. 593–617. Macmillan, New York.

Harborne, J. B. 1963. Distribution of anthocyanins in higher plants. *In* "Chemical Plant Taxonomy" (T. Swain, ed.), pp. 359–388. Academic Press, New York.

Harborne, J. B. 1964a. "Biochemistry of Phenolic Compounds," p. 618. Academic Press, New York.

Harborne, J. B. 1964b. Plant phenols. XI. The structure of acylated anthocyanins. *Phytochemistry* **3**, 151–160.

Harborne, J. B. 1965. Plant phenols. XIV. Characterization of flavonoid glycosides by acidic and enzymic hydrolyses. *Phytochemistry* **4**, 107–120.

Harborne, J. B. 1967. "Comparative Biochemistry of the Flavonoids," p. 383. Academic Press, New York.

Harmon, F. N. 1956. White Emperor virus in Cardinal and Red Malaga grapes. *Proc. Am. Soc. Hort. Sci.* **67**, 302–303.

Harmon, F. N., and Snyder, E. 1946. Investigations on the occurrence, transmission, spread, and effect of "white" fruit color in the Emperor grape. *Proc. Am. Soc. Hort. Sci.* **47**, 190–194.

Harper, K. A., and Chandler, B. V. 1967a. Structural changes of flavylium salts. I. Polarographic and spectrophotometric examination of 7,4′-dihydroxyflavylium perchlorate. *Australian J. Chem.* **20**, 731–744.

Harper, K. A., and Chandler, B. V. 1967b. Structural changes of flavylium salts. II. Polarographic and spectrometric investigation of 4′-hydroxyflavylium perchlorate. *Australian J. Chem.* **20**, 745–756.

Harris, G., and Ricketts, R. W. 1962. Metabolism of phenolic compounds by yeasts. *Nature* **195**, 473–474.

Hartmann, B. G., and Tolman, L. M. 1918. Concord grape juice: manufacture and chemical composition. *U. S. Dept. Agr., Bur. Chem. Bull.* **656**, 1–27.

Harvalia, A. 1961. Detection of hybrid wines with paper chromatography (trans). *Chem. Chronika* **26**, 180–187.

Harvalia, A. 1965. Contribution à l'étude du comportement de la couleur des vins rouges. *Chem. Chronika* **30A**, 155–159.

Haslam, E. 1966. "Chemistry of Vegetable Tannins," p. 179. Academic Press, New York.

Hassall, C. H., and Scott, A. I. 1961. Some recent studies relating to the intramolecular free radical coupling of phenols. *In* "Recent Developments in the Chemistry of Natural Phenolic Compounds" (W. D. Ollis, ed.), pp. 119–133. Pergamon Press, New York.

Hathway, D. E. 1957a. The transformation of gallates into ellagate. *Biochem. J.* **67**, 445–450.

Hathway, D. E. 1957b. Autoxidation of polyphenols. I. Autoxidation of methyl gallate and its methoxy ethers in aqueous ammonia. *J. Chem. Soc.* **1957**, 519–523.

Hathway, D. E. 1958. Autoxidation of polyphenols. IV. Oxidative degradation of the catechin autoxidative polymer. *J. Chem. Soc.* **1958**, 520–521.

Hathway, D. E., and Seakins, J. W. T. 1955. Autoxidation of catechin. *Nature* **176**, 218.

Hathway, D. E., and Seakins, J. W. T. 1957a. Autoxidation of polyphenols. III. Autoxidation in neutral aqueous solution of flavans related to catechin. *J. Chem. Soc.* **1957**, 1562–1566.

Hathway, D. E., and Seakins, J. W. T. 1957b. Enzymic oxidation of catechin to a polymer structurally related to some phlobatannins. *Biochem. J.* **67**, 239–245.

Hedrick, W. P., and Anthony, R. D. 1915. Inheritance of certain characters of grapes. *J. Agr. Res.* **4**, 315–330.

Heimann, W., and Andler, S. 1965. Über die hydroxylierende Wirkung der Phenoloxylase. II. Untersuchungen des Systems Sauerstoff-Phenolase-Brenzcatechin bei Anwesenheit von substituierten Hydrochinonen. *Z. Lebensm.-Untersuch. Forsch.* **128**, 154–158.

Heintze, K. 1965. Untersuchungen über die Vitamin P—wirksame Flavonoidverbindungen und ihre Bedeutung in Pflanzlichen Lebensmitteln. *Deut. Lebensm. Rundshau* **61**, 309–311.

Henke, O. 1960. Biochemische und morphologische Untersuchungen an Vitis—Artbastarden. *Züchter* **30**, 216–219.

Henke, O. 1963. Uber den Stoffwechsel reblausanfälliger und -unanfälliger Reben. *Phytopathol. Z.* **47**, 314–326.

Hennig, K. 1962. Weitere technische Anwendungsmöglichkeiten des Pyrokohlensäurediäthylesters (Diäthylpyrokarbonats) in der Kellerei. *Deut. Wein-Ztg.* **98**, 552, 554, 556.

Hennig, K., and Burkhardt, R. 1957a. Über die Gerbstoffe und Polyphenole der Weine. *Naturwissenschaften* **44**, 328–329.

Hennig, K., and Burkhardt, R. 1957b. Über die Farb-und Gerbstoffe, sowie Polyphenole und ihre Veränderungen im Wein. *Weinberg Keller* **4**, 374–387.

Hennig, K., and Burkhardt, R. 1958. Der Nachweis phenolartiger Verbindungen und hydroaromatischer Oxycarbonsäuren in Trauben. *Weinberg Keller* **5**, 542–552. 593–600; *Am. J. Enol. Viticult.* **11**, 64–79 (1960).

Hennig, K., and Burkhardt, R. 1960. Vorkommen und Nachweis von Quercitrin und Myricitrin in Trauben und Wein. *Weinberg Keller* **7**, 1–3.

Hennig, K., and Burkhardt, R. 1962. Chromatographische Trennung von Eichenholzauszügen und deren Nachweis in Weinbränden. *Weinberg Keller* **9**, 223–231.

Hennig, K., and Villforth, F. 1942. Die Aromastoffe der Weine. Ein Beitrag zur Untersuchung der Duft-oder Bukettstoffe der Weine. I. Isolierung der Aromastoffe. Aldehyde. Ketone. *Vorratspflege Lebensmittelforsch.* **5**, 181–199.

Herrmann, K. 1958. Über oxydationsfermente und phenolische Substrate im Gemüsse und Obst. III. Catechine, Oxyzimtsäuren und *o*-polyphenoloxydase im Obst. *Z. Lebensm.-Untersuch. Forsch.* **108**, 152–157.

Herrmann, K. 1959. Über Katechine und Katechin-Gerbstoffe und ihre Bedeutung in Lebensmitteln. *Z. Lebensm.-Untersuch. Forsch.* **109**, 487–507.

Herrmann, K. 1963a. Über die phenolischen Inhaltsstoffe der Trauben und des Weines (Flavonoide, Phenolkarbonsäuren, Farbstoffe, Gerbstoffe). *Weinberg Keller* **10**, 154–165, 208–220.

Herrmann, K. 1963b. Über die Adsorption phenolischer Inhaltsstoffe durch Polyamid. *Weinberg Keller* **10**, 322–327.

Heyns, K., Grützmacher, H.-F., Mayer, W., Merger, F., and Frank, G. 1964. Über Kondensationsprodukte der Catechine. VII. Massenspektrometrische Strukturanalyse der Kondensationsprodukte von Catechin mit Phloroglucin bzw. C-Äthylphloroglucin. *Ann. Chem. Liebigs* **675**, 134–141.

Hidalgo Sanchez, S. 1961. Détermination de l'intensité colorante du vin. *Bull. Offic. Intern. Vigne Vin.* **34**(366), 47–49.

Hilgard, E. W. 1896. The composition and classification of grapes, musts, and wines. *Calif. Expt. Sta. Rept. Viticult. Work 1887–1893*, 3–16.

Hilger, H., and Gross, L. 1887. Die Bestandteile einzelner Organe des Weinstockes. II. *Landwirtsch. Vers. Sta.* **33**, 170–196.

Hillis, W. E. 1962. "Wood Extractives and Their Significance to the Pulp and Paper Industries," p. 513. Academic Press, New York.

Hillis, W. E., and Isoi, K. 1965. Biosynthesis of polyphenols in Eucalyptus species. *Phytochemistry* **4**, 905–918.

Hillis, W. E., and Urbach, G. 1958. Leucoanthocyanins as the possible precursors of tannins. *Nature* **182**, 657–658.

Hillis, W. E., and Urbach, G. 1959. The reaction of (+)-catechin with formaldehyde. *J. Appl. Chem.* (*London*) **9**, 474–482.

Hinreiner, E., Filipello, F., Webb, A. D., and Berg, H. W. 1955a. Evaluation of thresholds and minimum difference concentrations for various constituents of wines. III. Ethyl alcohol, glycerol and acidity in aqueous solution. *Food Technol.* **9**, 351–353.

Hinreiner, E., Filipello, F., Berg, H. W., and Webb, A. D. 1955b. Evaluation of thresholds and minimum difference concentrations for various constituents of wines. IV. Detectable differences in wine. *Food Technol.* **9**, 489–490.

Hodge, J. E. 1953. Browning reactions in model systems. *J. Agr. Food Chem.* **1**, 928–943.

Hodge, J. E. 1967. Origin of flavor in foods. Nonenzymic browning reactions. *In* "The Chemistry and Physiology of Flavors" (H. W. Schultz, E. A. Day, and L. M. Libby, eds.), pp. 465–491. Avi, Westport, Connecticut.

Holm, T. 1911. Medicinal plants of North America, *Ampelopsis quinquefolia. Merck's Rept.* **20**, 309–311.

Holz, G., and Wilharm, G. 1950. Beitrag zur Kenntnis des Acroleins in Obstbränden, Maischen, und Mosten. I. Verhalten und Vorschläge zu seiner Beseitigung. *Z. Lebensm.-Untersuch. Forsch.* **91**, 256–240.

Hörhammer, L., and Wagner, H. 1963. Neue Ergebnisse auf dem Gebiet der Isolierung, Analytik, Pharmakologie und Phytochemie von Flavonen und Isoflavonen. *Pharm. Deltion, Sci. Ed.* **3**, 18–36.

Horwitt, M. K. 1933. Observations on behavior of the anthocyan pigment from Concord grapes in the animal body. *Proc. Soc. Exptl. Biol. Med.* **30**, 949–951.

Huang, H. T. 1955. Decolorization of anthocyanins by fungal enzymes. *J. Agr. Food Chem.* **3**, 141–146.

Huang, H. T. 1956. The kinetics of the decolorization of anthocyanins by fungal "anthocyanase." *J. Am. Chem. Soc.* **78**, 2390–2393.

Hulme, A. C. 1958a. Some aspects of the biochemistry of apple and pear fruits. *Advan. Food Res.* **8**, 297–413.

Hulme, A. C. 1958b. Quinic and shikimic acids in fruits. *Qualitas Plant. Mater. Vegetabiles* **3-4**, 468–473.

Humeau, G. 1966. Pressure tests. *Chem. Chronika* **31**, 23–37; *Chem. Abstr.* **65**, 2964 (1966).

Hussein, A. A., and Cruess, W. V. 1940. Properties of the oxidizing enzymes of certain vinifera grapes. *Food Res.* **5**, 637–648.

Hussein, A. A., Mrak, E. M., and Cruess, W. V. 1942. The effect of pretreatment and subsequent drying on the activity of grape oxidase. *Hilgardia* **14**, 349–357.

Iankov, I. A. 1966. Chromatographic determination of tannic materials in stems. *Logarstvo i Vinarstvo* **15**(5), 42–45; *Bull. Offic. Intern. Vigne Vin* **40**, 1145 (1967).

Ibrahim, R. K., and Towers, G. H. N. 1960. Identification, by chromatography, of plant acids. *Arch. Biochem. Biophys.* **87**, 125–128.

Ibrahim, R. K., Towers, G. H. N., and Gibbs, R. D. 1962. Syringic and sinapic acids as indicators of differences between major groups of vascular plants. *J. Linnean Soc. London Botany* **58**, 223–230.

Ingalsbe, D. W., Neubert, A. M., and Carter, G. H. 1963. Condord grape pigments. *J. Agr. Food Chem.* **11**, 263–268.

Iñigo Leal, B., and Bravo Abad, F. 1960. Modernas orientaciones en la vinificación y crianza biológica de tintos. *Agricultura (Madrid)* (341), 487–491.

Ito, S., and Joslyn, M. A. 1964. Presence of several phenolic components in fruit proanthocyanidins. *Nature* **204**(4957), 475–476.

Ito, U. 1967a. The mechanisms of flocculation of flocculent yeast during wort fermentation. III. The action of tannin-protein complex in wort to accelerate yeast flocculation, and the mechanism of yeast flocculation during wort fermentation. *Mem. Coll. Sci. Univ. Kyoto, Ser. A.* **31**(2), 127–135.

Ito. U. 1967b. Differences between the flocculent yeast and the nonflocculent yeast. *Mem. Coll. Sci. Univ. Kyoto, Ser. A.* **31**(2), 137–151.

Ivanov, T. 1967. Sur l'oxydation du moût de raisin. I. Activité de la polyphénoloxydase du raisin des cépages "Muscat rouge," "Dimiat," "Riesling" et "Aligoté." *Ann. Technol. Agr. Paris* **16**, 35–39.

Ivanova, T. M., Davydova, M. A., and Rubin, B. A. 1965. O fungistatichnom deistvii fenolov i ikh roli v immunitete rastenii. *Dokl. Akad. Nauk. SSSR* **164**, 705–708.

Iwano, S. 1967. Treatment of white wine of browning color material with Nylon 66. *Vitis* **6**, 309–313.

Jacobs, M. B. 1944. "The Chemistry and Technology of Food and Food Products," Vol. 3, p. 890. Wiley (Interscience), New York.

Jacobs, M. B. 1959. "Manufacture and Analysis of Carbonated Beverages," p. 333. Chem. Publ. Co., New York.

Jaulmes, P. 1965. Les méthodes internationales et les nouvelles méthodes officielles d'-analyse des vins et des moûts. *In* "Mises au Point de Chimie Analytique, Organique, Pharmaceutique et Bromatologique" (J. A. Gautier and P. Malangeau, eds.), 14th ser., pp. 182–215. Masson, Paris.

Jayasankar, N. P., and Bhat, J. V. 1966. Isolation and properties of catechol-cleaving yeasts from coir rets. *Antonie van Leeuwenhoek J. Microbiol. Serol.* **32**, 124–134.

Jayle, G. E., and Aubert, L. 1964. Action of anthocyanin glucosides on scotopic and mesopic vision of normal subjects. *Therapie* **19**(1), 171–185; *Chem. Abstr.* **62**, 11045 (1965).

Jenkins, G. L., Hartung, W. H., Hamlin, K. E., Jr., and Data, J. B. 1957. "The Chemistry of Organic Medicinal Products," 4th ed., p. 569. Wiley, New York.

Johnson, J. R., Christian, S. D., and Affsprung, H. E. 1967. Self-association and hydration of phenol in organic solvents. II. *J. Chem. Soc.*, **A**(5), 764–768.

Joslyn, M. A. 1935. Preliminary observations on the mellowing and stabilization of wine. *Fruit Prods. J.* **15**, 10–12, 24.

Joslyn, M. A. 1950. "Methods in Food Analysis," p. 525. Academic Press, New York.

Joslyn, M. A., and Amerine, M. A. 1964. "Dessert, Appetizer and Related Flavored Wines, the Technology of Their Production," p. 483. Division of Agricultural Sciences, Univ. of California, Berkeley and Davis.

Joslyn, M. A., and Comar, C. L. 1941. The role of acetaldehyde in red wines. *Ind. Eng. Chem.* **33**, 919–928.

Joslyn, M. A., and Dittmar, H. F. K. 1967. Die Proanthocyanidine der Trauben. I. Die Proanthocyanidine der Schalen und Kerne von Trauben der Sorte Pinot blanc. *Mitt. (Klosterneuburg)* **17**(2), 92–108.

Joslyn, M. A., and Goldstein, J. L. 1964a. Astringency of fruits and fruit products in relation to phenolic content. *Advan. Food Res.* **13**, 179–217.

Joslyn, M. A., and Goldstein, J. L. 1964b. Conversion of leucoanthocyanins into the corresponding anthocyanidins. *Science* **143**, 954–955.

Joslyn, M. A., and Little, A. 1967. Relation of type and concentration of phenolics to the color and stability of rosé wines. *Am. J. Enol. Viticult.* **18**(3), 138–148.

Joslyn, M. A., and Ponting, J. D. 1951. Enzyme-catalyzed oxidative browning of fruit products. *Advan. Food Res.* **3**, 1–44.

Joslyn, M. A., and Smit, C. J. B. 1954. Trennung der Tannine und verwandter Polyphenole des Obstes auf einer Kolonne von Silicagel. *Mitt. (Klosterneuburg)* **4B**, 141–147.

Joslyn, M. A., Farley, H. B., and Reed, H. M. 1929. Effect of temperature and time of heating on extraction of color from red-juice grapes. *Ind. Eng. Chem.* **21**, 1135–1137.

Jovanović, V., Supica, M., Koncar, L., Vapa, M., Milovanović, M., and Knezević, N. 1963. Untersuchung über den Einfluss des Traubensaftes von der Hybridrebe auf den Tierorganismus. *Ann. Wiss. Arbeiten Landw. Fakultat Novi Sad.* No. 7; *Mitt. (Klosterneuburg)* **15A**, 270 (1965).

Jurd, L. 1962. Spectral properties of flavonoid compounds. *In* "The Chemistry of Flavonoid Compounds" (T. A. Geissman, ed.), pp. 107–155. Macmillan, New York.

Jurd, L. 1963. Anthocyanins and related compounds. I. Structural transformation of flavylium salts in acidic solutions. *J. Org. Chem.* **28**, 987–991.

Jurd, L. 1964a. Reactions involved in sulfite bleaching of anthocyanins. *J. Food. Sci.* **29**(1), 16–19.

Jurd, L. 1964b. Some benzopyrylium compounds potentially useful as color additives for fruit drinks and juices. *Food Technol.* **18**(4), 157–159.

Jurd, L. 1964c. Anthocyanins and related compounds. III. Oxidation of substituted flavylium salts to 2-phenylbenzofurans. *J. Org. Chem.* **29**(9), 2602–2605.

Jurd, L. 1964d. Reactions involved in the discoloration of fruit pigments. *Hornblower (Inst. Food Technol., N. Calif.)* **16**(1), 13–14.

Jurd, L. 1966a. Quinone oxidation of flavenes and flavan-3,4-diols. *Chem. & Ind.* **1966**, 1683–1684.

Jurd, L. 1966b. Anthocyanidins and related compounds. 10. Peroxide oxidation products of 3-alkylflavylium salts. *Tetrahedron* **22**(8), 2913–2921.

Jurd. L. 1967. Anthocyanidins and related compounds. XI. Catechin-flavylium salt condensation reactions. *Tetrahedron* **23**(3), 1057–1064.

Jurd, L., and Asen, S. 1966. The formation of metal and "co-pigment" complexes of cyanidin 3-glucoside. *Phytochemistry* **5**(6), 1263–1271.

Jurd, L., and Geissman, T. A. 1963. Anthocyanins and related compounds. II. Structural transformations of some anhydro bases. *J. Org. Chem.* **28**(9), 2394–2397.

Jurd, L., and Lundin, R. 1968. Anthocyanidins and related compounds. XII. Tetramethylleucocyanidin-phloroglucinol and resorcinol condensation products. *Tetrahedron* **24** (6), 2653–2661.

Jurics, Eva W. 1967a. Zur Analytik der in Früchten am häufigsten vorkommenden Hydroxyzimtsäuren und Catechine. *Ernährungsforschung* **12**, 427–433.

Jurics, Eva W. 1967b. Determination of the contents of (+)-catechol, (−)-epicatechol in Hungarian fruits by means of paper chromatography. *Élelmiszervizsgálati Közlemények* **13**(3), 158–66; *Chem. Abstr.* **68**, 76955 (1968).

Kabiev, O. K., and Vermenichev, S. M. 1966. Antitumor activity of leucoanthocyanidins and catechins. *Vopr. Onkol.* **12**(4), 61–64; *Chem. Abstr.* **65**, 2875 (1966).

Kain, W. 1967. Untersuchung und Beurteilung von Bentoniten zur Weinbehandlung. II. Über die Bestimmung der Wirksamkeit auf Wein. *Mitt. (Klosterneuburg)* **17**, 201–222.

Karrer, P., and de Meuron, G. 1932. Pflanzenfarbstoffe. XL. Zur Kenntniss des oxidativen Abbaus der Anthocyane. Konstitution des Malvons. *Helv. Chim. Acta* **15**, 507–512.

Karrer, P., and Widmer, R. 1927. Untersuchungen über Pflanzenfarbstoffe. I. Über die Konstitution einiger Anthocyanidine. *Helv. Chim. Acta* **10**, 5–33.

Karrer, W. 1958. "Konstitution und Vorkommen den Organischen Pflanzenstoffen," p. 1207. Birkhaüser, Basel.

Kayser, R. 1897. Über den Einfluss des Lagerns auf die Zusammensetzung der Weine. *Z. Öffentl. Chem.* **3**, 95–97.

Kazumov, N. B. 1961. Istochniki obrazovaniya al'degidov v protsesse maderizatsii i ikh identifikatsiya. *Sadovodstvo, Vinogradarstvo i Vinodelie Moldavii* **16**(10), 31–33.

Kefford, J. F. 1959. The chemical constituents of citrus fruits. *Advan. Food Res.* **9**, 285–372.

Keith, E. S., and Powers, J. J. 1965. Polarographic measurement and thermal decomposition of anthocyanin compounds. *J. Agr. Food Chem.* **13**(6), 577–579.

Keith, E. S., and Powers, J. J. 1966. Gas chromatographic determination of anthocyanins and other flavonoids as silyl derivatives. *J. Food Sci.* **31**, 971–979.

Kelhofer, W. 1908. Untersuchung von Räuschlingtrauben in verschiedenen Reifestadien. *Schweiz. Z. Obst.-Weinbau* **17**, 65–68.

Keller, O., and Berger, L. 1946. A dimorphic form of *d*-catechin. *J. Am. Chem. Soc.* **68**, 145–146.

Kielhöfer, E. 1944. Die Farbmessung bei Rotweinen. *Wein Rebe.* **26**, 1–14.

Kim, C. H. 1939. Über die Esterase der Hefe. *Enzymologia* **6**, 183–185.

King, H. G. C. 1955. The constitution and synthesis of leucoanthocyanidins. *Sci. Proc. Roy. Dublin Soc.* **27**(6), 87–92.

King, H. G. C., and White, T. 1956. The quantitative determination of specific nuclei and components of vegetable tannin extracts. *In* "The Chemistry of Vegetable Tannins," pp. 31–56. Soc. Leather Trades' Chemists, London.

Kishkovskii, Z. N., and Potii, V. S. 1964. Izmenenie ul'trafioletovogo spektra pogloshcheniya vin pri nagrevanii. *Vinodelie i Vinogradarstvo SSSR* **24**(7), 10–12.

Kishkovskii, Z. N., Mekhuzla, N. A., Potii, V. S., and Sakharova, T. A. 1966. Isolation of the products of the sugar-amine reaction from wine with sephadex. *Prikl. Biokhim i Mikrobiol.* **2**(4), 447–451; *Chem. Abstr.* **65**, 16027 (1966).

Klenk, E., and Maurer, R. 1963. Beeinflussung der Rotweinfarbe durch Inhaltsstoffe und Kellerbehandlung. *Deut. Weinbau* **18**, 679–688.

Klenk, E., and Maurer, R. 1967. Über die Störung der Weinsteinkristallisation durch phenolische Substanzen im Wein. *Deut. Weinbau* **22**, 674–676.

Klenk, E., Villforth, F., and Fuhrman, N. 1954. Prüfung von Rotwein-Maische-Gärverfahren. *Mitt. (Klosterneuburg)* **4A**, 1–17.

Kliewe, H., and Anabtawi, A. 1964. Ein Vergleich von Hybridenweinen mit Weinen von europäischen Edelreben. *Wein-Wiss.* **19**, 113–126.

Kliewer, W. M. 1966. Sugars and organic acids of *Vitis vinifera*. *Plant Physiol.* **41**, 923–931.

Kloss, J., and Seifert, W. 1927. The reactions of gallic acid and the presence of the acid in fruit and grape wine. *Österr. Chemiker.-Ztg.* **30**, 117–120; *Chem. Abstr.* **21**, 3251(1927).

Knudson, L. 1913a. Tannic acid fermentation. I. *J. Biol. Chem.* **14**, 159–184.

Knudson, L. 1913b. Tannic acid fermentation. II. Effect of nutrition on the production of the enzyme tannase. *J. Biol. Chem.* **14**, 185–202.

Knypl, J. S. 1967. Growth retardants in relation to the germination of seeds. I. A paradoxical concentration effect of N,N- dimethyl-aminosuccinamic acid on the germination of kale seeds affected with coumarin. *Can. J. Botany* **45**, 903–913.

Koch, J., and Bretthauer, G. 1957. Zur Kenntniss der Eiweissstoffe des Weines. I. Chemische Zusammensetzung des Wärmetrubes kurzzeiterhitzter Weissweine und seine Beziehung zur Eiweisstrubung und zum Weineiweiss. *Z. Lebensm. Untersuch. Forsch.* **106**, 272–280.

Kocsis, E. A., Csokán, P., and Horvai, R. 1941. Weinfarbenuntersuchung mittels Capillar-Luminicenz-Analyse. *Z. Lebensm-Untersuch. Forsch.* **81**, 316–321.

Koeppen, B. H., and Basson, D. S. 1966. The anthocyanin pigments of Barlinka grapes. *Phytochemistry* **5**, 183–187.

Kohman, E. F., and Sanborn, N. H. 1931. Isolation of quinic acid from fruits. *Ind. Eng. Chem.* **23**, 126.

Kokurewicz, J. 1962. Tannin materials as fungicides. U. S. Patent 3,060,082.

Kolesnik, A. A., and Kremlev, M. M. 1957. Changes in the chemical composition of "Michurin" varieties of grapes during ripening. *Sbornik Nauch. Rabot, Moskov. Inst. Narod. Khoz. im. G. V. Plekhanova* (11), 197–219; *Chem. Abstr.* **52**, 12263 (1958).

Kondo, G. F., and Zvezdina, T. V. 1953. Soderzhanie tanidov v vinograde i vine. *Vinodelie i Vinogradarstvo SSSR* **13**(11), 13–18.

Konlechner, H., and Haushofer, H. 1956. Versuche mit Rotwein—Bereitungsverfahren. *Mitt. (Klosterneuburg)* **6A**, 158–173.

Konlechner, H., and Haushofer, H. 1958. Versuche zur Herstellung von Rotwein mittels Heissfermentierung der Maische. *Mitt. (Klosterneuburg)* **8A**, 169–174.

Konlechner, H., and Haushofer, H. 1961. Warmbehandlung, ein neues Rotweinbereitungsverfahren. Versuchsergebnisse mit Erwärmung und Fermentierung. *Mitt. (Klosterneuburg)* **11A**, 148–156.

Korotkevich, A. V. 1960. Fotometricheskii ekspres-metod opredeleniya dubil'nykh veshchestv v vine i kon'yake. *Vinodelie i Vinogradarstvo SSSR* **20**(4), 9–11.

Korotkevich, A. V., and Korotkevich, L. M. 1962. Opredelenie dubil'nykh i krasyashchikh vinogradnoi yagody i usloviya perekhoda ikh v vino. *Biokhim. Vinodeliya, Acad. Nauk*

Korotkevich, A. V., Gaivoronskaya, Z. I., and Kravchenko, A. P. 1950. Tannidyne zapasy vinogradnoi yagody i usloviya perekhoda ikh v vino. *Biokhim. Vinodeliya, Acad. Nauk SSSR* **3**, 26–42.

Korpassy, B., Horvai, R., and Koltay, M. 1951. On the absorption of tannic acid from the gastro-intestinal tract. *Arch. Intern. Pharmacodynamie* **88**, 368–377.

Korte, F., Barkemeyer, H., and Korte, I. 1959. Neuere Ergebnisse der Chemie pflanzlicher Bitterstoffe. *Progr. Chem. Org. Nat. Prods.* **17**, 124–182.

Kotake, M., and Kubota, T. 1940. Über Inhaltsstoffe von *Ampelopsis meliaefolia* Kudo (Haku-Tya). *Ann. Chem. Liebigs* **544**, 253–271.

Kràmszky, L. 1905. Bestimmung des Gerbstoffgehaltes der Weine. *Z. Anal. Chem.* **44**, 756–765.

Kuć, J. 1963. The role of phenolic compounds in disease resistance. *Conn. Univ. Agr. Expt. Sta. Bull.* **633**, 20–30.

Kul'nevich, V. G. 1957a. Rol' kisloroda v protsesse maderizatsii vina. *Biokhim. Vinodeliya* **5**, 164–179.

Kul'nevich, V. G. 1957b. Prevrashcheniya dubil'nykh veshchestv v protsesse maderizatsii vina. *Biokhim. Vinodeliya* **5**, 180–198.

Kushida, T. 1960. Effect of thermal treatment of musts on the quality of wines. I. Experimental fermentation of the must treated with high temperature. *Bull. Res. Inst. Ferment. Yamanashi Univ.* (7), 27–31.

Kushida, T., and Ito, K. 1963. Studies on the use of antioxidants in vinification. I. Effects of ascorbic, erythorbic, and sulfurous acids on the quality of wine. *Bull. Res. Inst. Ferment. Yamanashi Univ.* (10), 31–41.

Laborde, J. 1897. Sur l'absorption d'oxygene dans la casse du vin. *Compt. Rend.* **125**, 248–250.

Laborde, J. 1908a. Sur les transformations de la matière chromogène des raisins pendant la maturation. *Compt. Rend.* **147**, 753–755.

Laborde, J. 1908b. Sur l'origine de la matière colorante des raisins rouges et autres organes végétaux. *Compt. Rend.* **146**, 1411–1413.

Laborde, J. 1908–1910. Etude sur les matières tannoides du vin: matière colorante et oenotanin. *Rev. Viticult.* **30**, 169–173, 200–203, 533–537; 564–569; **31**, 7–10, 47–51, 178–184, 204–207; **32**, 453–457, 480–484, 561–567, 706–711, 733–741; **33**, 206–211, 238–242.

Laborde, J. 1909. Sur le mécanisme physiologique de la coloration des raisins rouges et de la coloration automnale des feuilles. *Compt. Rend.* **147**, 993–996.

Laborde, J. 1911. Sur la caractérisation des vins rosés et des vins blancs de raisins rouges. *Ann. Fals. Fraudes* **4**, 389–391.

Laborde, J. 1918. Recherches sur le viellissement du vin. *Rev. viticult.* **48**, 225–230, 241–244, 386–390.

Laffer, H. E. 1936. Tannin in wine making. *Wines & Vines* **17**(9), 22.

Laszlo, I., Lepădatu, V., Giosanu, T., Macici, M., and Taras, S. 1966. Prepararea vinurilor rosii prin macerarea carbonică. *Lucrari Stiintifice* (*Bucharest*) **7**, 859–868.

Lavollay, J., and Sevestre, J. 1944. Le vin, considéré comme un aliment riche en vitamin P. *Compt. Rend. Acad. Agr. France* **30**, 259–261.

Lavollay, J., Sevestre, J., and Dussy, J. 1944. Sur la présence de catechins dans un certain nombre d'espèces végétales alimentaires. *Compt. Rend.* **218**, 82–84.

Lawrence, W. J. C., Price, J. R., Robinson, G. M., and Robinson, R. 1938. A survey of anthocyanins. V. *Biochem. J.* **32**, 1661–1667.

Lawrence, W. J. C., Price, J. R., Robinson, G. M., and Robinson, R. 1939. Distribution of anthocyanins in flowers, fruits and leaves. *Phil. Trans. Roy. Soc.* (*London*) Ser. B **320**, 149–178.

Lazenby, W. R. 1903. Composition and waste of fruits and nuts. *Proc. Soc. Promotion Agr. Sci.* **24**, 101–108.

Lee, F. 1951. Fruits and nuts. *In* "The Chemistry and Technology of Food and Food Products" (M. B. Jacobs, ed.), Vol. 2, 2nd ed., pp. 1348–1589. Wiley (Interscience), New York.

Lee, S., and Aronoff, S. 1967. Boron in plants: a biochemical role. *Science* **158**, 798–799.

Lefèvre, P. M. 1966. Determinação do diglucosidomalvosido em vinhos. Estudo comparativo de alguns metodos de cromatografia em papel. *De Vinea et Vino Portugaliae Documenta Ser.* 2, **3**(1), 1–35.

Leglise, M., Naudin, R., and Prevost, J. 1967. Essais de traitement de raisins rouges entiers par la vapeur d'eau avant vinification. *Progr. Agr. Vit.* **167**, 309–315, 350–353.

Leibnitz, E., Behrens, U., and Seifert, H. 1960. Abbau von Phenolen durch Hefen. *Z. Allg. Mikrobiol.* **1**, 13–17.

le Roux, M. S. 1953. Color experiments with table grapes. *Farming in South Africa*, 1–3.

Levy, L. F., Posternack, T., and Robinson, R. 1931. Experiments on the synthesis of the anthocyanins. 8. A synthesis of oenin chloride. *J. Chem. Soc.* **1931**, 2701–2715.

Lichev, V. 1962. V'rkhu s'd'rzhanieto na rutin v conyachniya spirt. *Lozarstvo Vinarstvo Sofia* **11**(5), 28–30.

Liebmann, A. J., and Scherl, B. 1949. Changes in whisky while maturing. *Ind. Eng. Chem.* **41**, 534–543.

Lingens, F., Goebel, W., and Uesseler, H. 1967. Regulation der Biosynethese der aromatischen Aminosäuren in *Saccharomyces cerevisiae.* 2. Repression, Induktion und Aktivierung. *Eur. J. Biochem.* **1**, 363–374.

Lipis, B. V., Lyalikova, R. Yu., and Chernichuk, L. L. 1964. Spectrofotometricheskii metod opredeleniya dubil'nykh i krasyashchikh veshchestv v vinogradnom susle i vine. *Tr. Moldavsk. Nauchn.-Issled. Inst. Pishchevoi Prom.* **4**, 109–114.

Liuni, C. S., Calò, A., and Cappelleri, G. 1965. Contributo allo studio sui pigmenti antocianici di alcune specie del genere *Vitis* e di loro ibridi. *Accad. Ital. Vite. Vino, Siena, Atti,* **17**, 161–167.

Lloyd, F. E. 1911. The tannin-colloid complex in the fruit of the persimmon, Diospyros. *Biochem. Bull.* **1**, 7–41.

Lloyd, F. E. 1913. The artificial ripening of bitter fruits. *Science* **36**, 879–887.

Lloyd, F. E. 1922. The occurrence and functions of tannin in the living cell. *Trans. Roy Soc. Canada* **16**, 1–13.

Lolli, G. 1941. Marchiafava's disease. *Quart. J. Studies Alc.* **2**, 486–495.

Long, H. C. 1917. "Plants Poisonous to Live Stock," p. 119. Cambridge Univ. Press, London.

Loomis, N. H., and Lutz, J. M. 1937. The relationship of leaf area and leaf area fruit ratios to composition and flavor of Concord grapes. *Proc. Am. Soc. Hort. Sci.* **35**, 461–465.

Löwenthal, J. 1860. Versuch einer allgemeinen Maasanalyse für sämmtliche Farbstoffe, Gerbstoffe, usw. *J. Prakt. Chem.* **81**, 150.

Löwenthal, J. 1877. Ueber die Bestimmung des Gerbstoffs. *Z. Anal. Chem.* **16**, 33–48, 201.

Lucia, S. P. 1954. "Wine as Food and Medicine," p. 149. McGraw-Hill (Blakiston), New York.

Lucia, S. P. 1963. "A History of Wine as Therapy," p. 234. Lippincott, Philadelphia, Pennsylvania.

Lüers, H., and Mangele, J. 1926. Phytochemische Reduktion von Chinonen. *Biochem. Z.* **179**, 238–247.

Lukton, A., Chichester, C. O., and Mackinney, G. 1956. The breakdown of strawberry anthocyanin pigment. *Food Technol.* **10**, 427–432.

Lüthi, H. 1957. Ueber die Ursachen unvollständiger Vergärung von Obstweinen. *Mitt. Gebiete Lebensm. Hyg.* **48**, 201–217.

McFarlane, W. D. 1965. Tyrosine derived amines and phenols in wort and beer. *European Brewing Conv. Proc. Congr.* **10**, 387–395.

McFarlane, W. D., and Millingen, M. B. 1964. Aromatic amino acids in alcoholic fermentations. *Am. Soc. Brewing Chemists Proc.* **1964**, 41–48.

McFarlane, W. D., and Thompson, K. D. 1964. Colorimetric methods for determination of aromatic alcohols in beer. *J. Inst. Brewing* **70**(6), 497–502.

Mackinney, G., and Chichester, C. O. 1954. The color problem in foods. *Advan. Food Res.* **5**, 301–351.

Maier, V. P., Metzler, D. M., and Huber, A. F. 1964. Effects of heat processing on the properties of dates. *Date Growers Inst. Rept.* **41**, 8–9.

Maiorov, V. S., Shashilova, V. P., and Matasova, N. N. 1963. Application of a natural food pigment from grape residues in the production of fruit and berry wines. *Tr. Tsentr. Nauchn.-Issled. Inst. Pivovarennoi, Bezalkogol' noi i Vinnoi Prom.* (11), 61–66.

Mallory, G. E., and Love, R. F. 1945. Quantitative determination of caramel in wine, distilled spirits, vinegar, and vanilla extract. *Ind. Eng. Chem. Anal. Ed.* **17**, 631–637.

Malvezin, P. 1908. Sur l'origine de la couleur des raisins rouges. *Compt. Rend.* **147**, 384–386.

Manceau, E. 1895. Sur le dosage du tannin dans les vins. *Compt. Rend.* **121**, 646–647.

Mancini, M., Oriente, P., and D'Andrea, L. 1960. Hypocholesterolemic effects of quinic acid 1,4-dicaffeate in atherosclerotic patients. *Drugs Affecting Lipid Metab. Proc. Symp., Milan, 1960*, pp. 533–537.

Mapson, L. W., and Swain, T. 1961. Oxidation of ascorbic acid and phenolic constituents. *Soc. Chem. Ind. (London) Monograph* **1961** (11), 121–133.

Mar Monux, D. 1966. Estudia cromatografica de los flavonidos en los vinos blancos. *La Semana Vitivinic.* **21**(1051), 3497, 3499, 3501.

Maranville, F. L., and Goldschmid, O. 1954. Ultraviolet absorption spectra as a measure of phenolic hydroxyl group content in polyphenolic tanninlike materials. *Anal. Chem.* **26**, 1423–1427.

Marcilla Arrazola, J., Alas, G., and Feduchy, E. 1939. Contribucion al estudio de las levaduras que forman velo sobre ciertos vinos de elvado grado alcohólico. *Ann. Centro Invest. Vinic. (Madrid)* **1**, 1–230 (1936).

Mareca Cortes, I. 1962. Sur les phénomènes redox oenologiques. *Ind. Aliment. Agr. Paris* **79**, 829–838.

Mareca Cortes, I., and del Amo Gili, E. 1959. Evolution de la matière colorante des vins de Rioja au cours du viellissement. II. *Ind. Aliment. Agr. Paris* **76**, 601–606.

Mareca Cortes, I., and de Campos Salcedo, M. 1957a. Sur la combinaison de l'éthanal et des polyphénols dans les vins rouges. *Ind. Aliment. Agr. Paris* **74**, 103–106.

Mareca Cortes, I., and de Campos Salcedo, M. 1957b. Modificaciones en la composition de los vinos durante su añejamiento. *Rev Cienc. Apl. Madrid* **11**, 320–326.

Mareca Cortes, I., and Escudero, M. J. 1966. Quelques essais sur les propriétés collöidales des polyphénols des vins. *Compt. Rend. Acad. Agr. France* **52**, 143–152.

Mareca Cortes, I., and Gonzalez, A. 1964. Sur la composition de la matière colorante des vins rouges. *Ind. Aliment. Agr. Paris* **81**, 391–397.

Mareca Cortes, I., and Gonzalez, A. 1965. Contribucion al conocimiento de la evolucion de la materia colorante desde la uva al vino añejo. *Vitis* **5**, 201–211.

Mareca Cortes, I., and de Miguel, M. 1965. Contribution à l'étude polarographique des flavonoides. *Compt. Rend. Acad. Agr. France* **51**, 123–128.

Mareca Cortes, I., and de Miguel, M. 1966. Deuxième contribution a l'étude polarographique des flavonoïdes naturels. *Compt. Rend. Acad. Agr. France* **52**(12), 855–861.

Margheri, G. 1960. Determinazione dei tannini nel vino mediante spettrofotometria nell'ultravioletto. *Riv. Viticolt. Enol.* **13**(12), 401–406.

Martakov, A. A., Kolesnikov, V. A., and Bazarbaeva, S. G. 1967. Formirovaine madery na biokatalitlicheskoi osnove. *Vinodelie i Vinogradarstvo SSSR* **27**(4), 18–21.

Marteau, G. 1960. Tanins et souplesse. *Progr. Agr. Vit.* 77, 281–289.

Marteau, G., and Olivieri, C. 1966. Perspectives et données actuelles de la vinification en rouge par macération à chaud. *Progr. Agr. Vit.* **166**(17), 133–136; (18), 150–163; (19) 215–219.

Martin, G. E., Schmit, J. A., and Schoeneman, R. L. 1965. Various components of bourbon whiskey by thin-layer and gas-liquid chromatography. *J. Assoc. Offic. Agr. Chemists* **48**, 962–964.

Martin, G. J. 1955. Biochemistry of the bioflavonoids. *Ann. N.Y. Acad. Sci.* **61**, 646–651.

Martins, J., and Drews, B. 1966. Zum Metabolismus von Phenol in Hefe. *Chemiker-Ztg.* **90** (15), 531–534.

Marutyan, S. A. 1954. Ob otlichii aktivnosti peroksidazy u seyantsev vinograda razlichnogo pola. *Dokl. Akad. Nauk SSSR* **99**, 295–296.

Marutyan, S. A., and Mantashyan, E. A. 1961. Aktivnost okislitel'nykh fermentov razlichnykh tkanei vinograda v zavisimosti ot ego morozostoikosti. *Vinodelie i Vinogradarstvo. SSSR* **21**(2), 31–33.

Maruyama, C., and Kushida, T. 1965. Enological studies on the color of red wines. 6. Extraction and utilization of red color from grape skins. *Bull. Res. Inst. Ferment. Yamanashi Univ.* **40**(12), 41–46.

Mason, H. S. 1965. Oxidases. *Ann. Rev. Biochem.* **34**, 595–634.

Masquelier, J. 1956. Identification et dosage des facteurs vitaminiques P dans diverses boissons fermentées. *Bull. Soc. Chim. Biol.* **38**, 65–70.

Masquelier, J. 1957. Les propriétés antimicrobiennes des vins. *Congr. Intern. L'Etude Sci. Vin Raisin, Bordeaux, Proc.* 44–52.

Masquelier, J. 1958. Sur quelques propriétés biologiques des anthocyanes et des leucoanthocyanes de la vigne (*Vitis vinifera*). I. *Qualitas Plant. Mater. Vegatabiles* **3–4**, 481–490.

Masquelier, J. 1961a. Le vin dans l'alimentation humaine. *In* "Traité d'Oenologie" (J. Ribéreau-Gayon and E. Peynaud), Vol. 2, pp. 93–148. Lib. Polytechnique Ch. Beranger, Paris.

Masquelier, J. 1961b. Les constituants du vin présentant une action hypocholesterolemiante. *Congr. Med. Intern. Etude Sci. Vin Raisin, Bordeaux,* 137–142.

Masquelier, J. 1963. Dosage des leucoanthocyannes du vin blanc emploi de la poudre de polyamide. *Bull. Soc. Pharm. Bordeaux* **102**, 51–52.

Masquelier, J., and Delaunay, D. 1965. Action bactéricide des acides—phenols du vin. *Bull. Soc. Pharm. Bordeaux* **104**, 152–156.

Masquelier, J., and Jensen, H. 1953a. Récherches sur l'action vitaminique P du vin. *Bull. Soc. Pharm. Bordeaux* **91**, 20–23.

Masquelier, J., and Jensen, H. 1953b. Récherches sur l'action bactericide des vins rouges. I. *Bull. Soc. Pharm. Bordeaux* **91**, 24–29.

Masquelier, J., and Jensen, H. 1953c. Récherches sur l'action bactericide des vins rouges. II. *Bull. Soc. Pharm. Bordeaux* **91**, 105–109.

Masquelier, J., and Laparra, J. 1967. Action cholérétique des constituants cinnamiques du vin. *Rev. Franc. Oenol.* **7**(26), 33.

Masquelier, J., and Point, G. 1954. A propos du flavonoside des raisins blancs. *Bull. Soc. Pharm. Bordeaux* **92**, 33–35.

Masquelier, J., and Point, G. 1956. Le leucoanthocyane des cépages blancs de *Vitis vinifera*. *Bull. Soc. Pharm. Bordeaux* **95**, 6–11.

Masquelier, J., and Ricci, R. 1964. Chromatographie des dérivés cinnamiques du vin. *Qualitas Plant. Mater. Vegetabiles* **11**, 244–248.

Masquelier, J., Vitte, G., and Ortega, M. 1959. Dosage colorimétrique des leucoanthocyannes dans les vins rouges. *Bull. Soc. Pharm. Bordeaux* **98**, 145–148.

Masquelier, J., Michaud, J., and Triaud, J. 1965. Fractionnement des leucoanthocyannes du vin. *Bull. Soc. Pharm. Bordeaux* **104**, 81–85.

Masschelein, C. A. 1963. Qualité de la bière et transformations biochimiques au cours de la fermentation. *Fermentatio* (5), 201–210.

Masschelein, C. A., and Haboucha, J. 1965. Cell-surface adsorption phenomena during fermentation. *Soc. Chem. Ind. (London), Monograph* **19**, 202–213.

Masuda, H., and Muraki, H. 1963. Studies on the formation of the Madeira-type wine. *Bull. Res. Inst. Ferment. Yamanashi Univ.* (10), 23–30.

Mathieu, L. 1907. Sur les variations de volume et de degré des eaux-de-vie pendant le vieillissement. *Bull. Assoc. Chimistes Sucr. Dist.* **24**, 1687–1688.

Mathieu, L. 1919. Note on the reducing effect of yeasts in vinification. *Bull. Assoc. Chimistes Sucr. Dist.* **37**, 174–175; *Chem. Abstr.* **14**, 1869 (1920).

Mattick, L. R., Weirs, L. D., and Robinson, M. B. 1967. Detection of adulterated Concord grape juice with other anthocyanin-containing products. *J. Assoc. Offic Agr. Chem.* **50**(2), 299–303.

Maumené, E. 1874. "Traité Theorique et Pratique du Travail des Vins," 2nd ed., p. 680. Masson, Paris.

Maurié, A. 1942. Les composés phénoliques du vin. Matières colorantes et tannoides. *Bull. Offic Intern. Vigne Vin* **15**(153), 41–56.

Maveroff, A. 1949. La caseína en la clarificacion de los vinos blancos. *Anales Inst. Vino (Mendoza)* **1**(1), 7–44.

Mayer, P. 1914. Bildung von Saligenin aus Salicylaldehyd durch Hefe. *Biochem. Z.* **62**, 459–461.

Mayer, W., and Merger, F. 1961. Darstellung optisch aktiver Catechine durch Racemattrennung mit Hilfe der adsortionschromatographie an Cellulose. *Ann. Chem. Liebigs* **644**, 65–69.

Mehlitz, A., and Matzik, B. 1957. Papierchromatographische Untersuchungen über die in Obsäften vorkommenden nichtflüchtigen Säuren. *Ind. Obst.-Gemüseverwert.* **42**, 127–131.

Meisinger, M. A. P., Kuehl, F. A., Jr., Rickes, E. L., Brink, N. G., Folkers, K., Forbes, M., Zilliken, F., and György, P. 1959. The structure of a new product from yeast: 2(6-hydroxy 2-methoxy-3,4-methylenedioxyphenyl)benzofuran. *J. Am. Chem. Soc.* **81**, 4979–4982.

Melamed, N. 1962. Détermination des sucres résiduels des vins, leur relation avec la fermentation malolactique. I. *Ann. Technol. Agr. Paris* **11**(1), 5–31.

Merli, G. M. 1965. Coffee in the diet of the community. Biological effect of chlorogenic acid. *Minerva Dietol.* **5**(3), 129–134; *Chem. Abstr.* **63**, 15419 (1965).

Merory, J. 1960. "Food flavorings, composition, manufacture, and use," p. 381. Avi., Westport, Connecticut.

Mestre Artigas, C. 1942. Le froid industriel appliqué aux vins. *Bull. Offic. Intern. Vigne. Vin* **15**(150), 53–59.

Metcalfe, C. R., and Chalk, L. 1950. "Anatomy of the Dicotyledons," 2 Vols., p. 1500. Oxford Univ. Press (Clarendon), Oxford.

Meyer, L. 1918. Abänderung der Methode von Nessler und Barth zur Bestimmung des Gerbstoffes im Wein. *Mitt. Lebensm. Hyg.* **9**, 131–135.

Mikhailov, M. V., Kirillov, A. F., Skurtul, A. M., and Levit, T. K. 1965. Some characteristics of accumulation, localization, and conversion of reserve forms of transitory metabolites in individual plant parts and tissues of grapevines. *Vopr. Fiziol. Zimostoik. i Zasukhoustoich. Plodovykh i Vinograda*, 3–22; *Chem. Abstr.* **65**, 19004 (1966).

Milisavljevic, D. 1957. Action of *Penicillium expansum* on grapes and the quality of the wine. *Poljopr. Vojvodine* **5**(6), 40–47; *Chem. Abstr.* **54**, 25552 (1960).

Milisavljevic, D. 1967. Action postfermentaire des levures sur les composes phenoliques des vins rouges. *Symp. Intern. Oenol. Bordeaux-Cognac.*

Mohler, H., and Hämmerle, W. 1935. Chromatographischer und spektrophotometrischer Nachweis von künstlicher Färbung in Wein. *Z. Untersuch. Lebensm.* **70**, 193–195.

Mohler, H., and Hammerle, W. 1936. Über den Nachweis von Weisswein in Rotwein. *Z. Untersuch. Lebensm.* **71**, 186–189.

Molchanova, Z. Ya. 1966a. Rol' oksidazy v ustoichivosti vinograda k neblagopriyatnym usloviyam zimovki. *Vinodelie. i Vinogradarstvo SSSR* **26**(1), 21–23.

Molchanova, Z. Ya. 1966b. Activity of some oxidative enzymes of grape plants during the winter period (Trans.). *Fiziol. Rastenii Acad. Nauk SSSR* **13**, 494–500 (Trans. 445–451).

Molisch, H. 1916. Beiträge zur Mikrochemie der Pflanze. 3. Über den braunen Farbstoff "goldgelber" Weinbeeren. *Ber. Deut. Botan. Ges.* **34**, 69–72.

Mondy, N. I., Mobley, E. O., and Gedde-Dahl, S. B. 1967. Influence of potassium fertilization on enzymatic activity, phenolic content and discoloration of potatoes. *J. Food Sci.* **32**, 378–381.

Monin, J. 1967. Etude d'un inhibiteur existant chez les embryons dormants d' *Euonymus europaeus* L. *Compt. Rend.* **264**, 2367–2370.

Monties, B. 1966. Nature et propriétés chimiques des principaux polyphénols de la pomme. *Ann. Physiol. Végétale* **8**, 101–135.

Monties, B. 1967. Potentiel d'oxydo-réduction des polyphénols. Caractérisation potentiométrique des constituants phénoliques du jus de pomme et des boissons. *Ann. Technol. Agr. Paris* **16**, 251–258.

Monties, B., and Jacquin, P. 1966. Valeur technologique du dosage des tannins dans les jus de pomme. *Ann. Technol. Agr. Paris* **15**, 335–347.

Moreau, L. 1932. Rôle des matières de réserve de la vigne dans la mise à fruit des cépages et dans la véraison des raisins. *Ann. Agron.* **2**, 363–374.

Moreau, L., and Vinet, E. 1937. Le tanin dans les vins blancs. *Rev. Viticult.* **86**, 65–69.

Mosiashvili, G. I. 1958. Primenenie efiroobrazuyushchikh i pektinorazlagayushchikh drozhzhei v vinodelii. *Vinodelie i Vinogradarstvo SSSR* **18**(2), 6–9.

Motalev, S. V., Ponomareva, R. V., and Shakhsuvaryan, A. V. 1962. Effect of certain environmental factors on the conversion of wine to sherry. *Tr. Nauchn.-Issled. Inst. Sadovodstva, Vinogradarstva, Vinodeliya. Uz. SSR* **26**, 145–150; *Chem. Abstr.* **61**, 2438 (1964).

Mulder, G. J. 1857. "The Chemistry of Wine" (H. Bence Jones, transl.), p. 390. Churchill, London.

Müller-Thurgau, H. 1898. Abhängigkeit der Ausbildung der Traubenbeeren und einiger anderer Früchte von der Entwicklung der Samen. *Landwirtsch Jahrb. Schweiz.* **12**, 135–205.

Muraki, H., Shijo, N., Otsuka, K., and Masuda, H. 1960. Thermal treatment of alcoholic beverages. II. Influence of the length of the skin fermentation on the quality of baking sherry. *Bull. Res. Inst. Ferment. Yamanashi Univ.* (7), 61–64.

Myers, A. T., and Caldwell, J. S. 1939. The preparation of new types of unfermented grape juices by blending. *Fruit Prods. J.* **19**, 5–10, 36–40, 57, 69–72, 80–89.

Nachet, L. I. 1829. Ueber ein Ersazmittel der Eichenrinde für die Gerbereien. *J. Pharm.*, 412; *Dinglers Polytech. J.* **33**, 463 (1829).

Nachkov, D. 1967. V'rku s'd'rshanieto na leikantotsiani v nyakoi trapezni vina. *Lozartsvo Vinar.* (*Sofia*) **16**(1), 36–38.

Naito, R. 1964. Studies on coloration of grapes. V. Influence of light intensity on the coloration and pigmentation in some black and red grapes. *J. Japan. Soc. Hort. Sci.* **33**, 213–220.

Naito, R. 1966. Studies on the coloration of grapes. VII. Behavior of anthocyanins in the skin of some black and red grapes as affected by light intensity. *J. Japan. Soc. Hort. Sci.* **35**, 225–232.

Naito, R., Kyo, S., and Sumi, T. 1965. Studies on the coloration of grapes. VI. Effects of shading on the color and pigmentation of Muscat Bailey A. *J. Japan. Soc. Hort. Sci.* **34**, 145–151.

Nakabayashi, T. 1962. Photometric titration for the determination of tannins by Loewenthal's method. *Nippon Shokuhin Kogyo Gakkaishi* **9**(8), 313–317; *Chem. Abstr.* **59**, 14285 (1963).

Nanitashvili, S. T. 1957. Termincheskaya obrabotka kakhetinskogo vina. *Vinodelie i Vinogradarstvo SSSR* 17(1), 5–7.

Nassar, A. R., and Kliewer, W. M. 1966. Free amino acids in various parts of *Vitis vinifera* at different stages of development. *Proc. Am. Soc. Hort. Sci.* **89**, 281–294.

Nebesky, E. A., Esselen, W. B., Jr., McConnell, J. E. W., and Fellers, C. R. 1949. Stability of color in fruit juices. *Food Res.* **14**, 261–274.

Nègre, E. 1939a. Dosage de l'oenotannin. *Ann. Fals. Fraudes* **32**, 175–178.

Nègre, E. 1939b. Contribution oenologique a l'étude des matières tannoïdes. *Compt. Rend. Acad. Agr. France* **25**, 647–652.

Nègre, E. 1942–43. Les matières tannoïdes et la composition des vins. *Bull. Offic Intern. Vigne Vin* **15**(154), 20–52; **16** (155), 25–56.

Nègre, E. 1966. L'influence des phases prefermentatives de la vinification sur la qualité finale de vin. *Le Vigneron Champenoise* **87**(7–8), 281–295.

Negrul', A. M., and Lya, Yu. 1963. Izmenchivost i nasledstvennost okraski yagod vinograda. *Tr. Vses. Nauchn.-Issled. Inst. Vinodeliya i Vinogradarstva, Magarch.* **12**, 36–74.

Nelson, K. E., and Wheeler, D. H. 1939. Natural aging of wine. A chemical study. *Ind. Eng. Chem.* **31**, 1279–1281.

Nelson, K. E., and Ough, C. S. 1966. Chemical and sensory effects of microorganisms on grape musts and wine. *Am. J. Enol. Viticult.* **17**, 38–47.

Neubauer, A. 1944. Verfahren zum Freilegen des Farbstoffs roter Trauben. *Obst-Gemüsse-Verwertungs-Ind., Braunschweig. Konserven Ztg.* **31**, 69–70; *Chem. Zentr.* **1944** II, 1129.

Neubauer, C. 1871. Ueber die quantitative Bestimmung des Gerbstoffsgehalts der Eichenrinde. *Z. Anal. Chem.* **10**, 1–40.

Neubauer, C. 1872. Studien über den Rothwein. *Ann. Oenologie* **2**, 1–41.

Neubauer, C. 1873. Beitrage zur qualitativen Analyse des Weinlaubs *Z. Anal. Chem.* **12**, 39–48.

Neubauer, C. 1874. Beiträge zur Analyse des Weinlaubs und des im Frühjahr aus den Reben ausfliessenden Säfte (Rebthränen). *Ann. Oenologie* **4**, 102–116.

Neuberg, C., and Liebermann, L. 1921. Zur Kenntnis der Carboligase. II. *Biochem. Z.* **121**, 311–325.

Neuberg, C., and Nord, F. F. 1919. Die phytochemische Reduktion der Ketone. Biochemische Darstellung optisch-aktiven sekundärer Alkohole. *Ber. Deut. Chem. Ges.* **52b**, 2237–2248.

Neuberg, C., and Simon, E. 1926. Uber das Verhalten des *p*-Xylochinones zu Hefe. *Biochem. Z.* **171**, 256–260.

Nierenstein, M. 1925. Gallotannin. XIV. The action of yeast on gallotannin. *J. Am. Chem. Soc.* **47**, 1726–1728.

Niethammer, A. 1931. Weitere Studien über die Mikroskopie und Histochemie einiger Früchte und Blätter. *Z. Untersuch. Lebensm.* **61**, 217–221.

Nikandrova, V. N. 1958. Photocolorimetric determination of red pigments in wine. *Byul. Nauchn.-Tekhn. Inform., Moldavsk. Nauchn.-Issled. Inst. Sadovodstva, Vinograd. Vinodel.* (2), 26–27; *Chem. Abstr.* **57**, 1380 (1962).

Nilov, V. I., and Begunova, R. D. 1953. Prevrashchenie krasyashchikh veshchestv pri sozrevanii i pererabotke vinograda. *Biokhimiya* **18**, 275–278.

Nilov, V. I., and Ogorodnik, S. T. 1967. Nefermentativnye prevrashcheniya aminokislot pri vyderzhke vin. *Tr. Vses. Nauchn.-Issled. Inst. Vinodeliya i Vinogradarstva, Magarach* **16**, 246–260.

Nilov, V. V. 1964. O pogloshchenii kisloroda vinogradnym suslom v moment drobleniya yagody. *Tr. Vses. Nauchn.-Issled. Inst. Vinodelya i Vinogradarstva, Magarach* **13**, 57–59.

Nitsch, J. P. 1962. Basic physiological processes affecting fruit development. *In* "Proceedings Plant Science Symposium," pp. 5–23. Campbell Soup Co., Cambden, New Jersey.

Noyes, H. A., King, H. T., and Martsolf, J. H. 1922. Variations in the Concord grape during ripening. *J. Assoc. Offic. Agr. Chemists* **6**, 197–205.
Nutsubidze, N. N. 1958a. Khromatograficheskoe opredelenie katekhinov v vine. *Vinodelie i Vinogradarstvo SSSR* **18**(7), 11–14.
Nutsubidze, N. N. 1958b. Prevrashcheniya katekhinov pri maderižatsii vina. *Soobshcheniya Akad. Nauk Gruzin. SSSR* **21**(1), 41–56.
Nutsubidze, N. N., and Gulbani, D. I. 1964. Flavonols in the grapevine. *Soobshcheniya Akad. Nauk Gruzin. SSR* **33**(2), 345–352; *Chem. Abstr.* **62**, 9463 (1965).
Nykänen, L., Puputti, E., and Suomalainen, H. 1966. Gas chromatographic determination of tyrosol and tryptophol in wines and beers. *J. Inst. Brewing* **72**, 24–28.
Ogorodnik, S. T., and Nilov, V. I. 1966. Oxygen and $CO_2$ of air and oxidative decomposition of amino acids. *Prikl. Biokhim. i Mikrobiol.* **2**(5), 576–583; *Chem. Abstr.* **65**, 20796 (1966).
Oka, S. 1964. Mechanism of antimicrobial effect of various food preservatives. *Intern. Symp. Food Microbiol., 4th, Goteborg, Sweden*, 3–16.
Okolelov, I. N., and Kotlyarenko, M. R. 1951. Ispol'zovanie kholoda v vinodelii. *Vinodelie i Vinogradarstvo SSSR* **11**(12), 5–8.
Olmo, H. P. 1934. Empty-seededness in varieties of *Vitis vinifera*. *Proc. Am. Soc. Hort. Sci.* **32**, 376–380.
Olmo, H. P. 1942. The use of seed characters in the identification of grape varieties. *Proc. Am. Soc. Hort. Sci.* **40**, 305–309.
Olmo, H. P. 1946. Correlations between seed and berry development in some seeded varieties of *Vitis vinifera*. *Proc. Am. Soc. Hort. Sci.* **48**, 291–297.
Olmo, H. P. 1952. Wine grape varieties of the future. *Proc. Am. Soc. Enol.* **3**, 45–51.
Onslow, M. W. 1925. "The anthocyanin pigments of plants," 2nd ed., p. 314. Cambridge Univ. Press, Cambridge, England.
Oparin, A. I. 1963. Biochemie du vin. *Bull. Offic. Intern. Vigne Vin.* **36**(393), 1301–1310.
Orzechowski, G. 1962. Flavonoide in der Therapie. *Planta Med.* **10**, 403–411.
Otsuka, K., and Hara, S. 1967. Phenolic substances in alcoholic beverages. III. Phenolic substances in sake brewing. *Nippon Jôzô Kyôkai Zasshi* **62**(1), 78–81.
Otsuka, K., Morinaga, K., and Imai, S. 1963. Sugars in aged distilled spirits and oakwood (Trans.). *Nippon Jôzô Kyôkai Zasshi* **59**(5), 448–449.
Otsuka, K., Imai, S., and Morinaga, K. 1965. Studies on the mechanism of aging of distilled liquors. II. Distribution of phenolic compounds in aged distilled liquors. *Agr. Biol. Chem. (Tokyo)* **29**, 27–31.
Ough, C. S. 1960. Gelatin and poly(vinylpyrrolidone) compared for fining red wines. *Am. J. Enol. Viticult.* **11**, 170–173.
Ough, C. S., and Amerine, M. A. 1960. Experiments with controlled fermentations. IV. *Am. J. Enol. Viticult.* **11**, 5–14.
Ough, C. S., and Amerine, M. A. 1961. Studies on controlled fermentation. V. Effects on color, composition, and quality of red wines. *Am. J. Enol. Viticult.* **12**, 9–19.
Ough, C. S., and Amerine, M. A. 1962. Studies with controlled fermentations. VII. Effect of antefermentation blending of red must and white juice on color, tannins, and quality of Cabernet Sauvignon wine. *Am. J. Enol. Viticult.* **13**, 181–188.
Ough, C. S., and Amerine, M. A. 1966. Effects of temperature on wine making. *Calif. Agr. Expt. Sta. Bull.* **827**, 1–36.
Ough, C. S., and Amerine, M. A. 1967. Rosé wine color preference and preference stability by an experienced and an inexperienced panel. *J. Food Sci.* **32**, 706–711.
Ough, C. S., and Berg, H. W. 1959. Studies on various light sources concerning the evaluation and differentiation of red wine color. *Am. J. Enol. Viticult.* **10**, 159–163.

Ough, C. S., Berg, H. W., and Chichester, C. O. 1962. Approximation of per cent brightness and dominant wavelength and some blending application with red wines. *Am. J. Enol. Viticult.* **13**, 32–39.

Ough, C. S., Berg, H. W., and Loinger, C. 1967. Acid treatment of red table wine musts for color retention. *Am. J. Enol. Viticult.* **18**(4), 182–189.

Ournac, A. 1953. Recherches sur la variation de la teneur en acide ascorbique et en facteur P du jus de raisin en fonction de son mode de préparation. *Ann. Technol. Agr. Paris* **2** (2), 99–111.

Ournac, A. 1965. Etude du dosage de l'acide ascorbique dans les vins et dans les jus fortement colorés. *Ann. Technol. Agr. Paris* **14**(4), 341–347.

Ournac, A. 1966. Evolution de l'acide ascorbique pendent la fermentation alcoolique du jus de raisin frais et du jus de raisin d'ésulfité et la conservation de vins correspondants. *Ann. Technol. Agr. Paris* **15**(2), 181–191.

Ournac, A., and Poux, C. 1966. Acide ascorbique dans le raisin au cours de son developpement. *Ann. Technol. Agr. Paris* **15**(2), 193–202.

Ouyan, S.-Z. 1963. Physiology of frost resistance in grape vines. *Fiziol. Rastenii Acad. Nauk SSSR* **10**(3), 366–368; (trans.) 300–306.

Overcash, J. P. 1955. Some effects of certain growth-regulating substances on the ripening of Concord grapes. *Proc. Am. Soc. Hort. Sci.* **65**, 54–58.

Overholser, E. L. 1917. Color development and maturity of a few fruits as affected by light exclusion. *Proc. Am. Soc. Hort. Sci.* **14**, 73–85.

Pacottet, P. 1904. Rôle des pépins, des pellicules, des rafles en vinification. *Rev. Viticult.* **22** (568), 485–488; (571), 581–584.

Pallavicini, C. 1966. Determinatzione degli enzimi fenolasi fosfatasi proteinasi, e saccarasi in 20 vine da diversa natura. *Riv. Viticolt. Enol.* **19**, 492–496.

Pallavicini, C. 1967 Distribuzione degli enzimi fenolasi, fosfatasi, proteinasi e saccarasi nelle parti costituenti gli acini di 4 uve e nei vini provenienti dalle uve stesse. *Ind. Agrar. (Florence)* **5**, 603–606.

Pallavicini, C., Campanella, V., and Losito, F. 1963. Azione chiarificante e decolorante di due polimeri del polivinilpirrolidone su vini colpiti da rottura ossidasica. *Vini Ital.* **5**(26), 321–324.

Pallotta, U., Donati, A.-M., and Spallacci, P. 1966a. Sulla composizione chimica e caratteristiche chimico fisiche del vino Verdicchio. *Ind. Agrar. (Florence)* **4**, 1–4.

Pallotta, U., Donati, A.-M., and Spallacci, P. 1966b. Sulla composizione chimica e carattерstiche chimico-fisiche del vino Verdicchio. 4. *Ind. Agrar. (Florence)* **4**, 525–538.

Papakyriakopoulos, V. G., and Amerine, M. A. 1956. Sensory tests on two types of wine. *Am. J. Enol.* **7**, 98–104.

Parisi, E., and Bruni, C. 1926. The anthocyanin of the "Fogarina" grape. *Staz. Sper. Agrar. Ital.* **59**, 130–138; *Chem. Abstr.* **22**, 1791 (1928).

Pastena, B. 1964. Correlazioni tra il peso ed il numero dei vinaccioli dell'acino, e tra il peso ed il grado rifrattometrico della stessa bacca. *Riv. Viticolt. Enol.* **17**, 149–164.

Pataky, M. B. 1965. Sur les particularités du dosage spectrophotométrique des anthocyannes. *Ann. Technol. Agr. Paris* **14**(1), 79–85.

Patay, R., Masquelier, J., Danon, G., and Bernard, J. P. 1961. Sur la toxicité d'un tannoïde extrait des parties ligneuses de la grappe de raisin. *Compt. Rend. Congr. Natl. Soc. Savantes, Paris Dépts., Sect. Sci.* **86**, 69–71.

Paulus, W., and Lorke, D. 1967. Zur Kenntnis des Pyrokohlensäurediäthylesters. III. Herstellung und toxikologische Prüfung repräsentativer Umsetzungsproducte des Pyrokohlensäurediäthylesters. *Z. Lebensm.-Unters. Forsch.* **132**(6), 325–333.

Pauly, H., and Feuerstein, K. 1929. Hadromal, Lignin und Coniferylaldehyd; deren Darstellung und Nachweis. Höhere Alkali-Kondensate des Acetaldehydes. *Ber. Deut. Chem. Ges.* **62B**, 297–311.

Paupardin, C. 1967. Sur la rapport entre la prolifération et la formation d'acide gentisique dans les tissues de tubercule de Topinambour (*Helianthus tuberosus* L., var ≪Violet de Rennes≫) cultivés in vitro dans diverses conditions. *Compt. Rend. Ser. D.* **264**, 2103–2105.

Pederson, C. S., Goresline, H. E., Curl, A. L., and Beavens, E. A. 1941. The keeping quality of bottled wines. Effect of head space. *Ind. Eng. Chem.* **33**, 304–307.

Pellet, H. 1906. L'acide salicylique; propriétés, recherche et dosage; de la présence normale de l'acide salicylique dans le règne végétal; la question des vins portugais (1903); (translated and amplified by A. J. Ferreira da Silva) "O acido salicylico e a questão dos vinhos portugueses no Brazil em 1900," p. 520. Imprensa Univ., Coimbra.

Peri, C. 1967a. Acid degradation of leucoanthocyanidins. *Am. J. Enol. Viticult.* **18**(4), 168–174.

Peri, C. 1967b. Osservazioni sulla degradazione acida dei leucoantociani. *Agrochimica* **11**(4-5), 439–450.

Peri, C., and Cantarelli, C. 1963. Esame comparativo dei detannizzanti proteici e proteicosimili. *Vini Italia* **5**(26), 285–294.

Perkins,H. J., and Aronoff, S. 1956. Identification of the blue-fluorescent compounds in boron-deficient plants. *Arch. Biochem. Biophys.* **64**, 506–507.

Peterson, R. G., and Caputi, A., Jr. 1967. The browning problem in wines. II. Ion exchange effects. *Am. J. Enol. Viticult.* **18**, 105–112.

Petri, L. 1911. Tannins of the roots of *Vitis* in relation to resistance to phylloxera. *Atti Reale Acad. Lincei* **201**, 57–65; *Chem. Abstr.* **5**, 2667 (1911).

Peynaud, E. 1965. Le goût et l'odeur du vin. *Bull. Soc. Sci. Hyg. Aliment.* **53**(10,11,12); *Bull. Soc. Natl. Appell. Orig. Vins Eaux-de-Vie* (96), 33–47.

Peynaud, E., and Dupuy, P. 1964. Méthodes d'isolement, de culture, et de classification des bactéries malolactiques. *Bull. Offic. Intern. Vigne Vin* **37**(403), 908–922.

Peynaud, E., and Guimberteau, G. 1962. Modification de la composition des raisins au cours de leur fermentation propre en anaérobiose. *Ann. Physiol. Végétale* **4**, 161–167.

Peyrot, E. 1934. Sulla probabile quantità e qualità delle principali sostanze coloranti nei vini bianchi. Studio comparativo per curve de assorbimento. *Ann. Chim. Appl.* **24**, 512–519.

Phillips, J. D., Pollard, A., and Whiting, G. C. 1956. Organic acid metabolism in cider and perry fermentations. I. A preliminary study. *J. Sci. Food Agr.* **7**, 31, 40.

Picard, F. 1924. Contribution a l'étude du rôle physiologique des tanins. Leur importance dans l'aoûtement des sarments de la vigne. *Compt. Rend.* **179**, 778–780.

Pifferi, P. G. 1965. Thin-layer chromatography on silica gel G of some phenol carboxylic acids. *Vitis* **5**(1), 24–26.

Pifferi, P. G. 1966. Studi sui pigmenti naturali. V. Ricerche sui flavonoidi e sui tannini estraibili dai vini. *Ind. Agrar. (Florence)* **4**, 475–479.

Pilone, B. F., and Berg, H. W. 1965. Some factors affecting tartrate stability in wine. *Am. J. Enol. Viticult.* **16**, 195–212.

Pires Martins, G. 1963. A cor do vinho, sua avaliacão exacta por novo processo espectrofotometrico. *Vinea Vino Port. Doc. Ser. 2* **1**(4), 1–30.

Pirovano, M. A. 1967. Transmission par hérédité des caractères des parents déduite des travaux poursuivis pour l'obtention des varietés nouvelles par croisement. *Bull. Offic. Intern. Vigne Vin.* **40**(440), 1057–1070.

Pisarnitskii, A. F. 1965. Change of aldehyde composition in the process of wine aging. *Prinkl. Biokhim. i Mikrobiol.* 1(2), 144–149; *Chem. Abstr.* **63**, 10631 (1965).

Pokorný, J., Karvánek, M., and Davídek, J. 1966. Oxidimetrische Gerbstoffbestimmung im Wein. *Sb. Vyskum Sk. Chem. Technol. Praze, Potravin. Technol.* **10**, 13–19.

Politis, J. 1921. Sur l'origine mitochondriale des pigments anthocyaniques dans les fruits. *Compt. Rend.* **172**, 1061–1063.

Politis, J. 1948. Recherches cytologiques sur le mode de formation de l'acide chlorogénique. *Rev. Cytol. Cytophysiol. Végétales* **10**, 229–231.

Politova-Sovzenko, T. K. 1950. Znachenie dubil'nykh i krasyashchikh veshchestv v proizvodstve Madery. *Vinodelie i Vinogradarstvo. SSSR* **10**(10), 21–24.

Politova-Sovzenko, T. K. 1957. Vliyanie enotanina i drozhzhevogo avtolizata na buket vina. *Vinodelie i Vinogradarstvo SSSR* **17**(3), 8–11.

Ponte, A., and Gualdi, G. 1931. Tannin in wines and its estimation. *Boll. uffic. staz. sper. ind. pelli mat. concianti*, 391–407; *Chem. Abstr.* **26**, 3326 (1932).

Ponting, J. D., Sanshuck, D. W., and Brekke, J. E. 1960. Color measurement and deterioration in grape and berry juices and concentrates. *Food Res.* **25**, 471–478.

Popovskii, V. G., Gasyuk, G. H., Matov, B. M., and Levina, M. V. 1961. Vliyanie ul'trazuvka na vychod i tsvet vinogradnogo soka. *Konservaya i Ovoshchesushil'naya Prom.* (1), 4–6; *Weinberg Keller* **8**, 346 (1961).

Porter, W. L., and Schwartz, J. H. 1962. Isolation and description of the pectinase-inhibiting tannins of grape leaves. *J. Food Sci.* **27**, 416–418.

Porter, W. L., Schwartz, J. H., Bell, T. A., and Etchells, J. L. 1961. Probable identity of the pectinase inhibitor in grape leaves. *J. Food Sci.* **26**, 600–605.

Pourrat, H., Bastide, P., Dorier, P., Pourrat, A., and Tronche, P. 1967. Preparation and therapeutic activity of some anthocyanin glucosides. *Chim. Ther.* 2(1), 33–38; *Chem. Abstr.* **66**, 114375 (1967).

Poux, C. 1966. Polyphénoloxydase dans le raisin. *Ann. Technol. Agr. Paris* **15**, 149–158.

Powers, J.J. 1964. Action of anthocyanin and related compounds on bacterial cells. *Intern. Symp. Food Microbiol., 4th, Goteborg, Sweden*, 59–75.

Powers, J. J., Somaatmadja, D., Pratt, D. E., Hamdy, M. K. 1960. Anthocyanins. II. Action of anthocyanin pigments and related compounds on the growth of certain microorganisms. *Food Technol.* **14**, 626–632.

Pratt, D. E., Powers, J. J., and Somaatmadja, D. 1960. Anthocyanins. I. The influence of strawberry and grape anthocyanins on the growth of certain bacteria. *Food Res.* **25**, 26–32.

Pratt, R., Sah, P. P. T., Dufrenoy, J., and Pickering, V. L. 1948. Vitamin $K_5$ as an inhibitor of the growth of fungi and of fermentation by yeast. *Proc. Natl. Acad. Sci.* **34**, 323–328.

Prehoda, J. 1966. Rotweinbereitung mit und ohne Maischegärung. *Mitt.* (*Klosterneuburg*) **16**, 463–468.

Preobrazhenskii, A. A. 1963. Osobennosti maderizatsii v germetizirovannykh rezervuarakh *Vinodelie i Vinogradarstvo SSSR* **23**(6), 4–9.

Preobrazhenskii, A. A., and Kharin, Y. S. 1965. A rational method to prepare Madeira. *Pishchevaya Prom., Min. Vysshego i Srednego Spets. Obrazov. Ukr. SSR, Mezhvedomstv. Resp. Nauchn.-Tekkn. Sb.* **2**, 94–101; *Chem. Abstr.* **64**, 18357 (1966).

Pridham, J. B. 1960. "Phenolics in Plants in Health and Disease," p. 131. Pergamon Press, New York.

Pridham, J. B. 1963. "Enzyme Chemistry of Phenolic Compounds," *Proc. Plant Phenolics Group Symp., 1962,* p. 142. Macmillan, New York.

Pridham, J. B. 1964. "Methods in Polyphenol Chemistry," *Proc. Plant Phenolics Group Symp.*, 1963, p. 146. Macmillan, New York.

Prillinger, F. 1952. Gerbstoffe als Ursache von Trübungserscheinungen von Süssmost. *Mitt. (Klosterneuburg)* **2**, 197–198.

Prillinger, F. 1963. Protection des vins de l'oxydation par l'emploi de l'acide sulfureux et de l'acide ascorbique. *Ann. Technol. Agr. Paris Numero Hors Ser. I.* **12**(1), 159–169.

Prillinger, F. 1965. Akazienholzfarbstoffe im Wein. *Mitt. (Klosterneuburg)* **15A**, 21–24.

Pro, M. J. 1952. Reports on spectrophotometric determination of tannin in wines and whiskies. *J. Assoc. Offic. Agr. Chemists* **37**, 255–257.

Pro, M. J. 1955. Report on spectrophotometric determination of tannin in wines and spirits. *J. Assoc. Offic. Agr. Chemists* **38**, 757.

Protin, R. 1966. Situation de la viticulture dans la monde en 1965. *Bull. Offic. Intern. Vigne Vin* **39**(428), 1206–1247.

Puissant, A., and Léon, H. 1967. La matière colorante des grains de raisins de certains cépages cultivés en Anjou en 1965. *Ann. Technol. Agr. Paris* **16**, 217–225.

Quesnel, V. C. 1964. Suitability of sulfurous acid for hydrolysis of condensed tannins. *Tetrahedron Letters*, 3699–3702.

Quintanilla Iniesta, E., Bravo Abad, F., Iñigo Leal, B., and Garrido Marquez, J. M. 1966. Anthocyanogens, tannins, and nitrogen in wines from western Andalucia (Montilla-Aguilar-Moriles). *Rev. Cienc. Apl. (Madrid)* **20**(3), 206–210; *Chem. Abstr.* **66**, 27690 (1967).

Rabak, F. 1913. The utilization of waste raisin seeds. *U.S. Bur. Plant Ind. Bull.* **276**, 7–36.

Rabak, F., and Shrader, J. H. 1921. Commercial utilization of grape pomace and stems from the grape-juice industry. *U.S. Dept. Agr. Bull.* **952**, 1–22.

Radler, F. 1964. The prevention of browning during drying by the cold dipping treatment of Sultana grapes. *J. Sci. Food Agr.* **15**, 864–869.

Radoev, A., and Mladenov, A. 1959. The concentration of vitamin P substance in wild and cultivated vegetables and fruits. *Nauch. Trudove, Visshiya Inst. Khranitelna i Vkusova Prom.-Plovdiv* **5**, 375–378; *Chem. Abstr.* **55**, 5801 (1961).

Rafei, M. S. 1941. The anatomy of grape and related genera. Ph.D. thesis. Oregon State Univ., Corvallis, Oregon.

Rakcsányi, L. 1964. A szölöhéj cserzöanyagtartalmának kitermelése. *Borgazdasag* **12**, 5–15.

Ramos, M. daC., and de Oliveira, H. T. 1949. Estudo cromatográfico das matérias tintu-riais do vinho do porto, tendo como primeiro objectivo a pesquisa da baga. *Anais Inst. Vinho Porto* **10**(1), 11–24.

Ramos, M. daC., and Pereira, J. 1954. Matérias corantes no vinho do Porto, Reacções e aspectos cromatográficos. *Anais Inst. Vinho Porto* **15**(2), 1–22.

Rankine, B. C. 1962. Bentonite and wine fining. *Australian Wine, Brew. Spirit Rev.* **80**(3), 18, 20; **81**(2), 18, 20, 22.

Rankine, B. C. 1964. Heat extraction of colour from red grapes for wine. *Australian Wine, Brew. Spirit Rev.* **82**(6), 41, 42.

Rankine, B. C., Kepner, R. E., and Webb, A. D. 1958. Comparison of anthocyan pigments of vinifera grapes. *Am. J. Enol.* **9**, 105–110.

Rasulpuri, M. L., Anderson, A. W., and Yang, H. Y. 1965. Mode of action of vitamin $K_5$ on *Saccharomyces cerevisiae*. *J. Food Sci.* **30**, 160–165.

Rasuvaev, N. I. 1965. Potochnye skhemy kompleksnoi pererabotki otkhodov vinodeliya. *Vinodelie i Vinogradarstvo SSSR* **25**(4), 42–46.

Ravaz, L., Dupont, E., and Callaudaux, R. 1933. Recherches sur le rougeau de la vigne. *Ann. Agron.* **3**, 225–231.

Rebelein, H. 1965a. Beitrag zur Bestimmung des Catechingehaltes im Wein. *Deut. Lebensm. Rundschau* **61**, 182–183.

Rebelein, H. 1965b. Beitrag zum Catechin-und Methanolgehalt von Weinen. *Deut. Lebensm. Rundschau* **61**, 239–240.

Reichel, L. 1937. Anthocyane als biologische Wasserstoff-Acceptoren *Naturwissenschaften* **25**, 318.

Reichel, L., and Buckart, W. 1938. Über biogenetische Beziehungen der Anthocyanidine zu Flavonfarbstoffen und Catechinen. *Ann. Chem. Liebigs* **536**, 164–173.

Rentschler, H. 1963. Troubles métalliques des vins et leurs traitements. *Ann. Technol. Agr. Paris Numero Hors Ser. I.* **12**(1), 267–276.

Rentschler, H., and Schaeppi, R. 1957. Die Beeinflussung von Rotweinen durch Kälte. *Schweiz. Z. Obst-Weinbau* **66**, 437–439.

Rentschler, H., and Tanner, H. 1951. Das Bitterwerden der Rotweine. Beitrag zur Kenntnis des Vorkommens von Acroleins in Getränken und seine Bezihung zum Bitterwerden der Weine. *Mitt. Gebiete Lebensm. Hyg.* **42**, 463–475.

Rentschler, H., and Tanner, H. 1956. Über Polyphenole der Kernobst- und Traubensäfte. I. De Balavoine-Reaktion und ihre Ursache. *Mitt. Lebensm. Hyg.* **47**, 28–32.

Reuther, G. 1961. Genetisch-biochemische Untersuchungen an Rebenartbastarden. *Züchter* **31**, 319–328.

Reynolds, H., and Vaile, J. E. 1942. Chemical composition of Arkansas-grown American grapes. *Arkansas Univ. Agr. Expt. Sta. Bull.* **420**, 1–53.

Reynolds, T. M. 1965. Chemistry of nonenzymic browning. II. *Advan. Food Res.* **14**, 167–283.

Ribbons, D. W. 1965. The microbiological degradation of aromatic compounds. *Ann. Repts. Progr. Chem. (Chem. Soc., London)* **62**, 445–462.

Ribéreau-Gayon, J. 1933. "Contribution a l'Étude des Oxydations et Reductions dans les Vins; Application a l'Etude du Vieillissement et des Casses," p. 213. Delmas, Bordeaux.

Ribéreau-Gayon, J. 1936. Les phénomènes colloïdaux dans les vins. *Bull. Soc. Chim. France* **3**, 603–612.

Ribéreau-Gayon, J. 1939. Clarification et stabilisation spontanées des vins. *Rev. Viticult.* **91**, 73–78, 97–103.

Ribéreau-Gayon, P. 1953. Différenciation des matières colorantes des raisins et des vins des cépages français et hybrides. *Compt. Rend. Acad. Agr. France* **39**, 800–807.

Ribéreau-Gayon, P. 1957. Le leucocyanidol dans les vins rouges. *Compt. Rend. Acad. Agr. France* **43**, 197–199, 596–598, 821–823.

Ribéreau-Gayon, P. 1958a. Les anthocyannes des raisins. *Qualitas Plant. Mater. Vegetabiles* **3–4**, 491–499.

Ribéreau-Gayon, P. 1958b. Formation et évolution des anthocyannes au cours de la maturation du raisin. *Compt. Rend.* **246**, 1271–1273.

Ribéreau-Gayon, P. 1959. "Recherches sur les Anthocyannes des Végétaux. Application au Genre *Vitis*," p. 114. Librairie General Enseignement, Paris.

Ribéreau-Gayon, P. 1960a. Les anthocyannes au genre *Vitis*. Application à la différenciation des vins. *Compt. Rend.* **250**, 591–593.

Ribereau-Gayon, P. 1960b. Untersuchungen über die Farbstoffe roter Trauben. *Deut. Lebensm. Rundschau* **56**, 217–223.

Ribéreau-Gayon, P. 1963a. Les acides-phénols de *Vitis vinifera*. *Compt. Rend.* **256**, 4108–4111.

Ribéreau-Gayon, P. 1963b. La différenciation des vins par l'analyse chromatographique de leur matière colorante. *Ind. Aliment. Agr.* **80**, 1079–1084.

Ribéreau-Gayon, P. 1963c. Les anthocyannes des fruits. Méthodes d'identification et applications. *Mises Point Chim. Anal., Org., Pharm., Bromatol.* **11**, 189–220.

Ribéreau-Gayon, P. 1963d. Evolution des anthocyannes au cours du vieillisement des vins rouges. *Ann. Technol. Agr. Paris Numero Hors Ser. I.* **12**(1), 264–265.

Ribéreau-Gayon, P. 1964a. Les composés phénoliques du raisin et du vin. *Ann. Physiol. Végétale* **6**, 119–147, 211–242, 259–282.

Ribéreau-Gayon, P. 1964b. Les flavonosides de la baie dans le genre *Vitis. Compt. Rend.* **258**, 1335–1337.

Ribéreau-Gayon, P. 1964c. La differenciation des vins par l'analyse chromatographique de leur matière colorante. *Bull. Inst. Natl. Appell. Orig. Vins Eaux-de-Vie* (88), 29–43.

Ribéreau-Gayon, P. 1965a. Phenolic compounds of grapes and wine. *Wines & Vines* **46**(5), 26–27.

Ribéreau-Gayon, P. 1965b. Nouvelles observations sur la différenciation des vins de *V. vinifera* et d'hybrides. *Compt. Rend. Acad. Agr. France* **51**, 135–140.

Ribéreau-Gayon, P. 1965c. Identification d'esters des acides cinnamiques et l'acide tartique dans les limbes et les baies de *V. vinifera. Compt. Rend.* **260**, 341–343.

Ribéreau-Gayon, P. 1965d. La couleur des vins. *Aliment. Vie* **53**(10–12), 232–248.

Ribéreau-Gayon, J. 1966. Influence des températures de fermentation et de conservation du vin et des vins spéciaux sur leurs caractères chimiques, microbiologiques et organoleptiques. Rapport Francais. *Bull. Offic. Intern. Vigne Vin* **39**(424), 728–736.

Ribéreau-Gayon, P. 1968. "Les Composés Phénoliques des Végétaux," p. 254. Dunod, Paris.

Ribéreau-Gayon, J., and Gardrat, J. 1956. Titrage potentiométrique des anthocyannes du raisin. *Compt. Rend.* **243**, 788–790.

Ribéreau-Gayon, J., and Gardrat, J. 1957. Application du titrage potentiométrique à l'étude du vin. *Ann. Technol. Agr. Paris* **6**, 185–216.

Ribéreau-Gayon, J., and Maurié, A. 1942. Les composés phénoliques du vin (matières colorantes et tanoïdes). *Bull. Offic Intern. Vigne Vin* **15**(150), 60–76.

Ribéreau-Gayon, P., and Nedeltchev, N. 1965. Discussion et application des méthode modernes de dosage des anthocyanes et des tanins dans les vins. *Ann. Technol. Agr. Paris* **14**, 321–330.

Ribéreau-Gayon, J., and Peynaud, E. 1934–35. Etudes sur le collage des vins. *Rev. Viticult.* **81**, 5–11, 37–44, 53–59, 117–124, 165–171, 201–205, 310–315, 341–346, 361–365, 389–397, 405–411; **82**, 8–13.

Ribéreau-Gayon, J., and Peynaud, E. 1935. Formations et précipitations de colloïdes dans les vins rouges. *Compt. Rend. Acad. Agr. France* **21**, 720–725.

Ribéreau-Gayon, J., and Peynaud, E. 1952. Action inhibitrice sur les levures de la vitamine $K_5$ et de quelques antibiotiques. *Compt. Rend. Acad. Agr. France* **38**, 479–481.

Ribéreau-Gayon, J., and Peynaud, E. 1960–61. "Traite d'Oenologie," 2 Vol., p. 753, 1065. Librairie Polytech. Beranger, Paris.

Ribéreau-Gayon, J., and Ribéreau-Gayon, P. 1953. Etude expérimentale de la vinification en rouge. II. Expérimentation 1952. *Chim & Ind. Paris* **70**, 694–700.

Ribéreau-Gayon, J., and Ribéreau-Gayon, P. 1954. Etude expérimentale de la vinification en rouge. III. Expérimentation 1953. *Chim. & Ind. Paris* **72**, 922–930.

Ribéreau-Gayon, J., and Ribéreau-Gayon, P. 1958. The anthocyans and leucoanthocyans of grapes and wines. *Am. J. Enol.* **9**, 1–9.

Ribéreau-Gayon, G., and Ribereau-Gayon, P. 1963. Utilisation de $^{14}CO_2$ et de glucose-U-$^{14}C$ pour l'etude du metabolisme des acides organiques de *Vitis vinifera* L. *Compt. Rend.* **257**, 778–780.

Ribéreau-Gayon, P., and Sapis, J. C. 1965. Sur la présence dans le vin de tyrosol, de tryptophol, d'alcool phényléthylique et de γ-butyrolactone, produits secondaires de la fermentation alcoolique. *Compt. Rend.* **261**, 1915–1916.

Ribéreau-Gayon, P., and Stonestreet, E. 1964. La constitution des tanins du raisin et du vin. *Compt. Rend. Acad. Agr. France* **50**, 662–671.

Ribéreau-Gayon, P., and Stonestreet, E. 1965. Le dosage des anthocyanes dans le vin rouge. *Bull. Soc. Chim. France*, 2649–2652.

Ribéreau-Gayon, P., and Stonestreet, E. 1966a. Vorkommen und Bedeutung der Catechine, Leukoanthocyanidine und der Gerbstoffe in Rotweinen. *Deut. Lebensm. Rundschau* **62**, 1–5.

Ribéreau-Gayon, P., and Stonestreet, E. 1966b. Dosage des tanins du vin rouge et détermination de leur structure. *Chim. Anal.* **48**, 188–196.

Ribéreau-Gayon, P., and Sudraud, P. 1957. Les anthocyannes de la baie dans le genre *Vitis*. *Compt. Rend.* **244**, 233–235.

Ribéreau-Gayon, J., and Sudraud, P. 1958. Etude expérimentale de la vinification en rouge. IV. Expérimentation 1954. *Chim. & Ind. Paris* **79**, 302–312.

Ribéreau-Gayon, P., Sudraud, P., and Durquety, P. M. 1955. Relations entre génétique et nature chimique des pigments anthocyaniques de la baie dans le genre *Vitis*. *Rev. Gen. Botan.* **62**, 667–674.

Rice, A. C. 1964. Identification of grape varieties. *J. Assoc. Offic. Agr. Chemists* **47**, 671–676.

Rice, A. C. 1965. Identification of grape varieties. *J. Assoc. Offic. Agr. Chemists* **48**, 525–530.

Rice, A. C., Ferguson, J. W., and Belscher, R. S. 1968. Residual sugars in New York State wines. *Am. J. Enol. Viticult.* **19**, 1–5.

Rieche, A., Hilgetag, G., Martini, A., and Lorenz, M. 1962. Continuous production of yeast from phenol-containing substrates. *Continuous Cultivation Microorganisms Symp. Prague, 1962* 2, 293–297.

Riffer, R., and Anderson, A. B. 1967. Chemistry of the genus *Sequoia*. IV. The structures of the $C_{17}$ phenols from *Sequoia sempervirens*. *Phytochemistry* **6**, 1557–1562.

Rinderknecht, H., and Jurd, L. 1958. A novel non-enzymic browning reaction. *Nature* **181**, 1268–1269.

Roberts, E. A. H. 1956a. Conformation of catechins and gallocatechins. *Chem. & Ind.* **1956**, 737.

Roberts, E. A. H. 1956b. Paper chromatography as an aid to the elucidation of the structure of polyphenols occurring in tea. *In* "The Chemistry of Vegetable Tannins," pp. 87–97. Soc. Leather Trades' Chemists, Croydon, England.

Roberts, E. A. H. 1959. The interaction of flavanol orthoquinones with cysteine and glutathione. *Chem. & Ind.* **1959**, 995.

Roberts, E. A. H. 1962. Economic importance of flavonoid substances: tea fermentation. *In* "The Chemistry of Flavonoid Compounds" (T. A. Geissman, ed.), pp. 468–512. Macmillan, New York.

Roberts, E. A. H., and Myers, M. M. 1960. Phenolic substances of manufactured tea. VIII. Enzymic oxidations of polyphenolic mixture. *J. Sci. Food Agr.* **11**, 153–157.

Roberts, E. A. H., and Wood, D. J. 1953. Separation of tea polyphenols on paper chromatograms. *Biochem. J.* **53**, 332–336.

Robinson, G. M., and Robinson, R. 1932. A survey of anthocyanins. II. *Biochem. J.* **26**, 1647–1664.

Robinson, G. M., and Robinson, R. 1933. Survey of anthocyanins. III. Notes on the distribution of leuco-anthocyanins. *Biochem. J.* **27**, 206–212.

Robinson, G. M., and Robinson, R. 1934. A survey of anthocyanin. IV. *Biochem. J.* **28**, 1712–1720.

Robinson, W. B., Shaulis, N. J., and Pederson, C. S. 1949. Ripening studies of grapes grown in 1948 for juice manufacture. *Fruit Prods. J.* **29**, 36–37, 54, 62.

Robinson, W. B., Bertino, J. J., and Whitcombe, J. E. 1966a. Objective measurement and specification of color in red wines. *Am. J. Enol. Viticult.* **17**, 118–125.

Robinson, W. B., Weirs, L. D., Bertino, J. J., and Mattick, L. R. 1966b. The relation of anthocyanin composition to color stability of New York State wines. *Am. J. Enol. Viticult.* **17**, 178–184.

Rodopulo, A. K. 1950. Rol' dubil'nykh veshchestv v okislenii susla i vina. *Vinodelie i Vinogradarstvo SSSR* **10**(9), 20–21.

Rodopulo, A. K. 1953. O fermentativnykh i nefermentativnykh okislitel'nykh sistemakh susla i vina. *Biokhim. Vinodeliya* **4**, 211–225.

Rodopulo. A. K. 1964. O roli produktov prevrashcheniya aminokislot v obrazovanii buketa shampanskogo. *Vinodelie i Vinogradarstvo SSSR* **24**(1), 6–9.

Roelofsen, P. A. 1958. Fermentation, drying, and storage of cacao beans. *Advan. Food Res.* **8**, 225–296.

Róna, E. 1914. I. Über die Reduktion des Zimtaldehyds durch Hefe. II. Vergärung von Benzylbrenztraubensäure. *Biochem. Z.* **67**, 137–142.

Rosenheim, O. 1920a. Note on the use of butyl alcohol as a solvent for anthocyanins. *Biochem. J.* **14**, 73–74.

Rosenheim, O. 1920b. Observations on anthocyanins. I. The anthocyanins of the young leaves of the grape vine. *Biochem. J.* **14**, 178–188.

Rosenstiehl, A. 1897. De la solubilité de la matière colorante rouge du raisin et de la sterilisation des moûts de fruits. *Compt. Rend.* **124**, 566–569.

Rosenstiehl, A. 1902. De l'action des tannins et des matières colorantes sur l'activité des levures. *Compt. Rend.* **134**, 119–122.

Roson, J.-P. 1967. Recherches sur la vinification des vines rosés. *Ann. Technol. Agr. Paris* **16**, 227–239.

Ross, S. L. 1942. Detannating wines for pharmaceutical use. *Am. J. Pharm.* **114**, 200–204.

Rossi, J. A., Jr., and Singleton, V. L. 1966a. Contributions of grape phenols to oxygen absorption and browning of wines. *Am. J. Enol. Viticult.* **17**, 231–239.

Rossi, J. A., Jr., and Singleton, V. L. 1966b. Flavor effects and adsorptive properties of purified fractions of grape-seed phenols. *Am. J. Enol. Viticult.* **17**, 240–246.

Rousseaux, E. 1909a. Durée du cuvage et dècuvage. *Rev. Viticult.* **32**, 324–326.

Rousseaux, E. 1909b. Vinification en blanc des raisins rouges. *Rev. Viticult.* **32**, 382–384.

Roux, D. G., Maihs, E. A., and Paulus, E. 1961. Chromatographic behavior of some stereochemically interrelated flavonoid compounds in aqueous medium. *J. Chromatography* **5**, 9–16.

Rubin, B. A., and Artsikhovskaya, Ye. V. 1963. "Biochemistry and Physiology of Plant Immunity" (H. Wareing, trans. from 1960 Russ. ed.), p. 358. Macmillan, New York.

Rudnev, N. M., and Leonov, B. I. 1961. Grape pigments, their chemical constitution and method of extraction. *Sadovodstvo, Vinogradarstvo i Vinodelie Moldavii* **16**(9), 38–41; *Chem. Abstr.* **56**, 15901 (1962).

Ruf, W. 1952. Über die Anwendung der Papierchromatographie zum Nachweis von Fremdfarbstoffen im Wein. *Z. Lebensm.-Untersuch. Forsch.* **94**, 190–194.

Rusznyak, S., and Szent-Györgyi, A. 1936. Vitamin P: Flavonols as vitamins. *Nature* **138**, 27.

Ryabchun, O. P. 1965. Carbohydrate and lignin content per cell in grape shoots. *Fiziol. Rastenii Acad. Nauk* **12**, 479–488, (trans. 415–423).

Ryugo, K. 1962. Occurrence of vanillin and structurally related aldehydes in the endocarp of some *Prunus* species. *Proc. Am. Soc. Hort. Sci.* **81**, 180–183.

Sadow, H. S. 1953. The anthocyan pigments of certain grapes and wine. Ph.D. Thesis. Univ. Connecticut, Storrs, Connecticut.

Sakamoto, H. 1959. The metabolism of flavonol. I. Products of metabolism of ampheloptin by a microorganism. *Bitamin (Kyoto)* **18**(3), 91–95; *Chem. Abstr.* **62**, 3109 (1965).

Sakamoto, H. 1960. The metabolism of flavonol. II. Products of metabolism of rutin and quercetin by *Penicillium brevi-compactum Bitamin* (*Kyto*) **19**(1), 155–160; *Chem. Abstr.* **62**, 3109 (1965).

Saller, W., and de Stefani, C. 1960. Holunderzusatz zu Traubenmost. *Weinberg Keller* 7, 45–49.

Salgues, R. 1960. Parasitisme chez la vigne et consequénces oenologiques. *Qualitas Plant. Mater. Vegetabiles* 7, 16–30.

Salgues, R. 1962. Nouvelles études chemiques et toxicologiques sur le genre *Rhamnus* L. (Rhamnacees). *Qualitas Plant. Mater. Vegetabiles* **9**, 14–32.

Samvelyan, A. M. 1959. Ismenenie krasyashchikh veshchestv pri vyderzhke vina. *Vinodelie i Vinogradarstvo SSSR* **19**(4), 6–8.

Santhanam, P. S., Ghosh, D., and Nayudamma, Y. 1965. Studies on the phenolic constituents of babul. I. Affinity studies. *Leather Sci.* **12**, 179–183.

Sastry, L. V. L., and Tischer, R. G. 1952a. Behavior of the anthocyanin pigments in Concord grapes during heat processing and storage. *Food Technol.* **6**, 82–86.

Sastry, L. V. L., and Tischer, R. G. 1952b. Stability of the anthocyanin pigments in Concord grape juice. *Food Technol.* **6**, 264–268.

Sauzeat, M. 1932. La recherche de la couleur en vinification. *Rev. Viticult.* 77, 125–126.

Saywell, L. G. 1934. The clarification of wine. *Ind. Eng. Chem.* **26**, 981–982.

Scarborough, H. 1945. Observations on the nature of vitamin P and the vitamin P potency of certain foodstuffs. *Biochem. J.* **39**, 271–278.

Scarpati, M. L., and D'Amico, A. 1960. Isolamento dell'acido monocaffeiltartarico dalla cicoria (*Chicorium intybus*). *Ricerca Sci.* **30**, 1746–1748.

Scarpati, M. L., and Oriente, G. 1958. Chicoric acid (dicaffeyltartic acid): its isolation from chicory (*Chicorium intybus*) and synthesis. *Tetrahedron* 4, 43–48.

Schanderl, H. 1951. Zum Lobe der Hefeschönung. *Deut. Wein-Ztg.* **87**, 78–79.

Schanderl, H. 1957. Über das Vorkommen und die Bildungsweise der Gerbstoffe und der Anthozyane in den Rebengewächsen. *Mitt.* (*Klosterneuburg*) **7A**, 229–245.

Schanderl, H. 1959. "Die Mikrobiologie des Mostes und Weines," 2nd ed., p. 321. Ulmer., Stuttgart.

Schanderl, H. 1962. Der Einfluss von Polyphenolen und Gerbstoffen auf die Physiologie der Weinhefe und der Wert des pH-7 Testes für die Auswahl von Sektgrundweinen. *Mitt.* (*Klosterneuburg*) **12A**, 265-274.

Schanderl, H. 1964. Phénomènes d'oxydoreduction au cours de la maturation et du vieillissement du vin et procédés pour le modifier. Rapport Allemand. *Bull. Offic. Intern. Vigne Vin* **37**(399), 490–497.

Schanderl, H., and Staudenmayer, T. 1961. Erfahrungen mit weissen Sektgrundweinen. *Deut. Wein-Ztg.* **97**, 92, 94.

Schmersahl, K. J. 1964. Uber die Wirkstoffe der diaphoretischen Drogen des DAB 6. *Naturwissenschaften* **51**(15), 361.

Schneyder, J. 1962. Einfaches Gerät zur Messung der Farbtiefe von Rotwein. *Mitt.* (*Klosterneuburg*) **12**, 131–133.

Schneyder, J., and Epp, F. 1959. Papierchromatographische Aufftrennung der Farbstoffkomponenten von Rotweinen, Direktträgerrotwein und vergorenen Beerensäften. *Mitt.* (*Klosterneuburg*) **9A**, 111–114.

Schuller, P. L., Ockhuizen, T., Werringloer, J., and Marquardt, P. 1967. Aflatoxin B, und Histamin in Wein. *Arzneimittel-Forsch.* **17**(17), 888–890.

Scott, A. I. 1965. Oxidative coupling of phenolic compounds. *Quart. Rev.* (*London*) **19**, 1–35.

Scurti, F., and Pavarino, G. L. 1934. Sulla scottatura dell'uva. Esperienze esequite sull'uva Regina. *Ann. Sper. Agrar. (Rome)* **15**, 19–22.

Seeger, P. G. 1967. Die Anthozyane von *Beta vulgaris* var. rubra (Rote Bete), *Vaccinum myrtillis* (Heidelbeere), *Vinum rubrum* (Rotwein), und ihre Bedeutung als Zellatmungsaktivatoren für die Krebsprophylaxe und Krebstherapie. *Ärztliche Forsch.* **21**(2), 68–78.

Segal, B., and Grager, W. 1967. Das Anwachsen der Bioflavone in Traubensäften durch enzymatische Maischebehandlung. *Nahrung* **11**(3), 223–228.

Serre, P. 1919. Utilisation des pépins de raisins en Californie. *Compt. Rend. Acad. Agr. France* **5**, 150–151.

Shaulis, N., and Robinson, W. 1953. The effect of season, pruning severity and trellising on some chemical characteristics of Concord and Fredonia grape juice. *Proc. Am. Soc. Hort. Sci.* **62**, 214–220.

Shewfelt, A. L. 1966. The nature of the anthocyanin pigments in South Carolina Concord grapes. *S. Carolina Agr. Expt. Sta. Tech. Bull.* **1025**, 1–14.

Shimazu, Y. 1950. Formation and decomposition of D-(—)-phenylacetylcarbinol by yeast. *J. Chem. Soc. Japan, Pure Chem. Sec.* **71**, 503–505; *Chem. Abstr.* **45**, 9004 (1951).

Shnaidman, L. O., Afanas'eva, V. S., Solodukhin, A. I., and Papanov, V. A. 1965. Ispol'zovanie otkhodov sokovogo i vinodel'cheskogo proizvodstva dlya polucheniya vitamina P i karotina. *Konservn. i Ovoshchesushil'n. Prom.* **20**(12), 31–33.

Shpritsman, E. M. 1961. The tannin substances of oak and their role in the process of aging of cognac spirits. *Tr. Armyansk. Nauchn.-Issled. Inst. Vinogradarstva, Vinodeliya i Plodovodstva, Min. Sel'sk. Khoz. Arm. SSR* No. 5, 208–218; *Chem. Abstr.* **48**, 9594 (1963).

Shriner, R., and Anderson, R. 1928. A contribution to chemistry of grape pigments. V. The anthocyans in Ives grapes. *J. Biol. Chem.* **80**, 743-752.

Shunk, E., Knecht, E., and Marchlewski, L. 1894. Über einen in den Rebenblättern vorkommenden Farbstoff. *Ber. Deut. Chem. Ges.* **27**, 487–488.

Sikovec, S. 1966a. Der Einfluss einiger Polyphenole auf die Physiologie von Weinhefen. I. Der Einfluss von Polyphenolen auf den Verlauf der alkoholischen Gärung insbesondere von Umgärungen. *Mitt. (Klosterneuburg)* **16**, 127–138.

Sikovec, S. 1966b. Der Einfluss von Polyphenolen auf die Physiologie von Weinhefen. II. Der Einfluss von Polyphenolen auf die Vermehrung und Atmung von Hefen. *Mitt. (Klosterneuburg)* **16**, 272–281.

Simpson, F. J., Narasimhachari, N., and Westlake, D. W. S. 1963. Degradation of rutin by *Aspergillus flavus*—the carbon monoxide producing system. *Can. J. Microbiol.* **9**, 15–25.

Singleton, V. L. 1961. An extraction technique for recovery of flavors, pigments, and other constituents from wines and other aqueous solutions. *Am. J. Enol. Viticult.* **12**, 1–8.

Singleton, V. L. 1963. Changes in quality and composition produced in wine by cobalt-60 gamma irradiation. *Food Technol.* **17**, 112–115.

Singleton, V. L. 1964. Application of charcoal in wine clarification. *Wines & Vines* **45**(3), 29–31.

Singleton, V. L. 1966. The total phenolic content of grape berries during the maturation of several varieties. *Am. J. Enol. Viticult.* **17**, 126–134.

Singleton, V. L. 1967a. Fining-phenolic relationships. *Wines & Vines* **48**(3), 23–26.

Singleton, V. L. 1967b. Adsorption of natural phenols from beer and wine. *Tech. Quart. Master Brewers Assoc. Am.* **4**(4), 245–253.

Singleton, V. L., and Draper, D. E. 1961. Wood chips and wine treatment; the nature of aqueous alcohol extracts. *Am. J. Enol. Viticult.* **12**, 152–158.

Singleton, V. L., and Draper, D. E. 1962. Adsorbents and wines. I. Selection of activated charcoals for treatment of wine. *Am. J. Enol. Viticult.* **13**, 114–125.

Singleton, V. L., and Draper, D. E. 1964. The transfer of polyphenolic compounds from grape seeds into wine. *Am. J. Enol. Viticult.* **15**, 34–40.

Singleton, V. L., and Guymon, J. F. 1963. A test of fractional addition of wine spirits to red and white port wines. *Am. J. Enol. Viticult.* **14**, 129–136.

Singleton, V. L., and Ough, C. S. 1962. Complexity of flavor and blending of wines. *J. Food Sci.* **27**, 189–196.

Singleton, V. L., and Rossi, J. A., Jr. 1965. Colorimetry of total phenolics with phosphomolybdic-phosphotungstic acid reagents. *Am. J. Enol. Viticult.* **16**, 144–158.

Singleton, V. L., Berg, H. W., and Guymon, J. F. 1964a. Anthocyanin color level in port-type wines as affected by the use of wine spirits containing aldehydes. *Am. J. Enol. Viticult.* **15**, 75–81.

Singleton, V. L., Ough, C. S., and Amerine, M. A. 1964b. Chemical and sensory effects of heating wines under different gases. *Am. J. Enol. Viticult.* **15**, 134–145.

Singleton, V. L., Draper, D. E., and Rossi, J. A., Jr. 1966a. Paper chromatography of phenolic compounds from grapes, particularly seeds, and some variety-ripeness relationships. *Am. J. Enol. Viticult.* **17**, 206–217.

Singleton, V. L., Marsh, V. L., and Coven, M. 1966b. Identification of ellagic acid as a precipitate from loganberry wine. *J. Agr. Food Chem.* **14**, 5–8.

Sisakyan, N. M., Egorov, I. A., and Afrikyan, B. L. 1947. Vozrastnye variatsii dubil'nykh veshchestv v sortakh vinograda. *Biokhim. Vinodeliya, Akad. Nauk SSSR* **1**, 158–169.

Sisakyan, N. M., Egorov, I. A., and Afrikyan, B. L. 1948. Biokhimicheskie osobennosti vinogradnoi lozy ee vliyanie na sort vina. *Biokhim. Vinodeliya, Akad. Nauk SSSR* **2**, 7–55.

Skurikhin, I. M. 1957. Khimicheskie protsessy pri vyderzhke spirtov v dubovykh bochkakh. *Tr. Vses. Nauchn.-Issled. Inst. Vinodeliya i Vinogradarstva, Magarach* **5**, 69–90.

Skurikhin, I. M. 1959. Issledovanie ekstrakta kon'yachnogo spirta (dubil'nych veshchestv, lignina i redutsiruyushchich sacharov). *Vinodelie i Vinogradarstvo SSSR* **19**(2), 28–36.

Skurikhin, I. M. 1962. Prevrashchenie lignina, dubil'nykh i redutsiruyushchikh veshchestv pri sozrevanii kon'yachnykh spirtov. *Vinodelie i Vinogradarstvo SSSR* **22**(2), 17–22.

Skurikhin, I. M. 1966. Vliyanie spirtuoznosti, temperatury, pH i vozrasta bochki na sozrevanie kon'yachnykh spirtov. *Vinodelie i Vinogradarstvo SSSR* **26**(6), 10–15.

Skurikhin, I. M. 1967. Aromatic aldehydes of sherry and madeira. *Prikl. Biokhim. Mikrobiol.* **3**(6), 674–679; *Chem. Abstr.* **68**, 76923 (1968).

Smit, C. J. B. 1953. The tannins and related polyphenols of fruits. Ph.D. thesis. Univ. of California, Berkeley, California.

Smit, C. J. B., Joslyn, M. A., and Lukton, A. 1955. Determination of tannins and related polyphenols in foods, comparison of Loewenthal and Pro methods. *Anal. Chem.* **27**, 1159–1162.

Smith, R. M., and Luh, B. S. 1965. Anthocyanin pigments in the hybrid grape variety Rubired. *J. Food Sci.* **30**, 995–1005.

Smul'skaya, O. P. 1961. Rol' nekotorykh veshchestv vinogradnykh semyan v osvetlenii vin. *Vinodelie i Vinogradarstvo* **21**(5), 17–21.

Snell, F. D., and Snell, C. T. 1953. "Colorimetric Methods of Analysis," 3rd ed., Vol. 3, p. 606. Van Nostrand, New York.

Snell, F. D., Snell, C. T., and Snell, C. A. 1961. "Colorimetric Methods of Analysis," Vol. 3A, p. 576. Van Nostrand, Princeton, New Jersey.

Somaatmadja, D., Powers, J. J., and Hamdy, M. K. 1964. Anthocyanins. VI. Chelation studies on anthocyanins and other related compounds. *J. Food Sci.* **29**, 655–660.

Somaatmadja, D., Powers, J. J., and Wheeler, R. 1965. Action of leucoanthocyanins of Cabernet grapes on reproduction and respiration of certain bacteria. *Am. J. Enol. Viticult.* **16**, 54–61.

Somers, T. C. 1966a. Wine tannins. Isolation of condensed flavonoid pigments by gel filtration. *Nature* **209**, 368–370.

Somers, T. C., 1966b. Grape phenolics, the anthocyanins of *Vitis vinifera* variety Shiraz. *J. Sci. Food Agr.* 17, 215–219.

Somers, T. C. 1967a. Resolution and analysis of total phenolic constituents of grape pigments. *J. Sci. Food Agr.* **18**, 193–196.

Somers, T. C. 1967b. Pigments and tannins of wine. *Australian Wine Brewing Spirit Rev.* **85** (11), 38, 40.

Sondheimer, E. 1958. On the distribution of caffeic acid and chlorogenic acid isomers in plants. *Arch. Biochem. Biophys.* **74**, 131–138.

Sondheimer, E., and Kertesz, Z. I. 1948. The anthocyanin of strawberries. *J. Am. Chem. Soc.* **70**, 3476–3479.

Sondheimer, E., and Kertesz, Z. I. 1952. Kinetics of the oxidation of strawberry anthocyanin by hydrogen peroxide. *Food Res.* 17, 288–298.

Sone, K., Kagami, M., Omura, S., and Takayama, M. 1964. Effect of pectinase on the depth of color of red wine. *Bull. Res. Inst. Ferment. Yamanashi Univ.* **11**, 45–51.

Sostegni, L. 1897. Sulle materie coloranti delle uve rosse. *Gazz. Chim. Ital.* **27**(II), 475–485.

Spaeth, E. C., and Rosenblatt, D. H. 1950. Partition chromatography of synthetic anthocyanidin mixtures. *Anal. Chem.* **22**, 1321–1326.

Stadtman, E. R. 1948. Nonenzymatic browning in fruit products. *Advan. Food Res.* **1**, 325–372.

Stahl, W. H. 1962. The chemistry of tea and tea manufacturing. *Advan. Food Res.* **11**, 201–262.

Stanier, R. Y., Doudoroff, M., and Adelberg, E. A. 1963. "The Microbial World," 2nd ed., p. 753. Prentice-Hall, Englewood Cliffs, New Jersey.

Stanko, S. A., and Bardinskaya, M. S. 1962. Anthocyanins of callus tissue of *Parthenocissus tricuspidata. Dokl. Akad. Nauk. SSSR* **146**, 956–959, trans. 1152–1155 (1963).

Stanley, R. L., and Kirk, P. L. 1963. Systematic identification of artificial food colors permitted in the United States. *J. Agr. Food Chem.* **11**(6), 492–495.

Stanley, W. L. 1963. Recent developments in coumarin chemistry. *Plant Phenolics Group N. Am., Proc. 3rd Symp.*, 79–103.

Starkov, Yu. M. 1958. Deistvie 2-metil-1,4-naftokhinona (vitamina $K_3$) na mikrofloru vinogradnogo vina i soka. *Tr. Vses. Nauch.-Issled. Inst. Vinodeliya i Vinogradarstva, Magarach.* **6**(2), 146–163.

Stasunas, V. J. 1955. The anthocyan pigments of Zinfandel grapes and wine. Ph.D. thesis. Univ. Connecticut, Storrs, Connecticut.

Steinke, R. D., and Paulson, M. C. 1964. The production of steam-volatile phenols during the cooking and alcoholic fermentation of grain. *J. Agr. Food Chem.* **12**, 381–387.

Stella, C. 1962. Indice chimico-fisico atto a definire lo stato di maturazione dei vini rossi e rosati. *Accad. Ital. Vite e Vino, Siena, Atti* **14**, 97–108.

Stephen, J. 1860. "Treatise on the Manufacture, Imitation, Adulteration and Reduction of Foreign Wines, Brandies, Gins, Rums . . . Based on the French System." p. 207. Philadelphia, Pennsylvania.

Stevens, K. L., Bomben, J. L., and McFadden, W. H. 1967. Volatiles from grapes. *Vitis vinifera* (Linn.) *cultivar* Grenache. *J. Agr. Food Chem.* **15**, 378–380.

Stevens, R. 1961. Formation of phenethyl alcohol and tyrosol during fermentation of a synthetic medium lacking amino-acids. *Nature* **191**, 913–914.

Stevenson, P. E. 1965. Effects of chemical substitution of the electronic spectra of aromatic compounds. IV. A general theory of substituent effects and its application to the spectra of the flower pigments. *J. Mol. Spectry.* **18**(1), 51–58.

Sudario, E. 1953. Le sostanze coloranti nelle uve degli ibridi produttori diretti. *Ann. Sper. Agrar. (Rome)* 7, 157–163.

Sudraud, P. 1958. Interprétation des courbes d'absorption des vins rouges. *Ann. Technol. Agr. Paris* 7, 203–208.

Sudraud, P., and Cassignard, R. 1958. Influence de certaines conditions dans la vinification en rouge. *Ann. Technol. Agr. Paris* 7, 209–216.

Sutidize, K. 1962. Effect of *Botrytis cinerea* on grape harvest, must yield, and wine quality. *Tr. Inst. Sadovodstva, Vinogradarstva, i Vinodeliya Gruz. SSR* **14**, 143–148; *Chem. Abstr.* **62**, 5850 (1965).

Swain, T. 1963. "Chemical Plant Taxonomy," p. 543. Academic Press, New York.

Swain, T. 1966. "Comparative Phytochemistry," p. 360. Academic Press, New York.

Swain, T., and Goldstein, J. L. 1964. The quantitative analysis of phenolic compounds. *In* "Methods in Polyphenol Chemistry" (J. B. Pridham, ed.), pp. 131–146. Macmillan, New York.

Swain, T., and Hillis, W. E. 1959. Phenolic constituents of *Prunus domestica.* I. Quantitative analysis of phenolic constituents. *J. Sci. Food Agr.* **10**, 63–68.

Szechenyi, L. 1964. Formation des complexes métalliques des anthocyanes dans les jus de fruits et leur influence sur la couleur. *Ind. Aliment. Agr.* **81**(4), 309–317.

Tagunkov, Yu. D. 1966. Snizhenie aktivnosti katekholoksidazy susla adsorbentami. *Vindelie i Vinogradarstvo SSSR* **26**(6), 20–23.

Takahashi, T., and Abe, G. 1913. On the chemical composition of sake. *J. Coll. Agr. Tokyo* **5**, 95–103.

Takino, Y., and Imagawa, H. 1963a. Studies on the oxidation of catechins by tea oxidase formation of a crystalline reddish orange pigment of benzotropolone nature. *Agr. Biol. Chem. (Tokyo)* **27**, 319–321.

Takino, Y., and Imagawa, H. 1963b. Studies on the oxidation of myricetin glycosides by tea oxidase. *Agr. Biol. Chem. (Tokyo)* **27**, 666–668.

Takino, Y., Imagawa, H., Horikawa, H., and Hirose, Y. 1964. Studies on the oxidation of tea leaf catechins. III. Formation of a reddish orange pigment and its spectral relationship to some benzotropolone derivatives. *Agr. Biol. Chem. (Tokyo)*. **28**, 64–71.

Tanner, H., and Rentschler, H. 1956. Über Polyphenole der Kernobst und Traubensäfte. *Fruchtsaft-Ind.* 1, 231–245.

Tanner, H., and Vetsch, U. 1956. How to characterize cloudiness in beverages. *Am. J. Enol.* 7, 142–149.

Tarantola, C. 1951. Costituenti polifenolici del vino. *Acad. Ital. Vite Vino, Siena, Atti* 3, 362–368.

Tarr, H. L. A., and Cooke, N. E. 1950. Endiols in fruit browning. *Food Technol.* **4**, 245–247.

Taylor, C. E., and Murant, A. F. 1966. Nematocidal activity of aqueous extracts from raspberry canes and roots. *Nematologica* **12**(4), 488–494.

Thomas, C. A., and Orellana, R. G. 1963. Biochemical tests indicative of reaction of castor bean to Botrytis. *Science* **139**, 334–335.

Thompson, G. M. 1956. "Alcoholism," p. 548. Thomas, Springfield, Illinois.

Thompson, R. H. 1957. "Naturally Occurring Quinones," p. 302. Academic Press, New York.

Thudichum, J. L. W., and Dupré, A. 1872. "A Treatise on the Origin, Nature, and Varieties of Wine: being a Complete Manual of Viticulture and Oenology," p. 760. Macmillan, New York.

Timberlake, C. F., and Bridle, P. 1966a. Effect of substituents on the ionisation of flavylium salts and anthocyanins and their reactions with sulphur dioxide. *Chem. & Ind.* 1965–1966.

Timberlake, C. F., and Bridle, P. 1966b. Spectral studies of anthocyanin and anthocyanidin equilibria in aqueous solution. *Nature* **212**, 158–159.

Timberlake, C. F., and Bridle, P. 1967a. Isosbestic points in the visible and ultra-violet spectra of three component systems. *Spectrochim. Acta,* **23A**, 313–319.

Timberlake, C. F., and Bridle, P. 1967b. Flavylium salts, anthocyanidins, and anthocyanins. I. Structural transformations in acid solutions. *J. Sci. Food Agr.* **18**, 473–478.

Timberlake, C. F., and Bridle, P. 1967c. Flavylium salts, anthocyanidins, and anthocyanins. II. Reactions with sulfur dioxide. *J. Sci. Food Agr.* **18**, 479–485.

Tinsley, I. J., and Bockian, A. H. 1960. Some effects of sugars on the breakdown of pelargonidin-3-glucoside in model systems at 90°. *Food Res.* **25**, 161–173.

Tolbert, N. E., Amerine, M. A., and Guymon, J. F. 1943. Studies with brandy. II. Tannin. *Food Res.* **8**, 231–236.

Tomaszewski, M. 1960. The occurrence of *p*-hydroxybenzoic acid and some other simple phenols in vascular plants. *Bull. Acad. Polon. Sci. Ser. Sci. Biol.* **8**, 61–65.

Traphagen, F. W., and Burke, E. 1903. Occurrence of salicylic acid in fruits. *J. Am. Chem. Soc.* **25**, 242–244.

Trautner, E. M., and Roberts, E. A. H. 1950. The chemical mechanism of the oxidative deamination of amino acids by catechol and polyphenolase. *Australian J. Sci. Res.* **3B**, 356–380.

Tressler, D. K., and Pederson, C. S. 1936. Preservation of grape juice. II. Factors controlling the rate of deterioration of bottled Concord juice. *Food Res.* **1**, 87–97.

Trieb, G., 1967. Untersuchungen über der Einfluss von Unterlagen auf Qualitätsmerkmale des Edelreises 1965. *Weinberg Keller* **14**, 459–471.

Trillat, A. 1909. Sur le mécanisme de la fixation du résidu aldéhydique à la matière colorant du vin. *Bull. Soc. Chim. France* **5**, 555–558.

Tronchet, J. 1961a. Paper-chromatographic study of the flavonoids of species resistant to the action of gibberellic acid. *Bull. Soc. Hist. Nat. Doubs* **63**(2), 39–43; *Chem. Abstr.* **58**, 6132 (1963).

Tronchet, J. 1961b. Quelques remarques sur le contenu en flavonoides de plantes "réfractaires" a un traitement gibberellique. *Bull. Soc. Franc. Physiol. Végétale* **7**, 109.

Tsagareli, K. K., and Shikhashvili, T. G. 1959. Ekspress-mikrometod opredeleniya tanidov. *Vinodelie i Vinogradarstvo SSSR* **19**(8), 9–11.

Tsakov, D. 1963. D'bilnite i bagrilnite veshchestva v sortovete G'mza i Zarchin. *Nauchni Tr., Nauchn. Izsled. Inst. po Vinarska Pivovarna Prom., Sofia* **6**, 5–30.

Tsetskhladze, T. V., Kipiani, R. Ya., and Lashkhi, A. D. 1957. Organoleptic and biochemical changes in wines irradiated with X-rays. *Soobshcheniya Akad. Nauk. Gruzin SSR* **18**(2), 183, 188; *Chem. Abstr.* **52**, 4099 (1958).

Turbovsky, M. W., Filipello, F., Cruess, W. V., and Esau, P. 1934. Observations on the use of tannin in wine making. *Fruit Prods. J.* **14**, 106, 121, 123.

Turković, Z. 1967. Beitrag zu Fragen der Sortenkunde. *Wein-Wissenschaft* **22**, 55–66.

Turney, T. A. 1965. "Oxidation Mechanisms," p. 208. Butterworth, Washington, D.C.

Underwood, J. C., and Filipic, V. J. 1964. Source of aromatic compounds in maple syrup flavor. *J. Food Sci.* **29**, 814–818.

Uritani, I. 1961. The role of plant phenolics in disease resistance and immunity. *Plant Phenolics Group N. Am., Proc. 1st. Symp.,* 98–124.

Valaer, P. 1939. California Brandy. III. Changes in storage. *Wine Rev.* **7**(10), 14–16.

Valaer, P. 1941. Report on whiskey and rum. Tannins in potable spirits. *J. Assoc. Offic. Agr. Chemists* **24**, 224–232.

Valaer, P. 1947. Methods of analysis of wine. *J. Assoc. Offic. Agr. Chemists* **30**, 327–331.

Valuiko, G. G. 1965a. Binding of tanning substances and pigments of wine by aldehydes. *Prikl. Biokhim. i Mikrobiol.* 1(2), 242–243; *Chem. Abstr.* **63**, 7622 (1965).

Valuiko, G. G. 1965b. Izmenenie soderzhaniya krasyashchikh i dubil'nykh veshchestv v protsesse vydershki krasnogo stolovogo vina. *Vinodelie i Vinogradarstvo SSSR* **25**(6), 8–10.

Valuiko, G. G. 1966. Biochemical and chemical changes in pigments during production of red wines. *Tekhnol. Pishch. Prod. Rast. Proiskhozhd.*, 72–88; *Chem. Abstr.* **67**, 89736 (1967).

Valuiko, G. G., and Godin, K. G. 1961. Novaya tekhnologicheskaya skhema prigotovleniya krasnykh stolovykh vin v potoke. *Vinodelie i Vinogradarstvo SSSR* **21**(5), 12–17.

Valuiko, G. G., and Ivanyutina, A. I. 1967. Izmenenie okraski krasnykh vin v khode sozrevaniya i stareniya. *Vinodelie i Vinogradarstvo SSSR* **27**(3), 21–25.

Valuiko, G. G., Godin, K. G., and Poznanskaya, M. N. 1964. Rezhimy termicheskoi obrabotki vinograda. *Tr. Vses. Nauchn.-Issled. Inst. Vinodeliya i Vinogradarstva, Magarach* **13**, 44–56.

van Bragt, J., Rohrbaugh, L. M., and Wender, S. H. 1965. The effect of 2,4-dichlorophenoxyacetic acid on the rutin content of tomato plants. *Phytochem.* **4**, 963–965.

van Sumere, C. F., Parmentier, F., and van Poucke, M. 1959. Interconversion of caffeic acid and esculetin. *Naturwissenschaften* **46**, 668.

Van Wyk, C. J., and Venter, P. J. 1964. An investigation into the occurrence of malvine in South African dry red wines. *S. African J. Agr. Sci.* 7, 731–738.

v. Brehmer, W. 1928. Hölzer. *In* "Die Rohstoffe des Pflanzenreiches" (J. v. Wiesner), Vol. 2, 4th ed., pp. 1123–1646. Engelmann, Leipzig.

Venezia, M., and Gentilini, L. 1934. La maturazione dell'uva. Richerche sperimentali sul succo e sulle bucce. *Ann. Staz. Sper. Viticolt. Enol. Conegliano* **4**, 207–218.

Vermorel, V., and Dantony, E. 1910. "Utilisation des Sous-Produits de la Vigne et du Vin," p. 166. Laveur, Paris.

Verona, O. 1952. Azione sui lieviti e, in particolare, lieviti della fermentazione vinaria, delle vitamin $K_3$ e $K_5$. *Accad. Ital. Vite Vino, Siena, Atti* 4, 345–357.

Vetsch, U., and Lüthi, H. 1964. Farbstoffverluste während des biologischen Säureabbaus von Rotweinen. *Mitt. Gebiete Lebensm. Hyg.* **55**, 93–98.

v. Fellenberg, T. 1912. Ueber eine Farbenreaktion des Weines. *Mitt. Lebensm. Hyg.* **3**, 228–231.

v. Fellenberg, T. 1913. Quercetinbestimmung in Wein. *Mitt. Lebensm. Hyg.* **4**, 1–14.

v. Fellenberg, T. 1944. Nachprüfung des Verfahens von A. Torricelli zum Nachweis von Tresterwein in Weisswein. *Mitt. Lebensm. Hyg.* **35**, 149–188.

Viard, E. 1892. "Traité Général de la Vigne et des Vins," 2d ed., pp 1–1131. J. Dujardin, Paris.

Villforth, F. 1958. Studien zum Farbwert und zur Farbanalyse von Rotweinen. *Wein-Wiss., Beih. Fachz. Deut. Weinbau,* 1–8, 11–14.

Vil'yams, V. V., and Begunova, R. D. 1955. K opredeleniyu krasyashchikh veshchestv v krasnykh sortakh vinograda i krasnykh stolovykh vinakh. *Vinodelie i Vinogradarstvo SSSR* **15**(7), 38–40.

Vil'yams, V. V., and Taranova, R. D. 1950. Kolichestvennyi metod opredeleniya krasyashchikh veshchestv krasnogo vina. *Vinodelie i Vinogradarstvo SSSR* **10**(4), 29–30.

Vil'yams, V. V., and Taranova, R. D. 1951. Khromatograficheskii metod razdeleniya krasyashchikh veshchestv vina. *Vinodelie i Vinogradarstvo SSSR* **11**(7), 16–18.

Violante, C. 1948. Ricerca spettrofotometriche sulla materia colorante delle uva. *Riv. Viticolt. Enol.* 1, 265–270, 298–304.

Visintini-Romanin, M. 1967. Ricerca sui pigmenti antocianici nella vite e nei vini. L'ossidazione della malvina. *Riv. Viticolt. Enol.* **20**, 79–88.

Vogin, E. E. 1960. A review of bioflavonoids. *Am. J. Pharm.* **132**, 363–382.

Vogt, E. 1958. "Handbuch der Kellerwirtschaft. III. Weinchemie und Weinanalyse," 2nd ed., p. 382. Ulmer, Stuttgart.

Voisenet, E. 1929a. Nouvelles recherches sur la nature de la substance qui, dans la maladie de l'amertume des vins, produit le goût amer. *Compt. Rend.* **188**, 941–943.

Voisenet, E. 1929b. Le divinylglycol considerée comme agent de la saveur amère, dans la maladie de l'amertume des vins. *Compt. Rend.* **188**, 1271–1273.

von Babo, U., and Mach, E. 1923. "Handbuch des Weinbaues und der Kellerwirtschaft," 4th ed., Vol. 1, Part 1, p. 626. Paul Parey, Berlin.

Vuataz, L, Brandenberger, H., and Egli, R. 1959. Plant phenols. I. Separation of the tea leaf polyphenols by cellulose column chromatography. *J. Chromatography* **2**, 173–187.

Wagner, H., Patel, J., Hörhammer, L., Yap, F., and Reichardt, A. 1967. Flavon-C-Glykoside in den Blättern von *Vitis cinerea* Darwin. *Z. Naturforsch.* **B22**, 988–989.

Wait, F. E. 1889. "Wines and Vines of California, a Treatise on the Ethics of Wine-Drinking," p. 215. Bancroft Co., San Francisco.

Wallen, L. L., Stodola, F. H., and Jackson, R. W. 1959. Type reactions in fermentation chemistry. *U. S. Dept. Agr., Agr. Res. Service, ARS-71-13*, 1–496.

Wardrop, A. B., and Cronshaw, C. D. 1962. Formation of phenolic substances in the ray parenchyma of angiosperms. *Nature* **193**, 90–92.

Wase, D. A. J., and Hough, J. S. 1966. Continous culture of yeast on phenol. *J. Gen. Microbiol.* **42**, 13–23.

Wasserman, A. E. 1966. Organoleptic evaluation of three phenols present in wood smoke. *J. Food Sci.* **31**, 1005–1010.

Weaver, R. J. 1952. Thinning and girdling of Red Malaga grapes in relation to size of berry, color, and percentage of total soluble solids of fruit. *Proc. Am. Soc. Hort. Sci.* **60**, 132–140.

Weaver, R. J., and McCune, S. B. 1960. Influence of light on color development in *Vitis vinifera* grapes. *Am. J. Enol. Viticult.* **11**, 179–184.

Weaver, R. J., and Pool, R. M. 1965. Relation of seededness and ringing to gibberellin-like activity in berries of *Vitis vinifera. Plant Physiol.* **40**, 770–776.

Weaver, R. J., McCune, S. B., and Hale, C. R. 1962. Effect of plant regulators on set and berry development in certain seedless and seeded varieties of *Vitis vinifera* L. *Vitis* **3**, 84–96.

Webb, A. D. 1962. Present knowledge of grape and wine flavors. *Food Technol.* **16**, 56–59.

Webb, A. D. 1964. Anthocyanins of grapes. *Plant Phenolics Group N. Am. Proc. 4th Symp.*, 21–39.

Webb, A. D. 1967. Wine flavor: volatile aroma compounds of wines. *In* "Symposium on Foods: Chemistry and Physiology of Flavors" (H. W. Schultz, E. A. Day, and L. M. Libbey, eds.), pp. 203–227. Avi, Westport, Connecticut.

Webb, A. D., Kepner, R. E., and Ikeda, R. M. 1952. Composition of a typical grape brandy fusel oil. *Anal. Chem.* **24**, 1944–1949.

Webster, J. E., and Cross, F. B. 1935. Chemical analyses of grape juices—variety comparisons. *Proc. Am. Soc. Hort. Sci.* **33**, 442–443.

Webster, J. E., and Cross, F. B. 1936. The composition of grape juice as affected by the method of vine training. *Proc. Am. Soc. Hort. Sci.* **34**, 405–407.

Webster, J. E., and Cross, F. B. 1942. The uneven ripening of Concord grapes: chemical and physiological studies. *Oklahoma Agr. Expt. Sta. Tech. Bull.* **T-13**, 5–48.

Webster, J. E., and Cross, F. B. 1944. Chemical composition of grape juices: a varietal study. *Proc. Oklahoma Acad. Sci* **24**, 92–95.

Webster, J. E., Anderson, E., and Cross, F. B. 1934. Chemical and enzymatic studies of the uneven ripening of Concord grapes. *Proc. Am. Soc. Hort. Sci.* **32**, 365–369.

Weger, B. 1965. Über ein abnormales Verhalten eines Rotweines bein Warmetest. *Weinberg Keller* **12**, 481–484.

Weinges, K. 1964. The occurrence of catechins in fruits. *Phytochemistry* **3**, 263–266.

Weinges, K., Kaltenhauser, W., Marx, H.-D., Nader, E., Nader, F., Perner, J., and Seiler, D. 1968. Zur Kenntnis der Proanthocyanidine. X. Procyanidine aus Früchten. *Ann. Chem. Liebigs* **711**, 184–204.

West, D.B., Lautenbach, A. F., and Brumsted, D. D. 1963. Phenolic characteristics in brewing. *Proc. Am. Soc. Brewing Chemists*, 194–199.

Westlake, D. W. S. 1963. Microbiological degradation of quercitrin. *Can. J. Microbiol.* **9**, 211–220.

Westlake, D. W. S., and Spencer, J. F. T. 1966. The utilization of flavonoid compounds by yeasts and yeast-like fungi. *Can. J. Microbiol.* **12**, 165–174.

Westlake, D. W. S., Roxburgh, J. M., and Talbot, G. 1961. Microbial production of carbon monoxide from flavonoids. *Nature* **189**, 510–511.

Weurman, C., and de Rooij, C. 1958. Chlorogenic acid isomers in black Alicante grapes. *Chem. & Ind. (London)*, 72.

Whalley, W. B. 1962. The stereochemistry of flavonoid compounds. *In* "The Chemistry of Flavonoid Compounds" (T. A. Geissman, ed.), pp. 441–467. Macmillan, New York.

Wheeler, O. R., Carpenter, J. A., Powers, J. J., and Hamdy, M. K. 1967. Inhibition of enzymes by the anthocyanin malvidin-3-monoglucoside. *Proc. Soc. Exptl. Biol. Med.* **125**, 651–657.

White, T. 1957. Tannins–their occurrence and significance. *J. Sci. Food Agr.* **8**, 377–385.

White, T., Kirby, K. S., and Knowles, E. 1952. Tannins. IV. The complexity of tannin extract composition. *J. Soc. Leather Trades Chemists* **36**, 148–155.

Whiting, G. C., and Carr, J. G. 1957. Chlorogenic acid metabolism in cider fermentation. *Nature* **180**, 1479.

Wiley, H. W. 1929. "The History of a Crime Against the Food Law." p. 413. Publ. by author.

Williams, A. H. 1952. The tannin of apple juice and cider: an interpretation of the permanganate titration method. *Long Ashton Res. Sta. Ann. Rept.* **1952**, 219–224.

Williams, B. L., and Wender, S. H. 1952a. The isolation and identification of quercetin and isoquercitrin from grapes (*Vitis vinifera*). *J. Am. Chem. Soc.* **74**, 4372–4373.

Williams, B. L., and Wender, S. H. 1952b. Isoquercitrin and quercetin in Concord grapes. *Proc. Oklahoma Acad. Sci.* **33**, 250–251.

Williams, W. O. 1946. California vineyard fertilizer experimentation. *Proc. Am. Soc. Hort. Sci.* **48**, 269–278.

Willstätter, R., and Everest, A. E. 1913. Untersuchungen über die Anthocyane. I. Über den Farbstoff der Kornblume. *Ann. Chem. Liebigs* **401**, 189–232.

Willstätter, R., and Nolan, T. J. 1915. Untersuchungen über die Anthocyane. II. Über den Färbstoff der Rose. *Ann. Chem. Liebigs* **408**, 1–14.

Willstätter, R., and Zollinger, E. H. 1915. Untersuchungen über die Anthocyane. VI. Über die Farbstoffe der Weintraube und der Heidelbeere. *Ann Chem. Liebigs* **408**, 83–109.

Willstätter, R., and Zollinger, E. H. 1916. Untersuchungen über die Anthocyane. XI. Über die Farbstoffe der Weintraube und der Heidelbeere. *Ann. Chem. Liebigs* **412**, 113–251.

Wilson, C. O., and Gisvold, O. 1956. "Textbook of Organic Medicinal and Pharmaceutical Chemistry," 3rd ed., p. 823. Lippincott, Philadelphia.

Winkler, A. J. 1930. Berry thinning of grapes. *Calif. Univ. Agr. Expt. Sta. Bull.* **492**, 3–22.

Winkler, A. J. 1936. Ripening changes in grapes. *Wine Rev.* **4**(7), 6–7.

Winkler, A. J. 1962. "General Viticulture," p. 633. Univ. of California Press, Berkeley, California.

Winkler, A. J., and Amerine, M. A. 1937. What climate does, the relation of weather to the composition of grapes and wine. *Wine Rev.* **5**(6), 9–11; (7), 9–11, 16.

Winkler, A. J., and Amerine, M. A. 1938a. Color in California wines. I. Methods for measurement of color. *Food Res.* **3**, 429–438.

Winkler, A. J., and Amerine, M. A. 1938b. Color in California wines. II. Preliminary comparisons of certain factors influencing color. *Food Res.* **3**, 439–447.

Winkler, A. J., and Amerine, M. A. 1943. Color in wine and influencing factors. *Wine Rev.* 11(10):10–12.

Winkler, A. J., and Williams, W. O. 1935. Effect of seed development on the growth of grapes. *Proc. Am. Soc. Hort. Sci.* **33**, 430–434.

Winton, A. L., and Winton, K. B. 1935. "The Structure and Composition of Foods," Vol. 2, p. 904. Wiley, New York.

Wobisch, F., and Schneyder, J. 1952. Kolorimetrischer Nachweis von Holunderbeerfarbstoff im Wein. *Monatsh.* **83**, 478–481.

Wobisch, F., and Schneyder, J. 1955. Die Bestimmung der Farbstoffe von Rotwein. *Mitt. (Klosterneuburg)* **5 A**, 49–52.

Wood, D. J., and Roberts, E. A. H. 1964. The chemical basis of quality in tea. III. Correlations of analytical results with tea tasters' reports and evaluations. *J. Sci. Food Agr.* **15**, 19–25.

Wright, H. E., Jr., Burton, W. W., and Berry, R. C., Jr. 1960. Soluble browning reaction pigments of aged burley tobacco. I. Nondialyzable fraction. *Arch. Biochem. Biophys.* **86**, 94–101.

Wucherpfennig, K., and Bretthauer, G. 1962a. Versuche zur Stabilisierung von Wein gegen oxydative Einflüsse durch Behandlung mit Polyamidpulver. *Weinberg Keller* **9**, 37–55.

Wucherpfennig, K., and Bretthauer, G. 1962b. Beitrag zur Stabilisierung von Fruchtsäften mit Polyamidpulver. *Fruchtsaft-Ind.* **7**, 40–54.

Wucherpfennig, K., and Franke, I. 1963. Auftrennung von Weininhaltsstoffen durch Gel-Filtration. *Z. Lebensm.-Untersuch. Forsch.* **124**, 22–35.

Wucherpfennig, K., and Franke, I. 1964. Beitrag zur Bestimmung einer Kennzahl für Polyphenole in Weinen durch Gel-Filtration. *Wein-Wiss.* **19**(8), 362–369.

Wucherpfennig, K., and Kleinknecht, E. M. 1965. Beitrag zur Veränderung der Farbe und der Polyphenole bei Abfüllung von Weisswein durch Einwirkung von Sauerstoff und Wärme. *Wein.-Wiss.* **20**(11), 489–514.

Wucherpfennig, K., and Ratzka, D. 1967. Über die Verzögerung der Weinsteinausscheidung durch polymere Substanzen des Weines. *Weinberg Keller* **14**(11), 499–509.

Wucherpfennig, K., Troost, G., and Fetter, K. 1964. Untersuchungen zur Maische-Warmfermentierung weisser und roter Weintrauben. *Deut. Wein-Ztg.* **100**, 688, 690, 692, 694, 696, 698, 699, 716, 718, 720.

Wüstenfeld, H., and Luckow, C. 1933. Untersuchung über die physikalischen und chemischen Vorgänge bei der Lagerung von Weindestillaten in Grossbetrieb. *Z. Untersuch. Lebensm.* **65**, 208–220.

Yakivchuk, A. P. 1966. Antotsiany kozhitsy yagod vinograda. *Vinodelie i Vinogradarstvo SSSR* **26**(7), 29–30.

Yamamoto, A. 1961. Flavors of sake, V. Separation and identification of some nonvolatile carbonyl compounds. *Nippon Nogeikagaku Kaishi* **35**, 819–823; *Chem. Abstr.* **60**, 3454 (1964).

Yamamoto, A., Sasaki, K., and Saruno, R. 1961. Flavors of sake. IV. Separation and identification of ferulic acid, vanillic acid, and vanillin. *Nippon Nogeikagaku Kaishi* **35**, 715–719; *Chem. Abstr.* **60**, 3454 (1964).

Yamamoto, A., Omori, T., and Yasui, H. 1966. Flavors in sake. IX. The precursor of *p*-hydroxybenzaldehyde and *p*-hydroxybenzoic acid in sake. *Nippon Nogeikagaku Kaishi* **40**, 152–160; *Chem. Abstr.* **64**, 18357 (1966).

Yang, H. Y. 1943. Spectrophotometric and chromatographic studies of the principal pigments in Evergreen blackberries and Alicante Bouschet grapes. Ph.D. thesis. Oregon State Univ., Corvallis, Oregon.

Yang, H. Y., and Steele, W. F. 1958. Removal of excessive anthocyanin pigment by enzyme. *Food Technol.* **12**, 517–519.

Yang, H. Y., Gomez, L. P., Rasulpuri, M. L., and Michalek, J. G. 1962. The effect of vitamin $K_5$ on gas production in fruit juices. *Food Technol.* **16**, 109–111.

Yankov, A. T., and Balkandzhiev, G. N. 1965. V'prkhu v'zmozhnostta za poluchavane na enotsianin ot dzhibrite na sorta Alikant Teras-20. *Lozarstvo Vinarstvo* (*Sofia*) **14**(7), 42–45.

Yap, F., and Reichardt, A. 1964. Vergleichende Untersuchungen der Flavonoide und Oxyzimtsäuren in den Blättern artreiner *Vitis*-Sorten und ihre Bastarde. *Züchter* **34**, 143–156.

Zamorani, A., and Pifferi, P. G. 1964. Contributo alla conoscenza della sostanza colorante dei vini. I. Identificazione e valutazione quantitativa degli antociani di vini da *Vitis vinifera* (Merlot) e da ibridi (Clinton e Baco). *Riv. Viticolt. Enol.* **17**, 85–93.

Zamorani, A., and Pifferi, P. G. 1966. Natural pigments. IV. Determination of the composition of some anthocyanins. *Boll. Sci. Fac. Chim. Ind. Bologna* **24**, 31–40; *Chem. Abstr.* **65**, 11297 (1966).

Zimmermann, R. 1958. Über phenolspaltende Hefen. *Naturwissenschaft* **7**, 165.

Zinchenko, V. I. 1964. Predel'noe soderzhanie azotistykh veshchestv v susle i vine, predotvrashchayushchee pereokislennost. *Tr. Vses. Nauchn.-Issled. Inst. Vinodeliya i Vinogradarstva, Magarach,* **13**, 24–43.

Zitko, V., and Rosik, J. 1962. Ein Beitrag zur Theorie der Tannin-Gelatine-Schönung von Obstsäften. *Nahrung* **6**, 561–577.

Zotov, V. V., and Sokolovskaya, T. I. 1959. Various forms of tannic substances in the roots of grape vines, not affected and affected with phylloxera (trans). *Biokhim. Plodov. i Ovoshchei* **5**, 195–203.

Zotov, V. V., Svelyakova, R. I., Sokolovskaya, T. I., Storozhak, E. M., and Kucher, A. A. 1966. Physiology of grape resistance to phylloxera. *Sel'skokhoz. Biol.* **1**(3), 410–420; *Chem. Abstr.* **66**, 551 (1967).

Zuman, P. 1953. Das polarographische Verhalten der Anthocyane. I. *Collection Czechoslov. Chem. Communs.* **18**, 36–42.

# Author Index

Numbers in italics refer to the pages on which the complete references are listed.

## C

## D

## E



## F

## G

## L



## M

## N

## O



## P

## S

## T



## U

## V



## W



## XYZ

# Subject Index

## D



## E



## F

## M



## N



## O



## P



## Q

## W



## Y